senior ncea

design and visual communication

elizabeth mchugh

Australia • Brazil • Japan • Korea • Mexico • Singapore • Spain • United Kingdom • United States

Design and Visual Communication
1st Edition
Elizabeth McHugh

Cover designer: Cheryl Rowe
Text designer: Macarn Design
Production controller: Siew Han Ong

Any URLs contained in this publication were checked for currency during the production process. Note, however, that the publisher cannot vouch for the ongoing currency of URLs.

First published in 2008 as Complete Graphics NCEA Level One by Cengage Learning Australia Pty Limited

Acknowledgements
I would like to thank the following people for their contribution to this textbook:
• Students from St Peter's College, Epsom Girls Grammar and Onslow College. Individual students have been identified within the text.
• Supporting teachers Megan Dunsmore (Epsom Girls Grammar School), Ron Van Musscher (Onslow College), Sunny Park (St Peter's College - Auckland) for contributing their students' exemplar work.
• Lesley Pearce (Team Solutions) and Motu Samaeli (St Kentigern College) for their contribution of ideas for the application of design work.
• Christian Burgos (Top Scholar 2009) who has not only contributed an enormous amount of exemplar material but also the images for the cover.
• My family for their ongoing support - Fabian McQueen, Lucinda McHugh, Judy and William McHugh.

For product information and technology assistance,
in Australia call **1300 790 853**;
in New Zealand call **0800 449 725**

For permission to use material from this text or product, please email **aust.permissions@cengage.com**

National Library of New Zealand Cataloguing-in-Publication Data
National Library of New Zealand Cataloguing-in-Publication Data

McHugh, Elizabeth.
Design and visual communication / Elizabeth McHugh.
ISBN 978-017023-327-9
1. Graphic arts. 2. Design.
I. Title.
740—dc 23

Cengage Learning Australia
Level 7, 80 Dorcas Street
South Melbourne, Victoria Australia 3205

Cengage Learning New Zealand
Unit 4B Rosedale Office Park
331 Rosedale Road, Albany, North Shore 0632, NZ

For learning solutions, visit **cengage.com.au**

Printed in Australia by Ligare Pty Ltd
3 4 5 6 7 8 9 20 19 18 17 16

contents

1 idea generation

Visual communication techniques can be used to generate design ideas (your responses to a design brief). These techniques include sketching, rendering, modelling/model making such as mockups and 3D constructions, collage and overlays or digital media.

Visual communication techniques can be used to explore the aesthetic and functional qualities when generating design ideas. The exploration of aesthetic qualities could include the colour, tone, texture, pattern, shape, balance and surface finish for an idea. The exploration of functional qualities could include the size, scale and proportion of an object, how an object may be constructed, the material value or joining of components and ergonomics.

The use of the work of an influential designer (Level 1) or the use of the characteristics of a design movement or era (Level 2) to inform design ideas is also an effective method to generate ideas.

ISBN: 9780170233279

Level 1

At Level 1, you are required to choose an architect or designer, analyse examples of their work, identifying and explaining the aesthetic and/or functional characteristics. You are then to generate initial ideas, explore and refine your ideas, using characteristics of your designers/architects works.

Research

- Select a designer or architect. A designer or architect may be selected for you or you may be given a list to choose from.
- Show examples of the designer / architect's work, using photographs or sketches to identify and explain the aesthetic and/or functional characteristics that are typical of their work.
- Analyse the features of the designer's product or spatial design. Consider the aesthetic (factors related to appearance) and/or functional (factors related to use) features of each work.

Initial ideas

- Using the designer's or architect's work, produce initial ideas for your given design brief. Communicate your ideas using design sketches and/or digital media for the product or spatial design brief. Using notes, explain how your ideas (shape, style, colour, material) relate to the designer's/ architect's work.

Develop ideas to a final solution

- Develop your ideas to a final outcome. Refer to *Chapter 4 – Developing Design Ideas* for further guidance. Throughout your development, communicate and explain how the chosen designer / architect has effectively informed your own ideas and thinking.

Level 2

At Level 2, you are required to choose a design movement or era and explain the elements of design and how they are used, describing social factors that influenced the movement or era. You will apply visual communication and design techniques and knowledge to explore, refine and review design ideas that integrate characteristics of the design movement or era with your own ideas for a given brief.

Research

- Select a design movement or era to research (Movements could be Modernism, De Stijl, Bauhaus, Deconstructivism, New Look or other. Eras could be Aztec, pre-European Maori, Shogun, Renaissance, Victorian, 1920s, 1960s or other). Explore the characteristics of the movement/era, as well as the historical/cultural context and the designers and architects who influenced it.
- Using the internet and books collate images of buildings, furniture or product designs that explain characteristics of the selected movement/era.
- Analyse and evaluate the spatial or product design explaining the aesthetics and function of the design. Refer to the elements of design (shape, form, line, rhythm, balance, colour, harmony and contrast, user friendliness, durability). Explain how the elements of these designs characterise the design movement/era. Also, describe the social factors e.g. what was happening at the time of the movement/era (cultural, historical, societal, and technological) that was an influence.

Initial ideas

- Using the movement/era research, produce initial ideas for your given design brief. Communicate your ideas using design sketches and/or digital media for the product or spatial design brief. Using notes, explain how your ideas (shape, style, colour, material) relate to the movement/era.

ISBN: 9780170233279

Develop ideas to a final solution

- Develop your ideas to a final outcome. Refer to *Chapter 4 – Developing Design Ideas* for further guidance. Throughout your development communicate and explain how the chosen movement/era has effectively informed your own ideas and thinking. Explore, refine and review your design ideas, explaining your choices. You should explore alternatives and consider features related to function and aesthetics.

Influential designer

At Level 1, students are to use the work of an influential designer to inform design ideas. Influential designers are those recognised in the context of their work as leading practitioners. The work of a designer could be an individual product or spatial design, or a body of work.

Robert Gorrie — St Peter's College

The following work by Robert Gorrie is a study of the product designer Philippe Starck. He has sketched Starck products to gain an understanding of the shapes and forms he uses in his work. Robert has investigated Starck's personal background and ideals for design. He has recognised the aesthetic and functional characteristics that are typical of Philippe Starck. Robert has then analysed various Starck products to gain further understanding of the shape, style and materials that he uses and why. Robert has used these charateristics to influence his own design for a shelter.

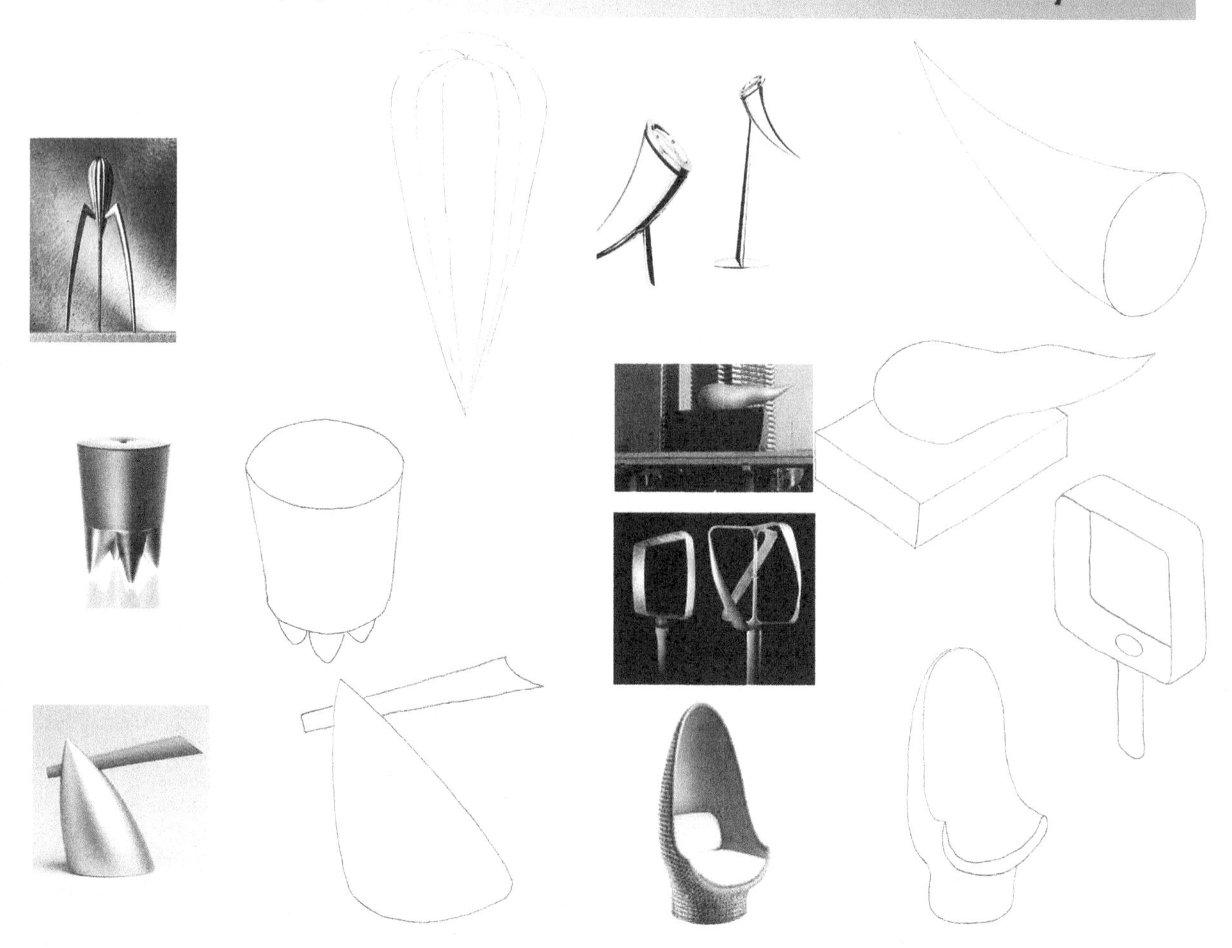

ISBN: 9780170233279

Philippe Starck

Personal Information

Philippe Patrick Starck (born January 18, 1949, Paris) is a French product designer and is one of the best known designers in the New Design style. He has designed pieces ranging from spectacular interior designs to mass produced consumer goods such as toothbrushes, chairs, and even houses. He was the son of an aeronautical engineer and attended the Ecole Nissim de Camondo in Paris from 1965 until 1967. In 1969 Starck became art director at Pierre Cardin. In the 1970s Philippe Starck embarked on a career as an interior decorator; stylish early interiors he designed include the "La Main Bleue" bar in Montreuil (1976) and "Les Bains Douches" in Paris (1978). In 1980 he founded Starck Products and, in 1985, a furniture-making firm, XO (with Gerard Mialet).). In 1982 Philippe Starck was one of the designers commissioned by François Mitterand to refurbish his private apartments in the Elysée Palace in Paris. The interior Philippe Starck designed for Café Costes in Paris (1984). One of the pieces of furniture he designed for the café was the elegant "Costes" chair (for Driade). This project made Philippe Starck famous worldwide.

Design Elements

For over 30 years, Philippe Starck has been defying the boundaries of imagination and creativity, going above and beyond to redefine shapes, textures, materials and the way we use space. Philippe is one of the leading figures of the New Design movement and so his name ranks along the likes of Raymond Loewy, Henry Dreyfus and Christian Dior of influential designers who helped shape the post-war era. He manipulates combination of sharp, straight lines and curvilinear, gentle shapes. His colour pallet is also surprising as he uses clean, light tones, contrasted with rich, florid hues. Starck differs from many New Design stylists as he has outgrown the artistic snobbery of creating shocking, one-off luxury items, but has instead developed into a prolific designer whose products range from exclusive and bespoke to everyday, mass-produced items. A true designer, Starck has created, shaped and moulded across the disciplines, applying his gifted touch to architecture, interior design, graphic art, fashion and his first love, industrial design.

Influence Starck has had on society

Philippe Starck's artistic philosophy gives people objects that could improve their live in aesthetical and quality range. He has designed noodles for Panzani, boats for Beneteau, mineral-water bottles for Glacier, kitchen appliances for Alessi, toothbrushes for Fluocaril, luggage for Vuitton, "Urban Fittings" for Decaux, office furniture for Vitra, as well as vehicles, computers, doorknobs, spectacle frames, etc. Starck's work has brought him numerous prizes and awards. During the 1980s Starck designed a number collections. He has created individual items of furniture that have been manufactured by firms all over the world. His style continues to evolve as his über-cool hotels, shops, restaurants, nightclubs and bespoke items spread across the continents, but again he finds time to pen designer corkscrews, bottles, labels and chairs – many of them stylish but also affordable. The difference is, Philippe Starck is also happy creating inexpensive products such as the Juicy Salif, a stylised yet inexpensive juicer that has since become a cult item. In the field of industrial design, he has been responsible for the creation of a wide variety of objects.

Societal influences on Starck

Much of Philippe Stark's work produced in the 1980's and 1990's was influenced by fashion and novelty. It has even been referred to by some as being 'overdesigned'. In the 21st century his approach to design seems to have changed. Although he has some ecological concerns other fascinations of Starck include high-technology and human intelligence. These factors have both had great influence on his design not only in case of hi-tech objects but also in connection with creation of furniture, clothes or home objects. Design infiltrates each aspect of our lives as a product for sale. Nowadays each of us have the opportunity to value the products created for mass by artist or designers. Among the influential brand names is Philippe Starck. Each of the pieces of Starck's objects are infiltrated by his imaginative nature, even oriental chic and sexy look. But, real gift of his dreams is present in his interior design realisations.

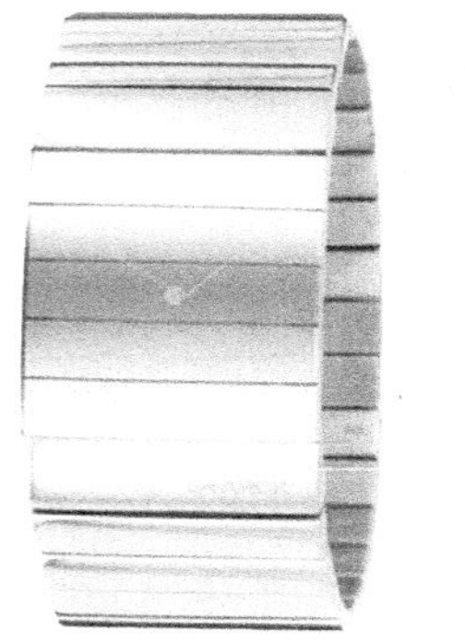

Robert Gorrie

ISBN: 9780170233279

Product Analysis

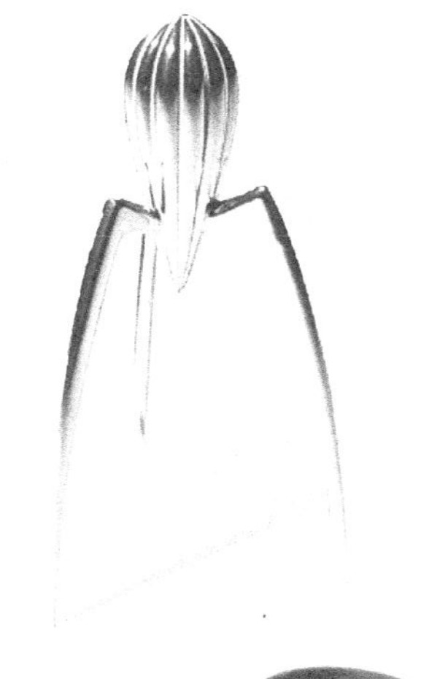

Philippe Starck is a revolutionary product designer and his Juicy Salif (orange squeezer) is one of his most famous pieces. The Juicy Salif was designed by Starck in 1990. It is one of Starcks most recognised pieces and an icon in industrial design. The orange squeezer is very aesthetically pleasing as it looks like and can be presented as a piece of art but all the while it is able to provide the function of squeezing fruit. This, like many of Starck's other designs, may look very expensive but is actually affordable which adds to its high appeal.

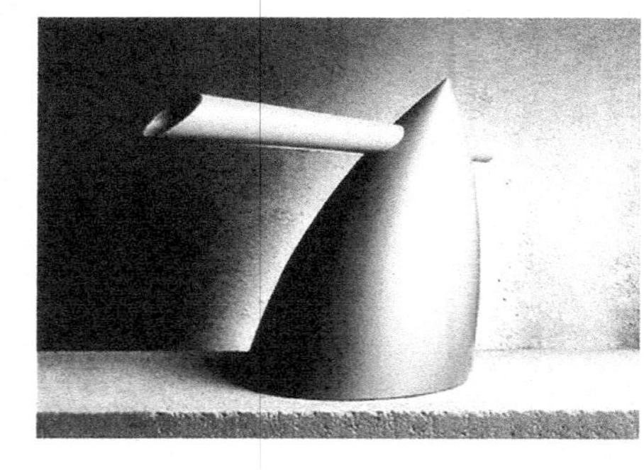

The Hot Bertaa is an aluminium kettle designed by Starck and launched in 1990. The design allows the water to enter the kettle, heat and leave through an opening that appears as an extension of the handle. This was one of Starcks first pieces produced by Alessi. Its amazing design allows for it to be a piece of art in the kitchen. It provides a modern kick all the while having the bonus attribute of actually being able to be used as a kettle. Although it may not be one of Starck's favourite designs, it is one of his most famous nonetheless.

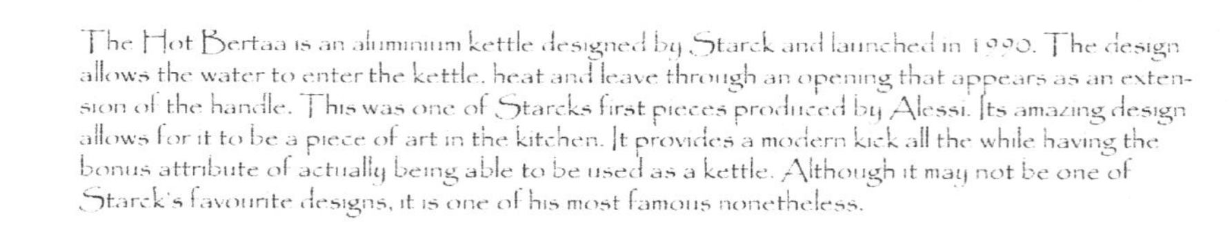

The Tooth Stool is another example of Starck's revolutionary design. This stool is simply shaped as a tooth. It can be used as both a seat and a side table allowing for multiple purposes. Starck designed this chair for the hotel lobby of the Saint Martin's Lane hotel in London. It is looked upon as a piece of jewellery for your home. This stool is made in rotation moulding polypropylene and comes in 3 colours: batch dyed silver as well as silver or gold lacquered. Although some may see this piece as being ugly or too radical, it shows off Starck's creativity and new era design.

Starck's Mac Gee Bookshelf was inspired by the novel written by Philip K. Dick. It is a collapsible bookshelf with 5 levels to place items on. The shape of it makes it look similar to the likes of a ladder or staircase. It is designed to be leaned and stabilised against a wall and stretches over two metres tall. The Mac Gee Bookshelf is yet another example of Starck's on-going creativity and design. This piece has steel sheets with epoxy finish in matt black, white and metallic silver making for a very modern, aesthetically pleasing bookshelf.

Robert Gorrie

ISBN: 9780170233279

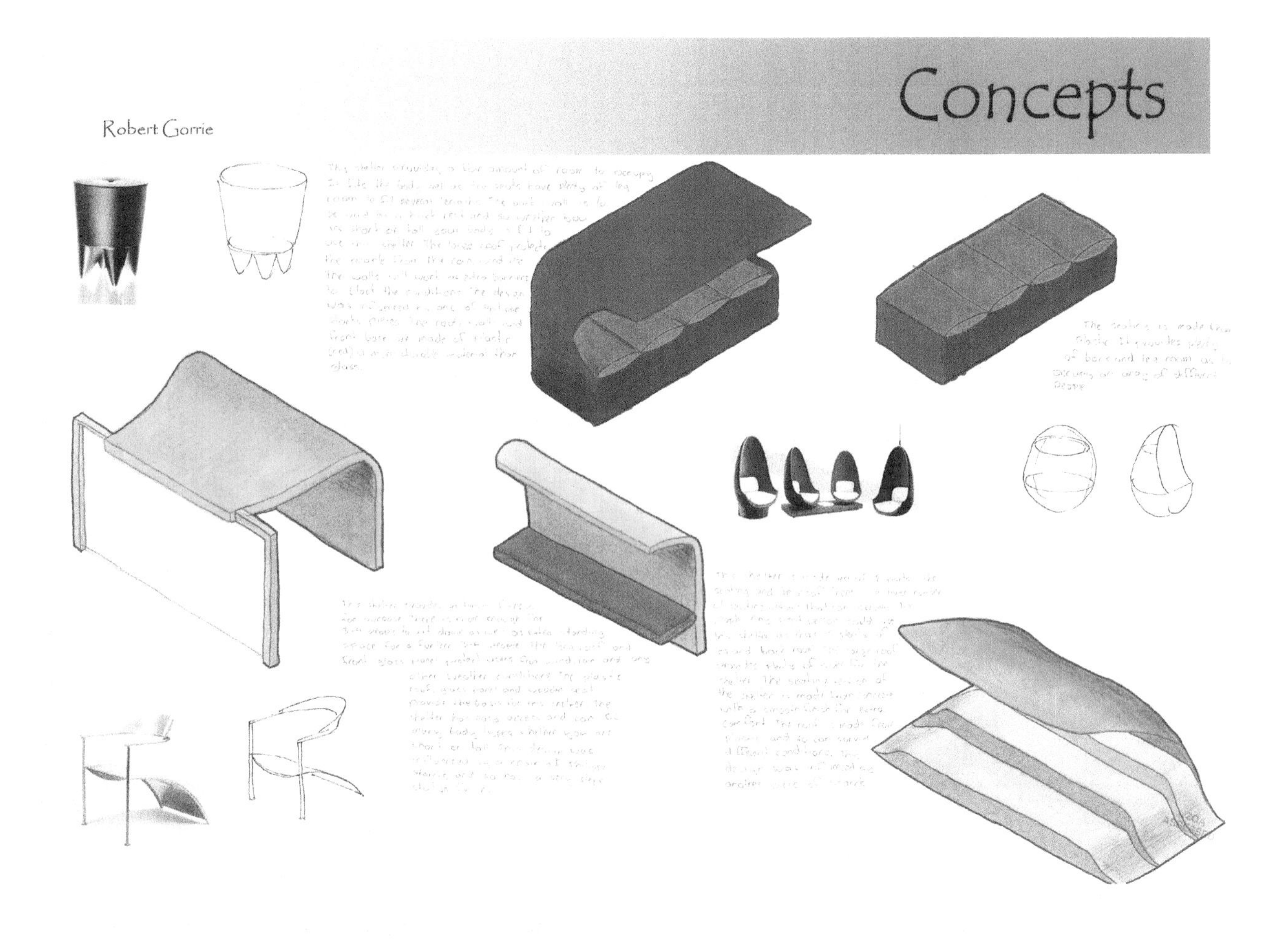

Influential design movement or era (Level 2)

At Level 2, students are to use the characteristics of a design movement or era to inform their design ideas. Possible design movements could be Modernism, De Stijl, Bauhaus, Deconstructivism, and New Look. Possible design eras could be Aztec, pre-European Maori, Shogun, Renaissance and Victorian.

Nathanael Hailemariam — St Peter's College

The following work by Nathanael Hailemariam demonstrates the study of the Modernist Movement. He has completed a detailed study of the architect Frank Lloyd Wright as part of his investigation of the Modernist Movement. Lloyd Wright was important to the history and development of this movement. Nathanael has analysed the key design principles of aesthetics and function, which include shape, form, rhythm, balance, proportion, colour, contrast and durability. He has also considered the societal factors such as cultural, historical, societal and technological. Nathanael has used Frank Lloyd Wright's ideals of the Modern Movement to inform his own design ideas. Nathanael has used the promotion of organic architecture and the simplification of form to influence his concept ideas for architectural design.

ISBN: 9780170233279

Modern Architecture

Nathanael Hailemariam

Modern Architecture is characterized by simplification of form and creation of ornament from the structure and theme of the building. Early modern architecture began at the turn of the 20th century with efforts to reconcile the principles underlying architectural design with rapid technological advancement and the modernization of society. It took the form of numerous movments, schools of design and Architectural styles.

The term Modernism describes the modernist movement in the arts, its set of cultural tendencies and associated cultural movements, originally arising from wide-scale and far-reaching changes to Western society in the late 19th and early 20th centuries. In particular the development of modern industrial societies and the rapid growth of cities, followed then by the horror of World War I, were among the factors that shaped Modernism.

In broad terms, the period was marked by sudden and unexpected breaks with traditional ways of viewing and interacting with the world. Experimentation and individualism became virtues, where in the past they were often heartily discouraged.

Architects important to the history and development of the modernist movement include Fank Lloyd Wright, Ludwig Mies van der Rohe, Le Corbusier, Oscar Niemeyer, Alvar Aalto, Walter Gropius and Louis I Kahn.

Wright's Larkin Building (1904) in Buffalo, New York, Unity Temple (1905) in Oak Park, Illinois, and the Robie House (1910) in Chicago, Illinois were some of the first examples of modern architecture in the United States. Frank Lloyd Wright was a major influence on European architects, including both Walter Gropius (founder of the Bauhaus) and Ludwig Mies van der Rohe, as well as on the whole of organic architecture. Gropius claimed that his "bible" for forming the Bauhaus was 100 Frank Lloyd Wright drawings that the architect shared with Germany over a decade prior to this point, the Wasmuth Portfolio. While Wright's career would parallel that of European architects, he refused to be categorized with them, claiming that they copied his ideas.[citation needed] Many architects in Germany[who?] believed that Wright's life would be wasted in the United States, since the US wasn't ready for his newer architecture.[citation needed] It would be several decades before the European architects would bring in turn their version to the United States. During the 1930s, Wright would experiment with his Usonian ideas for a uniquely U.S. American (ergo "US-onian") take on modernism.

Frank Lloyd Wright

Even though it's been fifty years from his death, Frank Lloyd Wright is still recognized today as the greatest 20th century architect.
To this day, Wright's influence on modernism remains uncontested and the reason for that is because Wright is the link that unites architecture of olden times to architecture of personal expression. A modern thought, character, or practice is what defines the word modernism and in addition, the term describes both a set of cultural inclination and a range of related cultural movements. Against the traditional standards of realism modernism was a revolution and the term includes the actions and output of those who felt the traditional appearance of art, architecture, literature, religious faith, social organisation and daily life and of a promising fully industrialized world they were becoming invalid in the new economic, social, and political situation.

Frank Lloyd Wright has influenced not only American architecture in a way that will always be seen but throughout the world, but the largest area of impact has been the American homes.
What Frank believed was that America should have its own architecture and that it should not be influenced by European styles therefore this led him to change the idea of the open floor plans and homes suitable to American lifestyles.
Wright also believed that the architecture should reflect individual needs of the client, the native materials available, and the nature of the site. This thinking led him to develop what he called organic architecture and it was this idea that was showed in many of his works.
Wright's ideas ranged throughout America and even the world through other architects. One of the ways that this happened was at Taliesin in Arizona which was Wrights school. Many young architects studied with Wright at Taliesin and were introduced to ideas such as organic architecture where his work has also been spread throughout the world. Other architects have studies Wright's work and ideas and incorporated them into their own work. This is how Frank's work and ideas have influenced America and been spread throughout the world.
Wright developed a language of architecture that did not look to Europe but was distinctive to the United States. As well as producing buildings which were fundamental in appearance, Wright had an unusual ability to put them together with the landscape that is stemming from his deep love and knowledge of nature. It was these gifts that marked him out from modern pioneers of modern architecture, such as Le Corbusier and Mies van der Rohe, and make his buildings seem in tune with our environmentally mindful period.

ISBN: 9780170233279

Modern Architecture - Frank Lloyd Wright

Nathanael Hailemariam

Fallingwater

The falling water was an exemplifier of Wright's ideas and it remains as one of his most astonishing and original work. The Fallingwater was built for the Kaufmann family for a weekend retreat. They wanted it on a site overseeing the Bear Run stream. The mix of natural and man-made forms is what the house is built of and Wright observed high rock ledge beside the stream therefore he used it to cantilever the house so that it stood above the stream and at the same time could become almost be part of it. The cantilevers each have a balcony which overlooks the stream, bringing in habitants to connect with water and nature.

Robie House.

Wright built this unusual residential home for wealthy bicycle and motorcycle manufacturer Frederick C. Robie in the first decade of the twentieth century The Robie House kindly moves away from the street in a series of horizontal overlapping planes, this exterior three-dimensional overlap is made up by an interior that is open to the outside, yet sheltered. This slight balance of private and exposed space which was wished by Robie himself, to shield his family from outsiders, but a constant theme in Wright's domestic architecture is amazing, as is the suitability of the house for modern living. The huge drop in ceiling heights, the guides at the Robie House emphasize that the low ceilings were not only modelled for forms of Wright's height or less (about 5' 7"), but that they expressed his commitment to democracy.

Wright's designs took his worry down to the smallest details with organic architecture. From his largest marketable works to the rather modest Usonian houses which were houses designed on a gridded concrete slab that joint the house's radiant heating system, and the house featured new approach to construction, which included sandwich walls that consisted of layers of wood siding, plywood cores and building paper, the important change from the usual framed walls. Usonian house normally contained flat roofs and were mostly constructed without basements, effecting the removal of roof space and basements from the house. Wright considered almost every aspect of both the external design and the internal fixtures, including furniture, carpets, windows, doors, tables and chairs, light fittings and decorative elements. He was one of the first architects to design and supply custom-made, purpose-built furniture and fittings that functioned as joined parts of the whole design, and he often returned to earlier contracts to redesign internal fittings. Some of the built-in furniture remains, while other renovations have built-in replacement pieces created using his plans. His Prairie houses use themed, matched the design elements that are repeated in windows, carpets and other fittings. He made original use of new building materials such as precast concrete blocks and glass bricks for his leadlight windows; instead of the traditional lead also he famously used Pyrex glass plumbing as a major element in the Johnson Wax Headquarters. Wright became the first architect to design and install custom-made electric light fittings, including some of the very first electric floor lamps, and his very early use of the then-novel spherical glass lampshade.

Wrights organic style had fully developed during the late 1920s and 1930s with the design of Fallingwater, Graycliff and Taliesin West. For the design of the Fallingwater, he designed it according to his desire which was to place the lodgers close to the natural surroundings which include the stream and waterfall running under the building. Balconies and walkways are what the construction of the building consist of and limestone is used for all verticals and concrete used for the horizontal. Then there was the masterpiece he was most recognized for and was working on for 16 years, which was the Solomon R.Guggenheim Museum. The interior of the building is very much similar to a seashell and the it has a unique central geometry which was intended to allow visitors to easily experience Guggenheim's collection of non-objective symmetrical paintings and to see them you would have to take an elevator to the top level and then from there you can view artworks by walking down the slowly downward central curved ramp which features a floor surrounded with circular shapes and triangular light fittings to match the geometric nature of the building.

Concepts

Nathanael Hailemariam

Hovering Planes
Influence of nature
Patterens and rythms found in nature
The branch of a tree (natural Cantilever)
Cascading Cantilevers (falling water)
"Snail" like Guggenheim

On the first floor of this building beneath, there will be the main entrance and Lounge. On the 2nd floor where the main cafe is placed is surrounded by glass and has one side that leads to a deck outside.

This building is made up of 3 Rectangular blocks with a glass area on the deck floor. This style of construction is based on cantilevers Frank Lloyd Wright based his buildings on. Getting the idea of cantilevers from nature, in this case, I have copied the features of the tree branches that are cantilevers. wood, Brick walls and glass will be used as materials.

A simple styled building with 3 decks cantilevering off the main building (main cafe). The cafe offers an outside dinning place with 3 decks. Then an indoor where the main cafe is placed

This building is formed in the eyes of a branch of a tree which is a natural Cantilever why Wright applied to his work alot. The materials used for this building consist of Timber, wood and some metal. The colour of the building copies the tree.

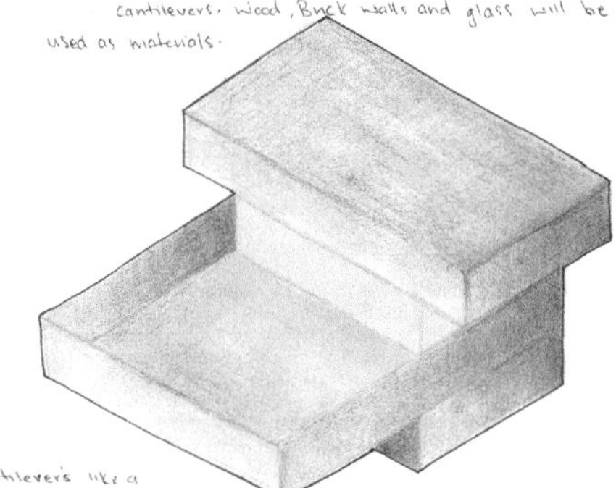

The use of looking at natural Cantilever's like a tree branch, Wright uses in his works for this building to the left, I have used this influence to apply to my design.

ISBN: 9780170233279

Concepts

Nathanael Hailemariam

Hovering Planes
Influence of nature
Patterens and rythms found in nature
The branch of a tree (natural Cantilever)
Cascading Cantilevers (falling water)
"Snail" like Guggenheim

The cafe on this building is placed on the first two floors (rectangular blocks) and the overlapping blocks creates space outside for a deck.

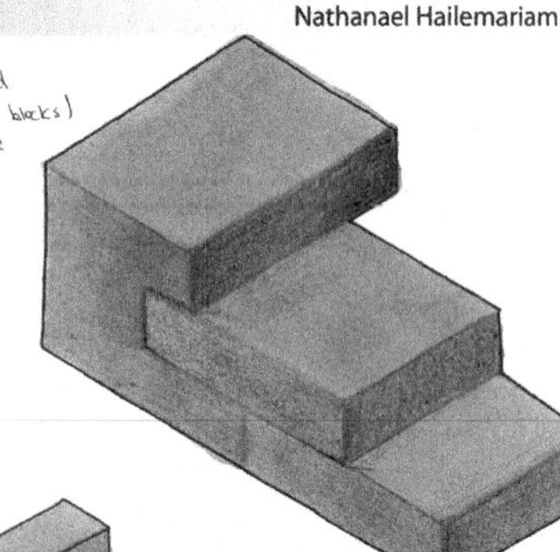

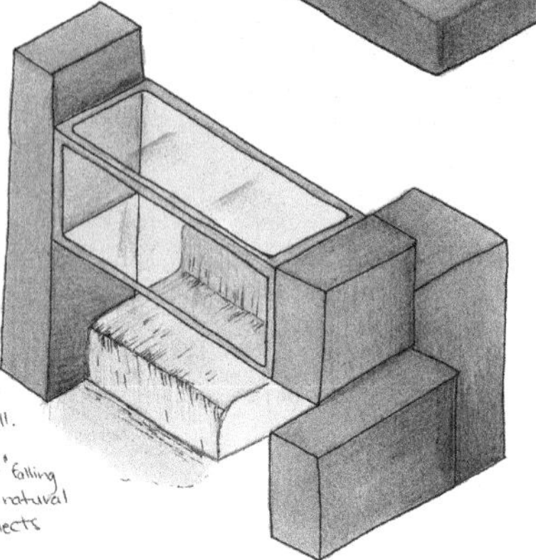

The Function for this building, there will be the waterfall flowing naturally in it's space and above it will be the main cafe surrounded by glass including the floor to provide a view above the waterfall when dining. This will be supported by cascading cantilevers on the side. The support on the right will be the main entrance and toilets.

The shape of the building will consist of solid rectangular blocks. The materials that would be used are obviously glass and wood which will give it that natural taste of nature. The form of the building is simply solid wood blocks supporting the glass building in the middle

Influence of nature is shown in this building as wright had in his work. The idea of combining the waterfall with the building is the work shown of the influence of nature.

Top right corner, the style of the building is structured to the levels of rocks formed in the waterfall. So the overlapping of rectangular blocks creates this style and formation. The cascading cantilevers of 'falling water' by wright influences on this design and the natural cantilevers created by the rocks in the waterfall connects to wrights ideas.

Mervin Tibay — St Peter's College

The following work by Mervin Tibay demonstrates the study of the Bauhaus. He has analysed the key design principles of aesthetics and function, which include shape, form, rhythm, balance, proportion, colour and contrast and durability. Mervin has also considered the societal factors such as cultural, historical, societal and technological. Mervin has used the shape and form of the Bauhaus to inform his own design ideas for lighting.

BAUHAUS

HISTORY AND INFLUENCE

In 1919 it was hard and challenging to try to start or found anything in Germany, especially an art school which ran along lines of radical, unthinkable and even revolutionary ideals. This tension and dificulty was caused by the presenting defeat, and the abdication of Kaiser, the nation was in terrible chaos, with hunger, unemployment and inflation. In such states of the country It was described chaotic, Bauhaus would be the inception. But this inception wasn't going to be easy because of the lack of unexceptable faculty, facilities, materials and even funding, to add to the weight of internal feuding followed by the relentless external political attacts only delayed its further progress and deminished its concepts of idealism through the use of realism. Even aganist these political and lacking of essentials reqired for growth Bauhaus still left an incrediable mark on the 20th century design, only even if it had a short history 1919-33. Art schools still owe some thanks to Bauhaus for the subjects on Art and crafts being taught in a certain way, the behavior of three dimensial design, materials, colour theory, and certain qualities that buildungs have on their occupants.Still

Bauhaus left a major mark and has contributed towards patterns also setting the standard for present day industrial design, it is also credited to have given birth to modern architecture Of the many societies which craft was a huge influence in Germany towards the end of the century, Munich had the most significant mark on the future design. In the approach towards 1900, it was now the time where design moved away from the idea of handcraftsmanship and onwards to serial production. Germany had come to realise the subject before the other influenced countries focusing on design (England and parts of Europe), the subject being on wheather the Morris tradition of individual craftsman being continued or should it face its superior to a form of industrialized manafacture on standard components. The argument was simply on fundamentalism concerning designers every where at the time, how to reconcile and accept changes of industrial design. In 1896 a special post at London embassy was created to serve as a place for Herman Muthesius to study and later report on British town planning and housing policies Seven years later Muthesins resarch led him to a publish a book which had influencial subjects. This book highlighted the details of his discoveries, and also suggested from his opinion on how English industrialisation shound best fit in Germany. Muthesius also had a blief of which mechanized production were irreconcilable. "What we expect from machine products is smooth form reduced to its essential function". It was an intension that students learned by making thier designs in workshop lessons not just designing on paper. On April 1919 Bauhaus became an establishment which took the form of a shcool, it had all the unique design ideals and concepts of previous influences of mechanized production. On that date all influences of a combination of artist-craftsmanship and mechanized serial production designs was now official and stood in the halls of Bauhuas school.

Mervin Tibay

ISBN: 9780170233279

BAUHAUS

MATERIALS

Due to the time era the materials were inflenced by the idea of cheap and machine assembled products. The major idea which influenced the materials to be used at the time, was to have a product that was what we expect from machines. Muthesius Quoted "what we epect from machine products is a smooth form ruduced to its essential function. This influenced the use of metal alloy, glass, wood, brass, procelain, silver, ivory, copper chrone because of these materals being easy to create and sculpt through machines also being easy to be placed into serial production. Another reason way these materials were used at the time was because they fulfilled the desired style of Bauhaus designers, in that time everything was designed and created by artists and craftsmen. Due to this most products were individually different causing different styles to be born. Bauhaus wanted differently they wanted to create a style which was recognisable and was uniformed, somthing radical and modern. Through the use of these materials they were able to create the desired style of Bauhas, a modern revolutionary style which consisted of attractive materials and well designed products and modern even revolutionary geometric shapes for simplicity.

MANUFACTURING

Because of the use of geometric shapes and various materials and also accompanying simplicity with in all products. Most of Bauhaus products were manufactured by machines, this production led away form handcraftsmanship and in to serial production. Because of materials such as metals it was seen more sufficient and cheap to use machines to produce and sculpt the design, also since there were raw elements of materials such as glass it would have fulfilled the desired style of Bauhaus. The manufacture was with in the style of function and essentials.

ASTHETICS

Bauhaus having its own concepts and style also using certain materials usually had a shiny and smooth apperance. The public also came to admire and recognise Bauhaus to use generally geometric shapes stripped down to its essentials and the sole function, accompanied with its modern dull materials made it quite subtle. Use of Metals meant that most products would be geometric shapes and reflect shiny dull colors. Use of glass was used to highlight and give the the room a sense of transparancy. Wood was used to create a contrast between the metal elements and provide a sense of unity between nature and technology. Reasons behind the Bauhaus asthetics are reasonably because of its simplicity of geometric shapes and its design is easy to maintain.

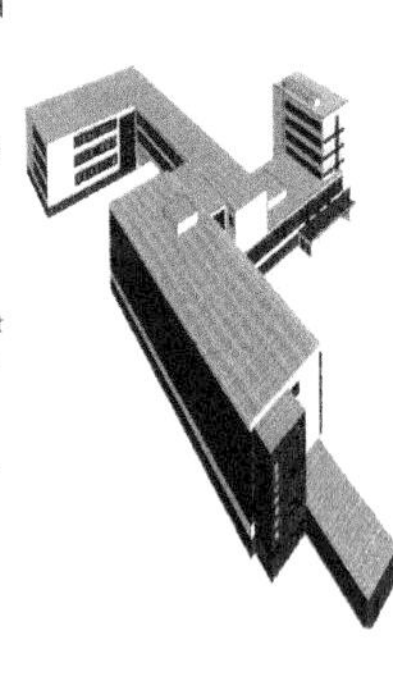

IMPACT

Bauhaus has had a huge influence on arts, crafts and architexture within western Europe, the United States and even Israel. Because of Bauhaus having such a unique design which is founded on essentials and basic fundamentalism design also using modern materials, its become quite popular in the public. Modern architecting owes a great amounts of gratitude. Many architechs have been influenced by Bauhaus to use certain materials and geometric shapes. The use of quite shiny and smooth streamline materials. Especially the use of metals such as silvier, copper and chrone to give the simple and clean feeling. A major base of impact for Bauhaus was the element of unity of art, craft and technology. Metals were considered a positive element, this became quite important for product design. Later it became a preliminary course which became a modern design thanks to Bauhaus. With Bauhaus help it later became a "Basic Design Course"which structured and was a foundation for architects and design schools across the world. Bauhaus is considered to be the founder of modern design.

Mervin Tibay.

BAUHAUS

ARCHITECTURAL EVALUATION

Baubuas architecture mainly focuses on geometric shapes and very attractive materials, another element to add is the concept of simplicity and function. The Bauhaus has mainly focused on simple shapes which combined with sleek and attractive materials. In the begining Bauhaus focused on its aesthetics, but later it fell towards funtionality. Bauhaus later focused on apartments buildings with the thought of function and not aesthetics. Bauhaus achitecture would then again be given the concept of essentils which would consist of buildings having essential items and function. Because of the attention to function and essentials, the modern architecture has been influenced to use Bauhaus style of design. Bauhuas has been considered one of the most impacting styles. A style which I admire for its now perfect balance between functionality, essentials and attractive simplicity of geometric shapes and material aesthetics.

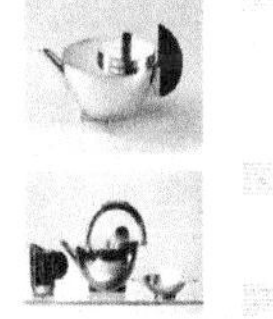

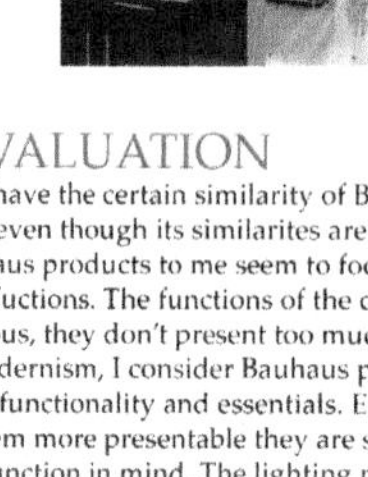

PRODUCT EVALUATION

Bauhaus products also have the certain similarity of Bauhus architecture. I consider even though its similarites are really obvious, in depth Bauhaus products to me seem to focus more towards essentials and fuctions. The functions of the chair and lighting are really obvious, they don't present too much aesthetics, style and modernism, I consider Bauhaus products are highly designed for functionality and essentials. Even though the products seem more presentable they are still deeply designed with function in mind. The lighting product is in its fullest essentils and serve its functions, only its materials and shape make it presentable. The chair is also at its most highest quality of function and essentials, even though there is materials which make it more presentable its main function seem to me to speak louder than its aesthetics

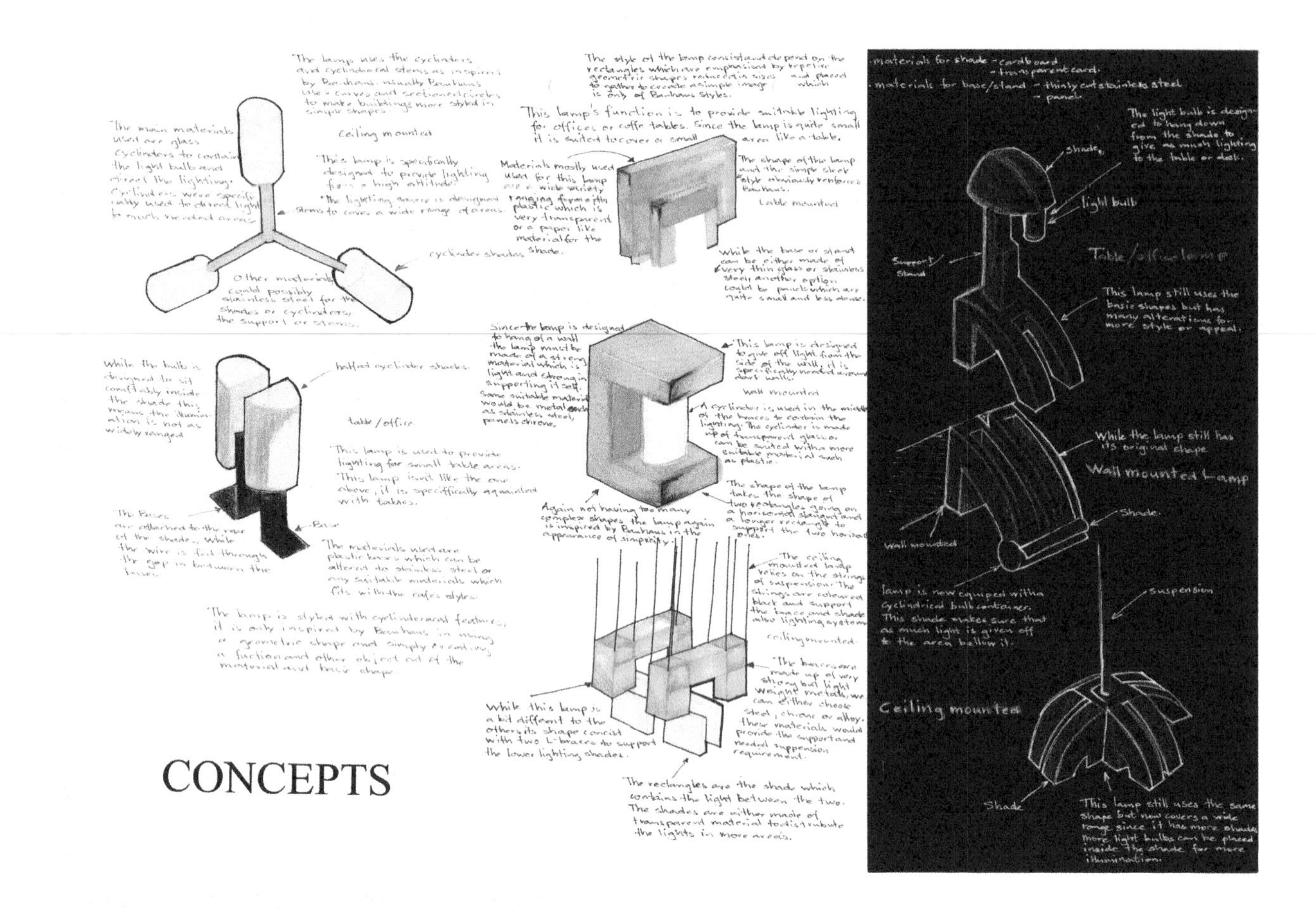

Using sketch models (manipulation of card and other materials) is a great way to generate ideas in response to a brief. Sketch models are a good starting point to produce ideas for the aesthetic qualities of an object.

Manipulation of card

The following photographs demonstrate the manipulation of card by bending, folding and gluing to create different forms. Not only plain card can be used for this technique but also translucent or clear plastic, and corrugated or perforated materials, to add texture to the sketch models.

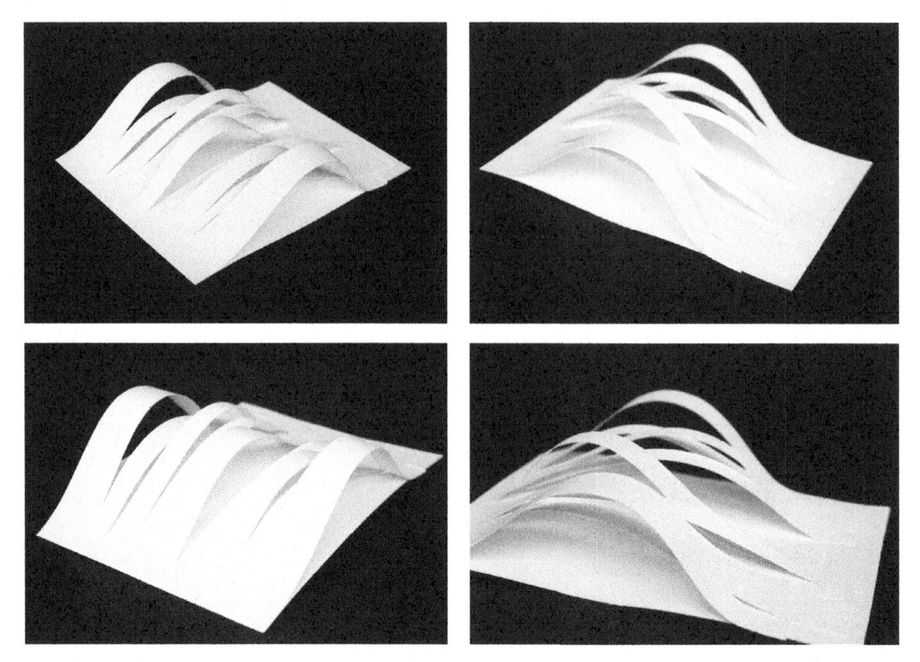

ISBN: 9780170233279

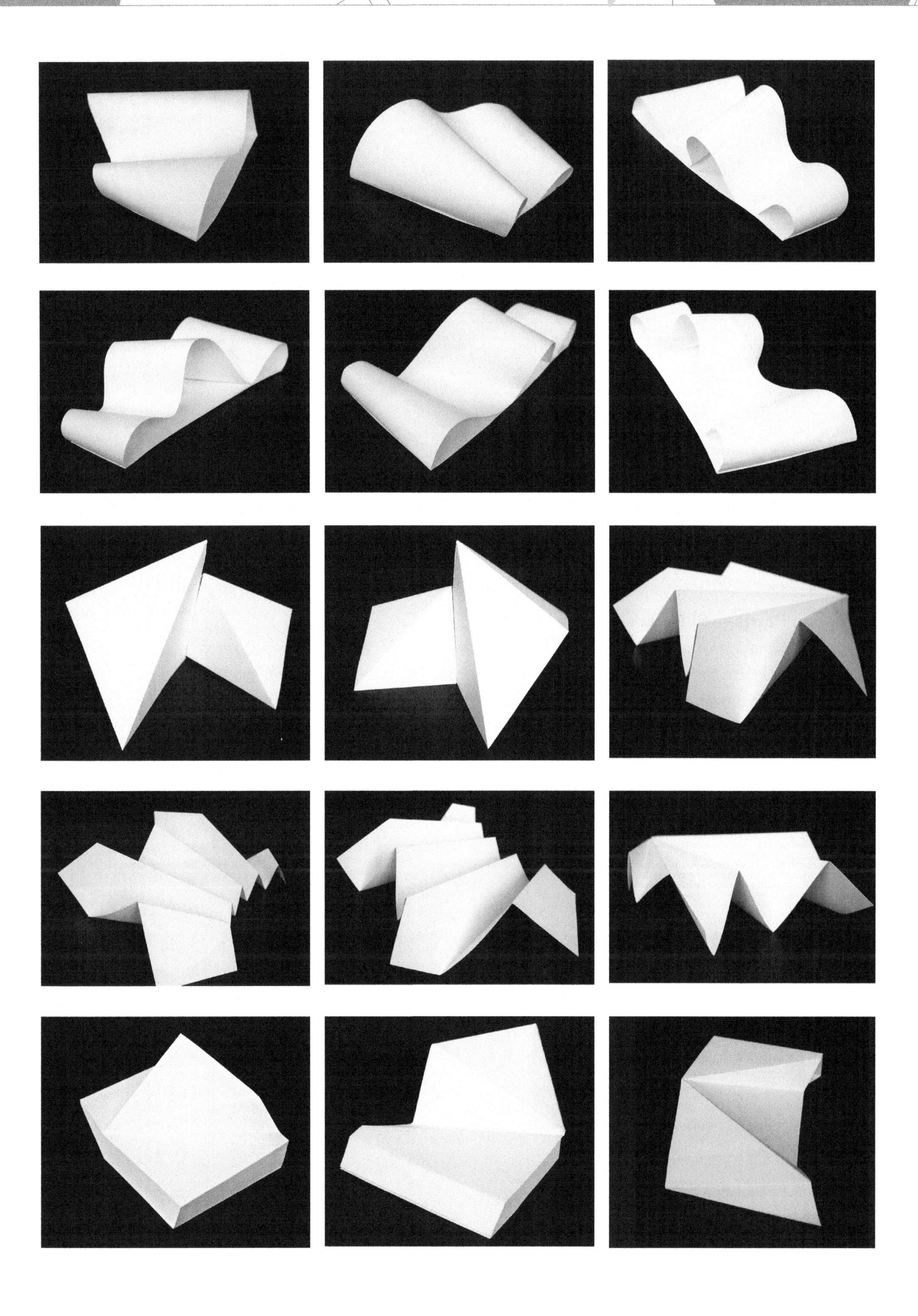

ISBN: 9780170233279

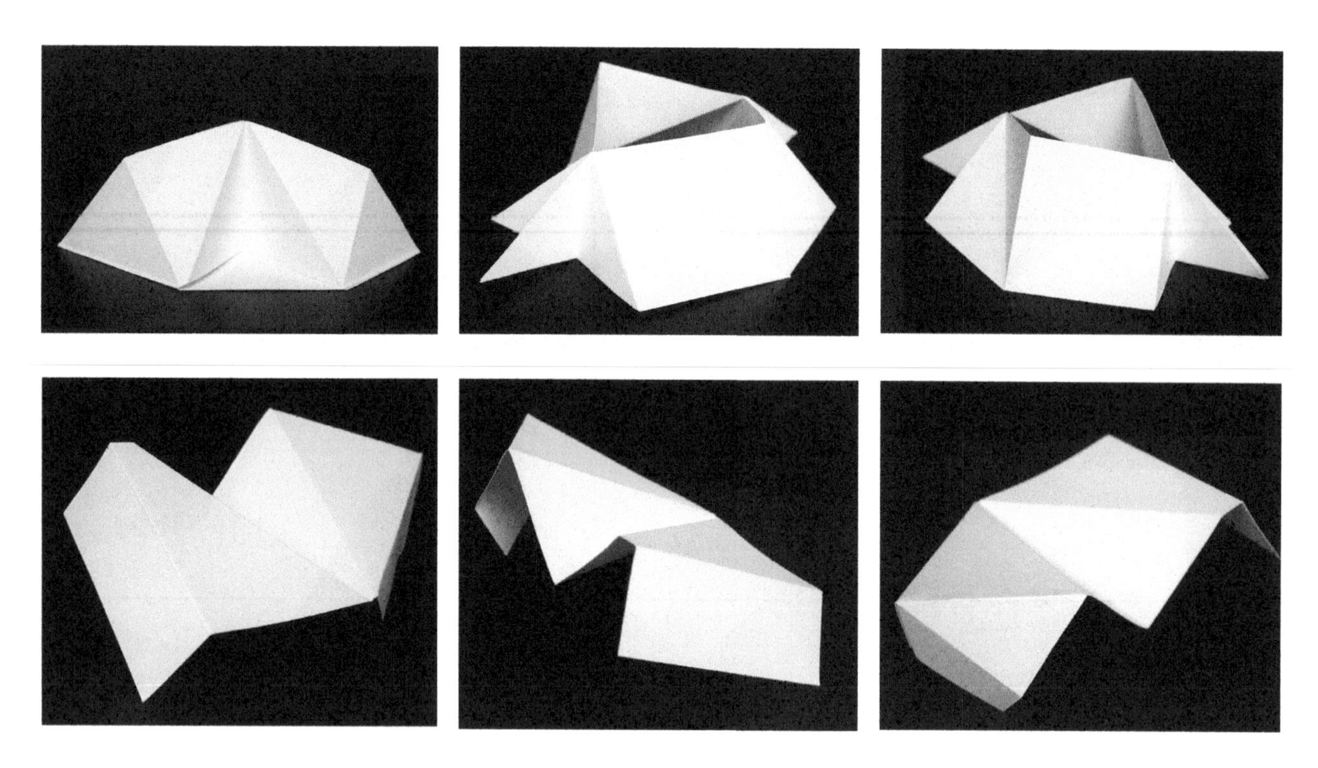

Models used to generate ideas

Simon Parris — St Peter's College

The following work produced by Simon Parris demonstrates the method of using sketch models to produce design ideas. He has used card and a glue gun to construct the models. Simon has then photographed each model from different directions. He has sketched and rendered the design idea in two and three dimensions.

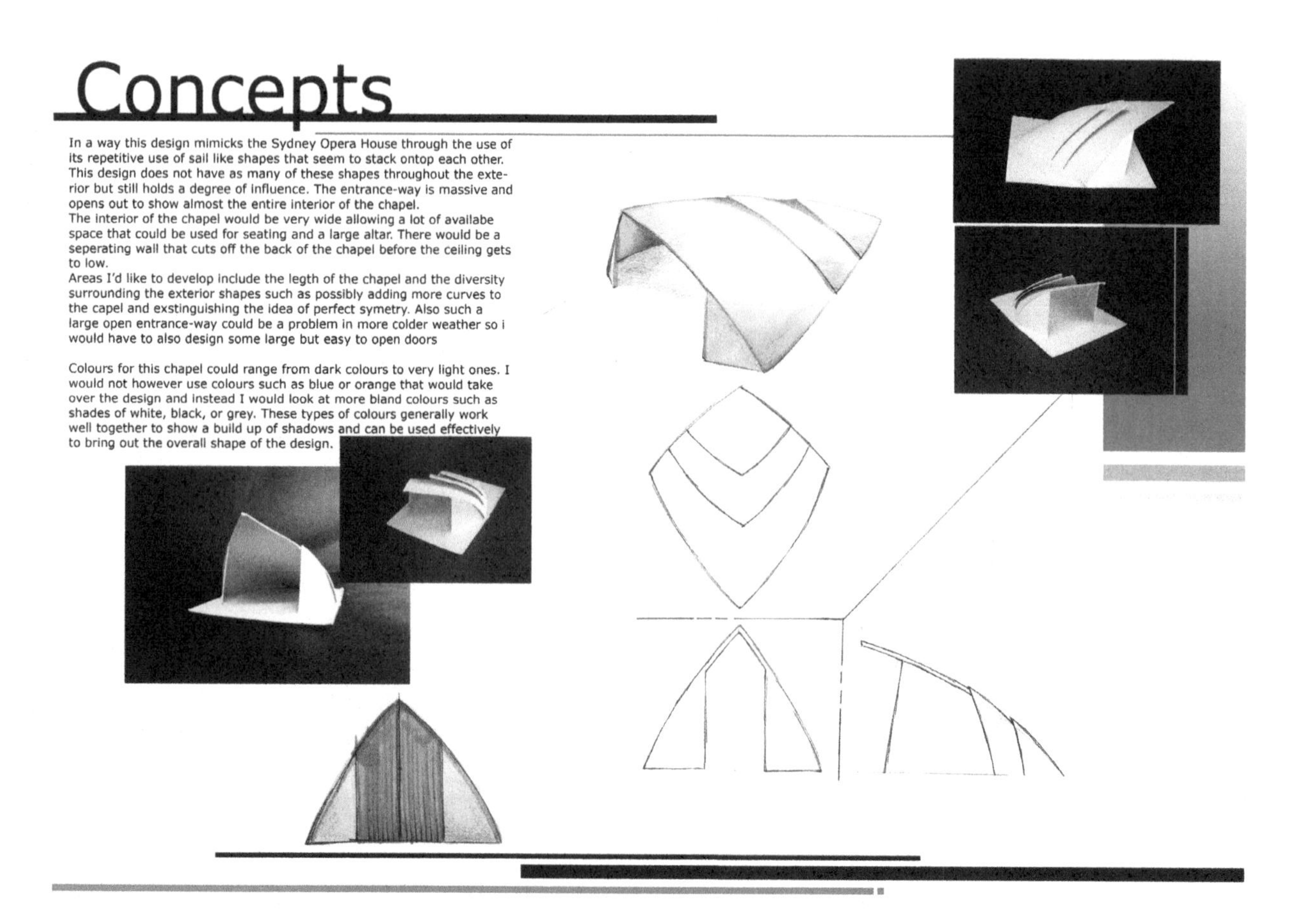

ISBN: 9780170233279

Concepts

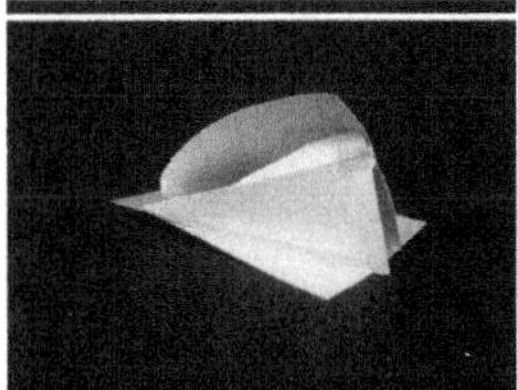

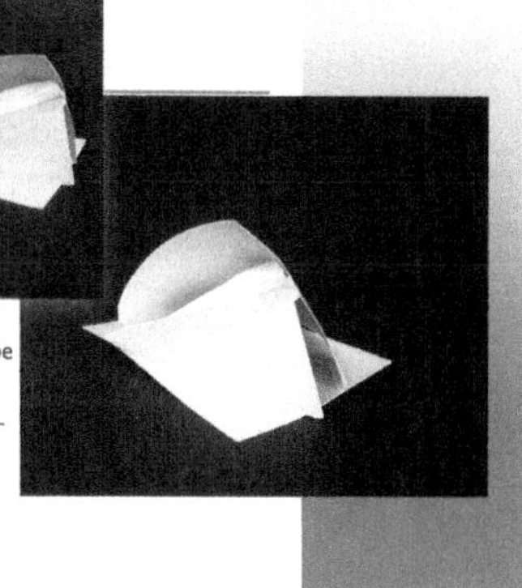

The design of this chapel shows a unique shape using a large curve that rolls over the more pointed side that generally consists of straight lines. The entranceway mimmicks that of oldstyle churches through the use of a pointed tip and in the way in how huge it is and that it extends to almost the maximum height of the chapel. The interior of the chapel mirrors the exterior with a straight wall on the left and a slanted ceiling that meets the east wall that curves down to the base of the chapel. The interior of the chapel ends however, when the ceiling of the chapel begings to get to low.

The colours I would incorporate with this design should not be too bright and would most likely be shades of grey and white.

I would like to further develop the body of the chapel in terms of adding more curves and adjusting the slants and lines to create a more appealing shape.
I would like to show a variety of symbolism including a large crucifix at the entrance-way, some sort of large 'Rock' in the courtyard illistating the underlineing theme of St Peter, statues, and a posibility of stained glass windows.

Concepts

This church design follows modern aesthetic techniques of most chapels we see today. The entraceway is very large, with a pointed tip and a triangular shape. The churches shape mirrors this idea through to the back of the church where it is also pointed triangular shape. The difference however that seperates this church from most is the way that both of these triangular shapes move up at opposite angles to each other and meet around the middle on the roofs lowest point. This has the effect that when someone walks through the entrance the building gradually gets skinnier and darker but then opens up to a grand main area with light pouring in from the huge windowed area which takes up the entire back wall.
For this design I would like to develop the idea of adding more sybolism, for example maybe a stained glass crusifix that could be incorporated into the huge glass area at the end of the chapel. Also I would like to possibly add further areas and shapes to the chapel to show it as a more complex and distinguished building.
Colours I would normally use for a chapel like this would consist of mostly whites and greys as to basically follow the guidelines in terms of more traditional chapels such as this one.

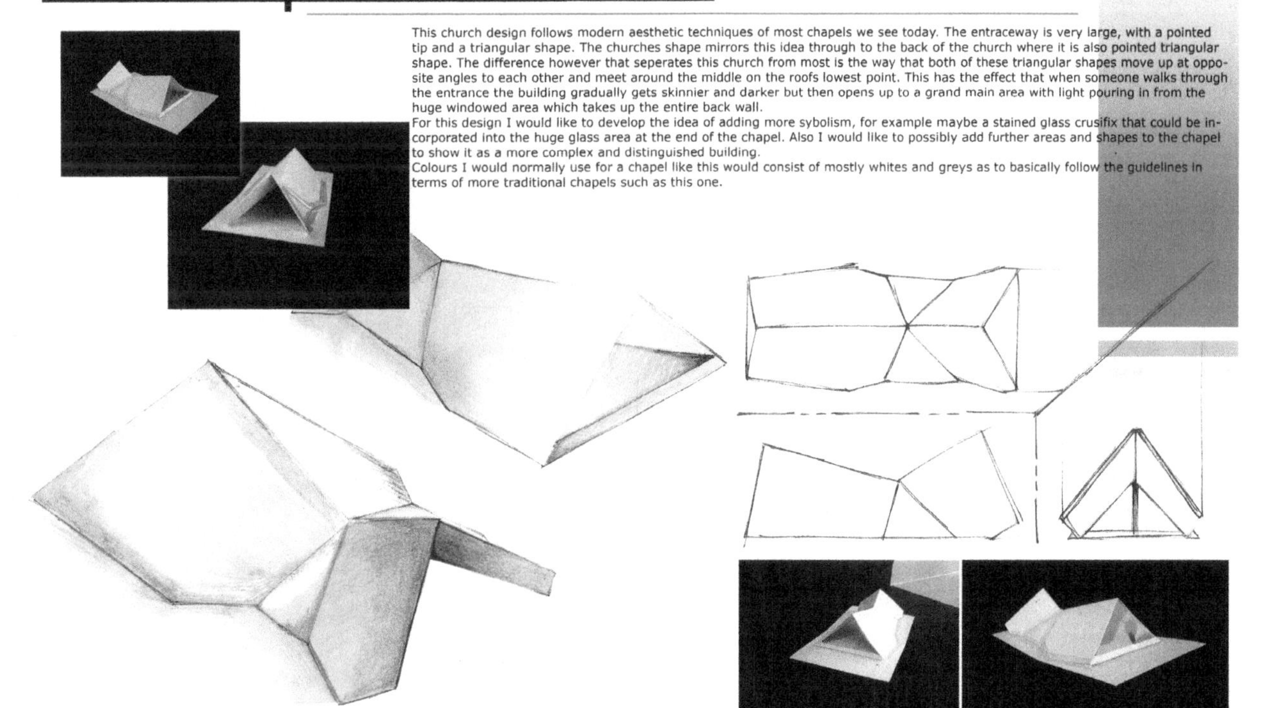

ISBN: 9780170233279

Paihere Tims — St Peter's College

The following work produced by Paihere Tims demonstrates the method of using sketch models to produce design ideas. He has used designers as a starting point to influence the structure of the card model. Paihere has photographed each model from different directions. He has sketched and rendered the design ideas in two and three dimensions. This gives Paihere a good starting point to continue the development of an idea or combination of ideas.

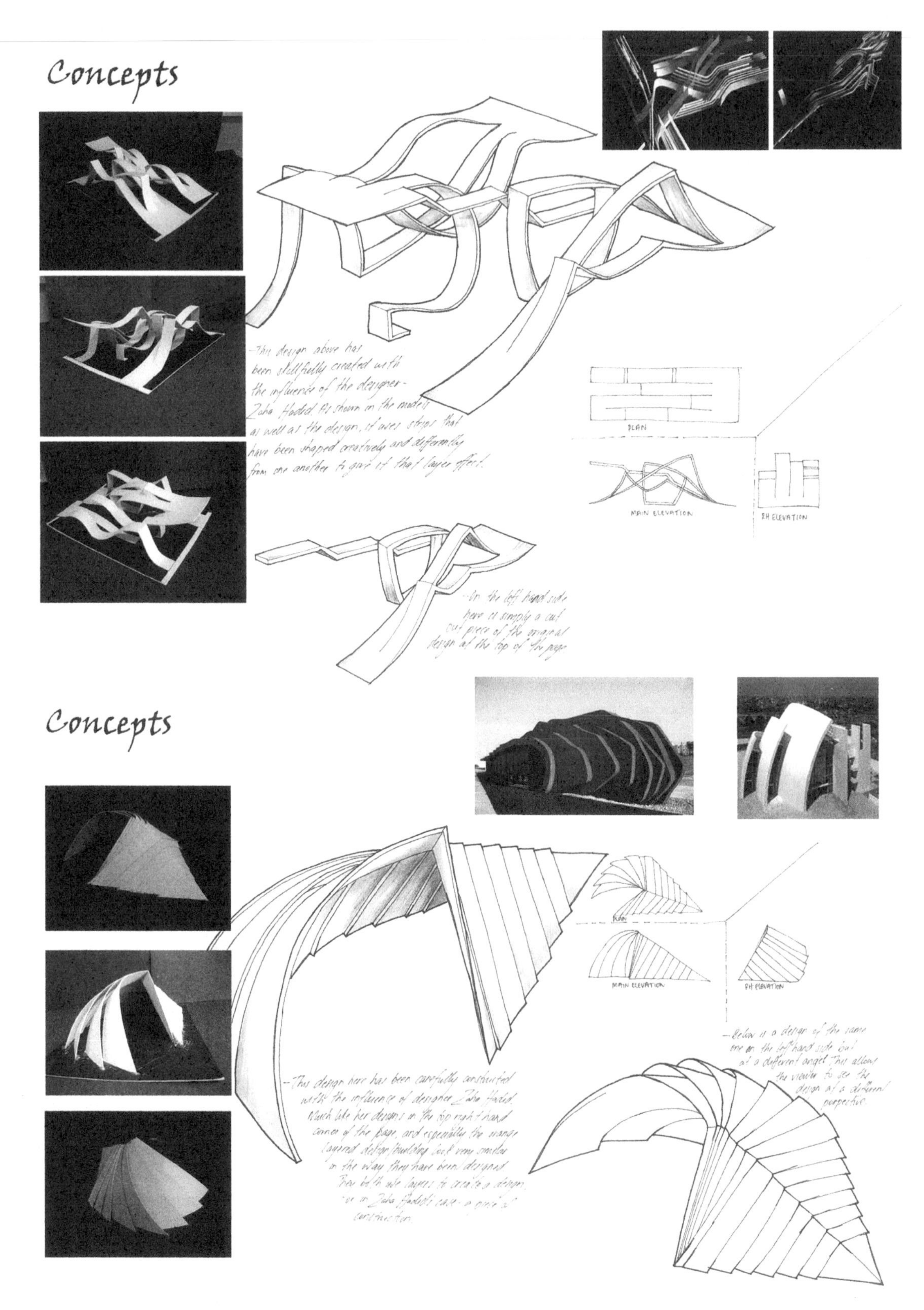

ISBN: 9780170233279

Olivia Wong — Epsom Girls Grammar School

The following work produced by Olivia Wong demonstrates the method of using sketch models to produce design ideas. She has used different materials (card and wire mesh) to explore the effect of different textures. Olivia has photographed each model and has sketched and rendered further exploration of these ideas.

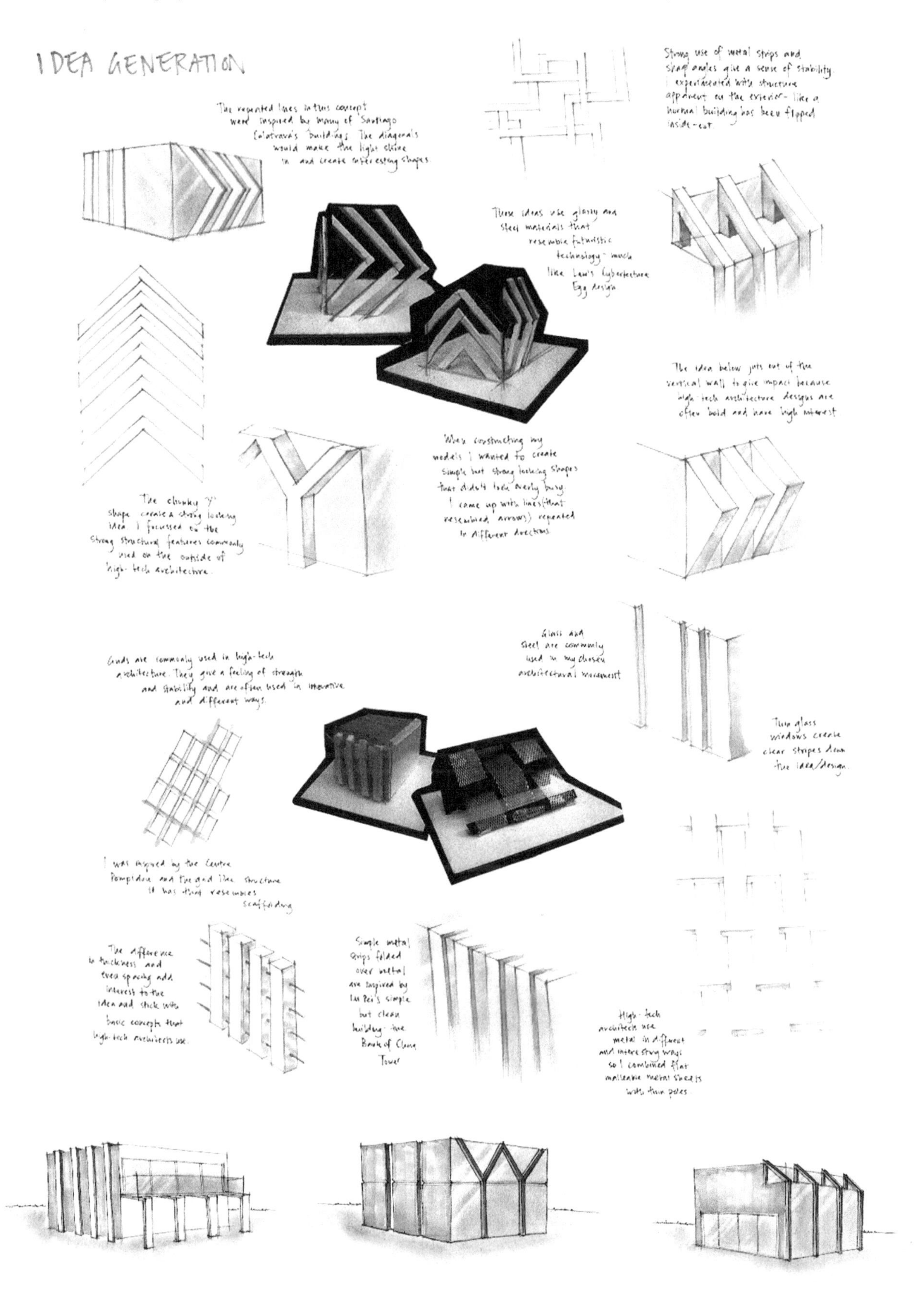

Sketch models for a brief

For the following brief, sketch models have been used to generate several design ideas. The advantage of using this technique is that the same components of the model can be reused to create several variations. This can be a faster way to explore the functional and aesthetic qualities of the design than sketching. It is also, then easier, to sketch these ideas from the model.

Shipping container brief

Situation/Problem

You are to design a 1 to 2 bedroom unit from 20 foot (6.1 m) standard shipping containers. The unit may be designed for any environment (city, hillside, beach or farmland). The design of the unit can be made up of a maximum of four containers stacked. You will need to consider efficiency of space. The unit will be self-contained but will only need the essentials such as a basic kitchen, bathroom, living and sleeping areas.

ISBN: 9780170233279

The following demonstrates the sketching of one of the generated ideas from different directions using isometric, oblique and perspective (refer to the *Chapter 2 – Freehand Sketching* for guidance using these techniques).

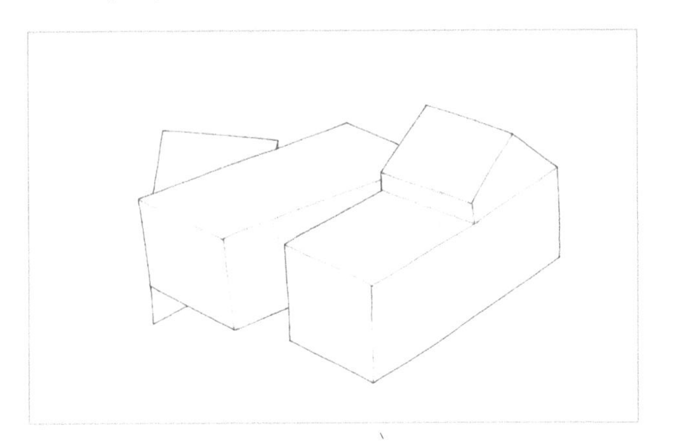

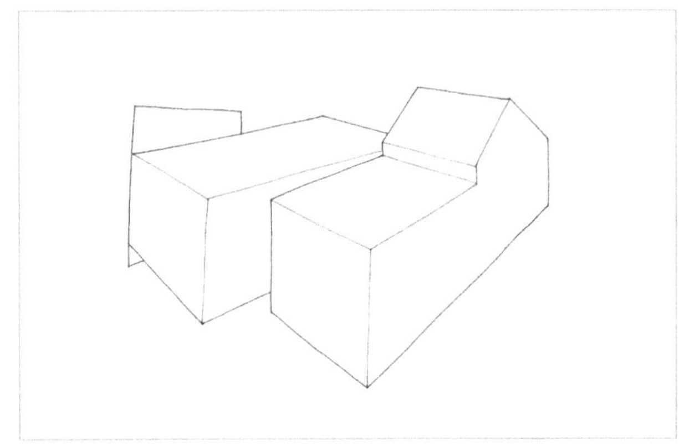

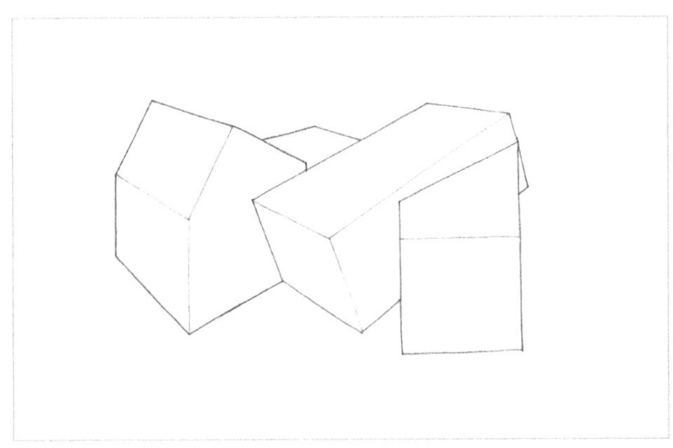

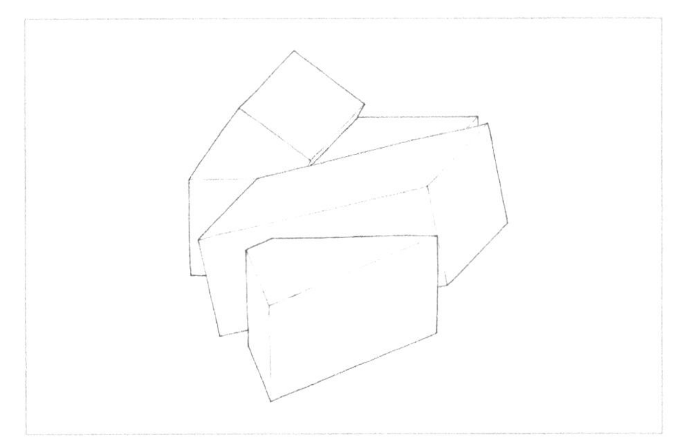

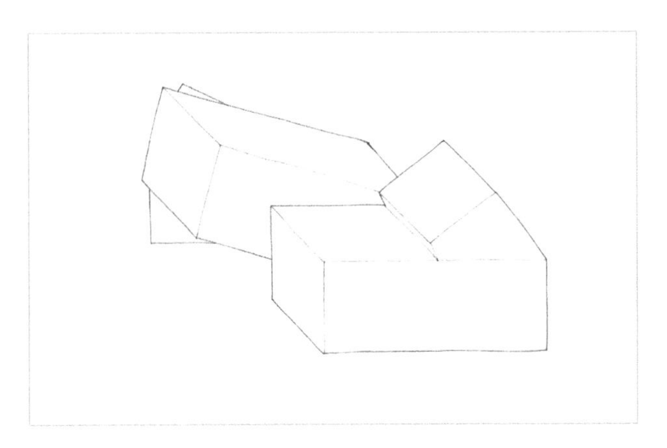

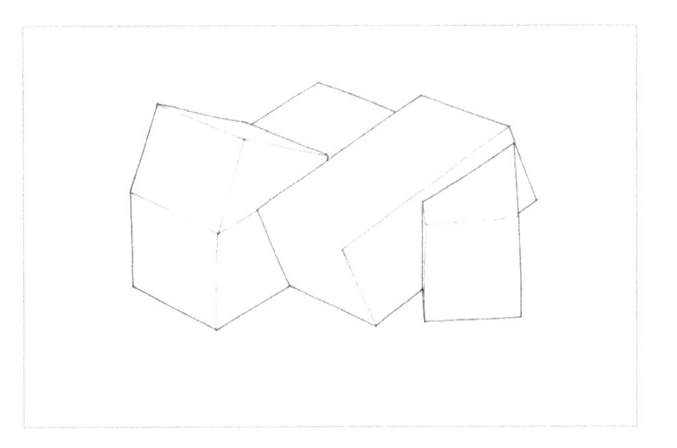

ISBN: 9780170233279

Mockups used to inform development

Mockups or sketch models may not necessarily be used at the initial idea generation stage. They may be used to help explore the functional and aesthetic qualities of the design further along in the development.

Nicholas Beel — St Peter's College

The following work produced by Nicholas Beel demonstrates his use of mockups, sketching and rendering to explore his design ideas for spatial design at the developmental stage. He has combined all three techniques using sketching and rendering on top of the photograph of the model to reflect on his idea. This is a good example of overlaying techniques.

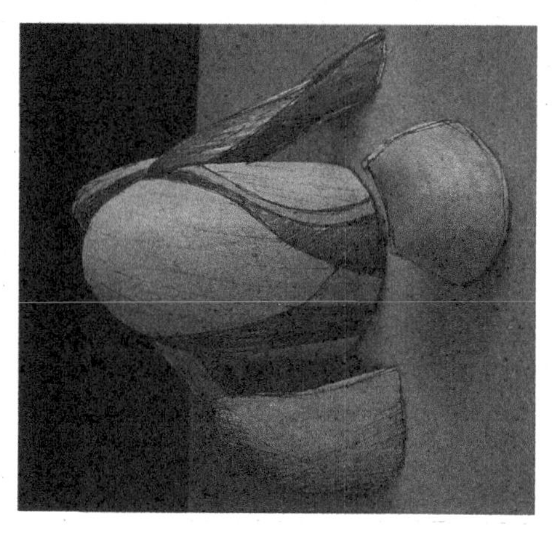

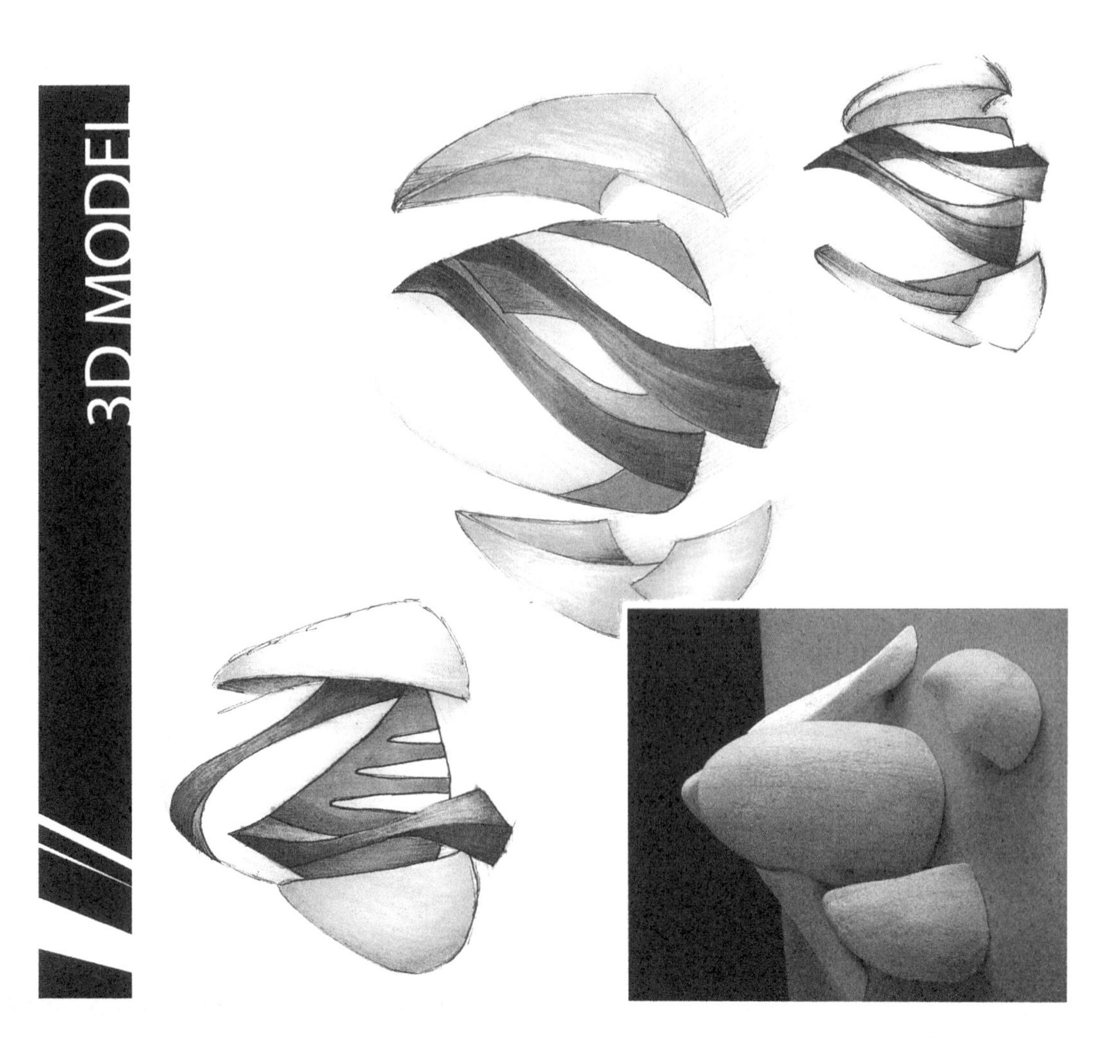

ISBN: 9780170233279

Zack Kite — St Peter's College

The following work produced by Zack Kite demonstrates his use of the influence of the architect Antoni Gaudi to inform ideas, sketches, rendering, a mockup and a final model. The work is a compiled from a larger body of work. Zack has used Antoni Gaudi as a starting point to generate ideas, sketching and rendering variations of form. As the shape is organic a mockup was made out of card and masking tape to help define the form. Sketches and renderings have been presented of this mockup. A final model has also been constructed to communicate the functional qualities of the design such as the texture of the exterior.

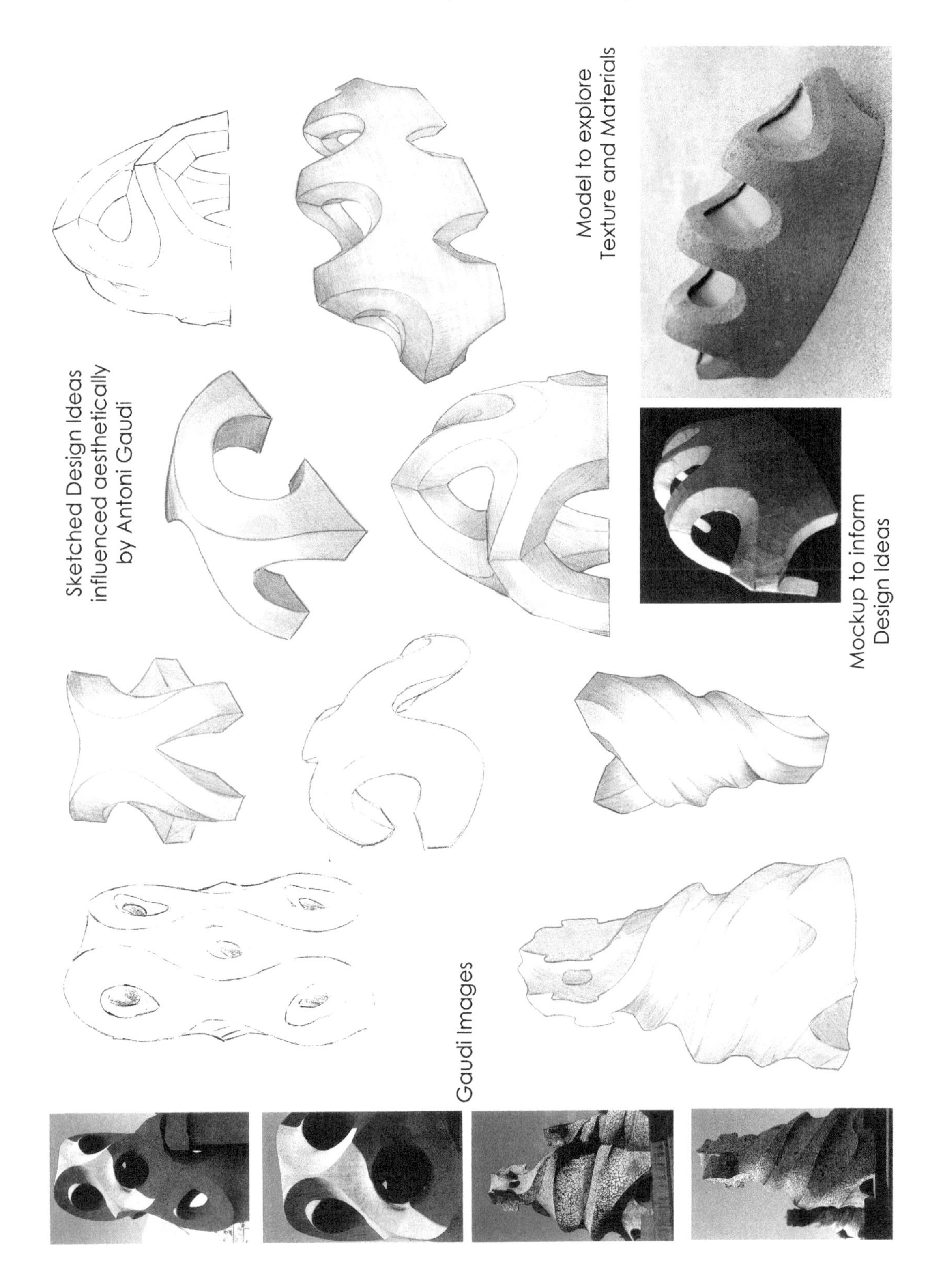

ISBN: 9780170233279

Christian Burgos — St Peter's College

The following work produced by Christian Burgos demonstrates the use of a mockup at the development stage to explore the functional qualities of the design. He has used the mockup to work out the components of thc work station and how the parts fit together. This exploration has informed his final idea. The final model communicates the materials and shape of the design clearly.

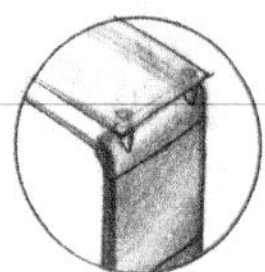

A major modification in the design is the right side of the desk, where the C-shaped leg has been replaced with a large rectangular block. It will be coloured glossy black and will hold the glass desktop by the two round cones as it normally was.

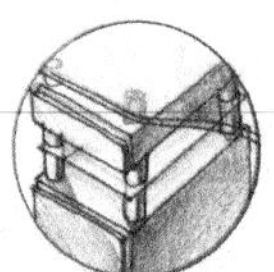

The most obvious change is the design of the drawers. The size of the drawers is unnecessarily big, and should be smaller. Instead of the two big drawers, there will be one drawer at the bottom and a shelf in the middle and a small stationary holder at the top of the shelf.

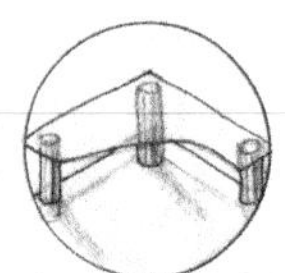

Another notable change is the top shelves, which are changed to one small one instead of two big ones. The two big shelves were not necessary and may not be stable since only a thin bar of steel holds it up. A small shelf is good enough and is much more reliable since it has three legs to hold it up.

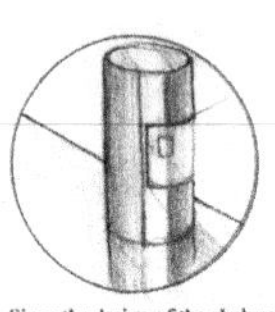

Since the design of the shelves has been changed, the light was needed to be placed somewhere else. The pole nearest to the drawing area is the best at possible place to put the light since it will illuminate the drawing well at close range.

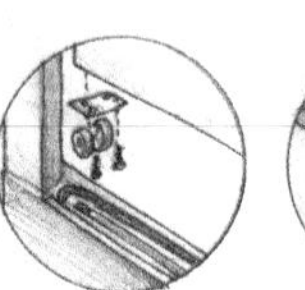

The main improvement in the design in a pull out drawer where the user designer can store the pages of design work neatly and easily. The pull out drawer opens like a computer table drawer for the keyboard and mouse. It has rollers stoppers and rails like the keyboard drawer has.

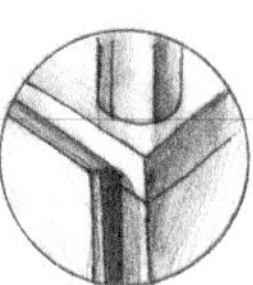

Another change in the design is the absence of drawer handles. It would look more stylish if the drawer handles were seamless. The edge of the drawer has an indent that runs around the whole door that lets you hold it and pull it out.

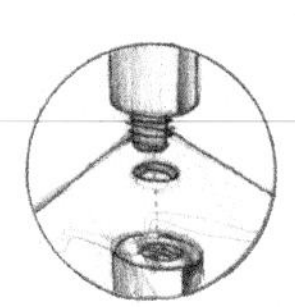

With the top shelf design changed, the poles stand on top of the glass to hold op the top shelf. It now holds the glass between the poles since the top pole attaches to the bottom pole by screwing them together. There is a hole in the glass so the poles can screw together.

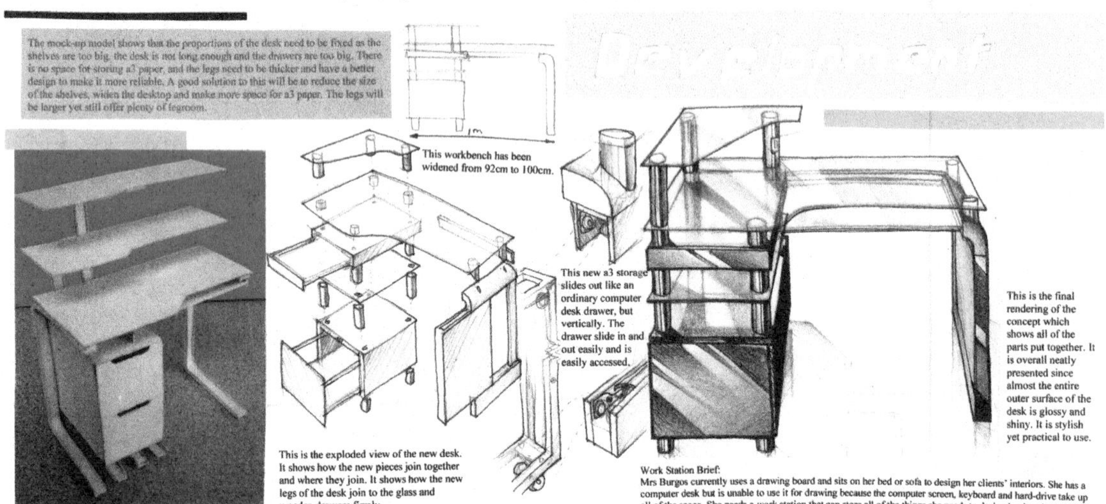

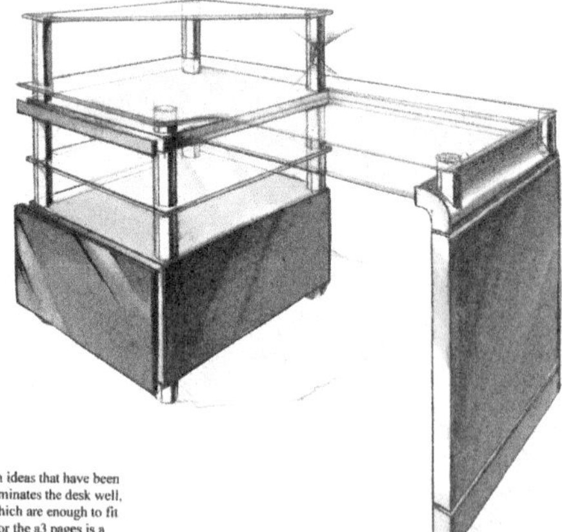

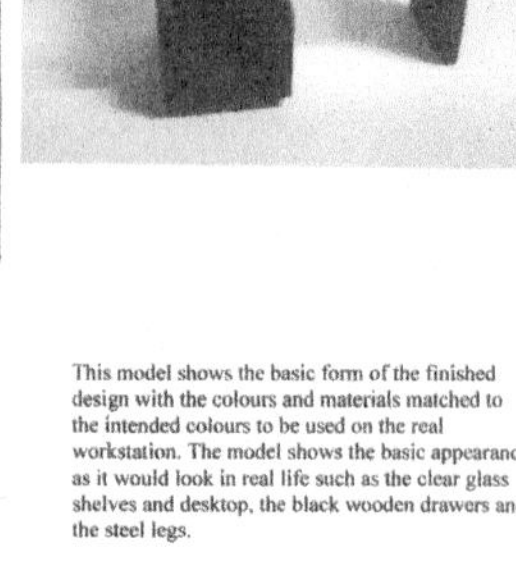

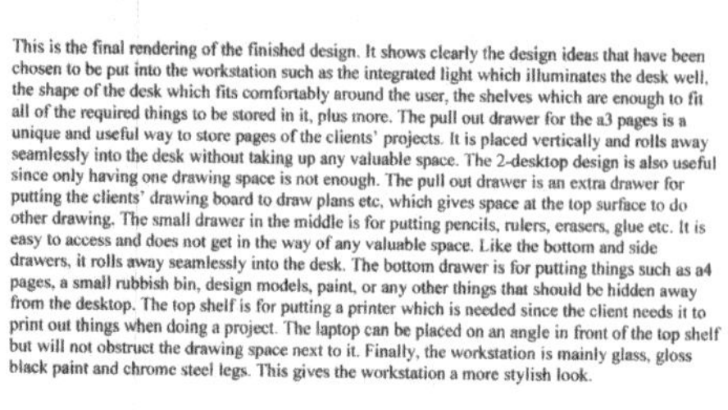

This is the final rendering of the finished design. It shows clearly the design ideas that have been chosen to be put into the workstation such as the integrated light which illuminates the desk well, the shape of the desk which fits comfortably around the user, the shelves which are enough to fit all of the required things to be stored in it, plus more. The pull out drawer for the a3 pages is a unique and useful way to store pages of the clients' projects. It is placed vertically and rolls away seamlessly into the desk without taking up any valuable space. The 2-desktop design is also useful since only having one drawing space is not enough. The pull out drawer is an extra drawer for putting the clients' drawing board to draw plans etc, which gives space at the top surface to do other drawing. The small drawer in the middle is for putting pencils, rulers, erasers, glue etc. It is easy to access and does not get in the way of any valuable space. Like the bottom and side drawers, it rolls away seamlessly into the desk. The bottom drawer is for putting things such as a4 pages, a small rubbish bin, design models, paint, or any other things that should be hidden away from the desktop. The top shelf is for putting a printer which is needed since the client needs it to print out things when doing a project. The laptop can be placed on an angle in front of the top shelf but will not obstruct the drawing space next to it. Finally, the workstation is mainly glass, gloss black paint and chrome steel legs. This gives the workstation a more stylish look.

This model shows the basic form of the finished design with the colours and materials matched to the intended colours to be used on the real workstation. The model shows the basic appearance as it would look in real life such as the clear glass shelves and desktop, the black wooden drawers and the steel legs.

Final

ISBN: 9780170233279

Students may also use other influences to inform and generate design ideas such as forms from nature. The following student work demonstrates this approach for idea generation.

Nicholas Beel — St Peter's College

The following work produced by Nicholas Beel demonstrates the use of plant forms to influence his design ideas. He has used the folds and overlapping of the flower form to inspire his initial ideas.

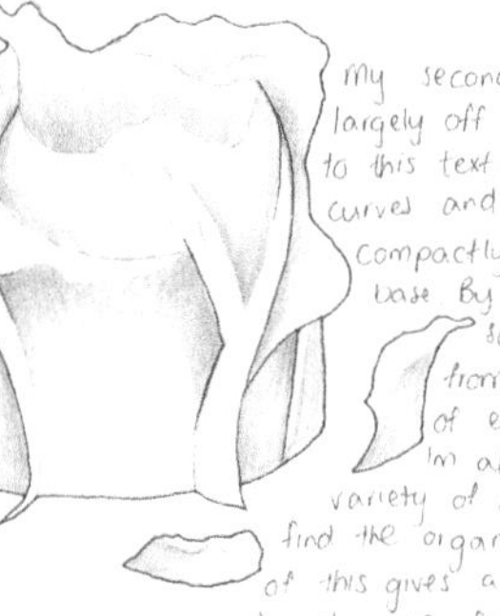

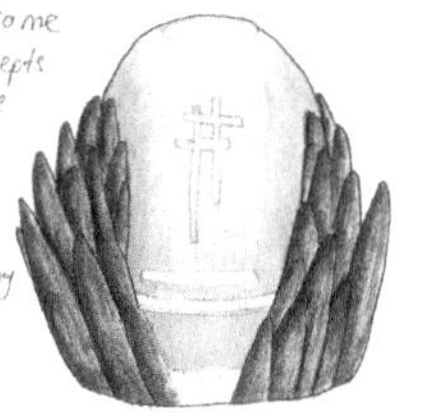

ISBN: 9780170233279

DEVELOPMENT

This development pushes on from the organic concepts. Ive taken the 'overlaping' theme and used it to create this development. Showing a circular structure with the petal like features portrayed over the chapel.

I took the concept of the 'Flower bud' and relooked at how I can use the centre theme and idea to make quite a simple and effective school chapel. Ive used the curving petals surrounding the main dome and reinforced that 'centred' theme again with a wooden ring around the main shape.

This development bridges into exploring the dome structure more. Pinching the top creates a very skinny cone like structure which seems ideal for the centre of the school. It also reinforces the organic theme of a flower petal by placing several curved structures around the dome.

This design is one of my favourite developments as it strongly reflects a centred 'heart' to the school aswell as the 'flower bud' symbol to give a very modern design appearance. The petal like structures curve around the glass dome aswell as one large concrete lip that hangs over the structure flooding light into the hall.

This development simply shows how to fit a building into the middle of the school to best fit the enviroment. A concrete dome with wooden petals surrounding the building and cross shaped windows running all around the design.

This development is extremely independant toward the others Ive started to break away from the common centred/circular shapes and started playing around with glass being the main material. Expermenting on how I can create buildings through this material while still following a natural and organic theme.

DEVELOPMENT

This development pushes on from the overlaping features of a flower. Three large concrete structures create the bulk of the design and stack up adjacent to one another. To contrast these more natural shapes Ive used an angled pyrimid for the main hall. With a full glass wall breaching the interior with light.

This design was developed from concepts that are revolved around the shape of a flower. It has the centered petal feature.

This design is very much designed from the small dome development on the previous page. I aimed to make it a much lighter and open building. The majority of the concept is a glass dome, flooding the interior with light.

I developed this drawing from Antoni Gaudis abstract shapes and features Ive also brought through the idea of flower petals.

This design links back to the concepts that were produced from a flower. It has a circle base and an angled roof. This abstract design is then featured with three concrete walls that are placed through the heart of the glass structure.

This development is ideal for the school chapel I think. It combines the concepts ideas of a dome and the flower petal features. The chapel is made mainly of a varnished wood aswell as a range of concrete or plaster. It also has three large windows flooding the interior with sunlight and portray a slightly abstract approach.

ISBN: 9780170233279

Melody Chen — Epsom Girls Grammar School

The following work produced by Melody Chen demonstrates the use of water in several forms to influence her ideas. She has used the repetition of water's movement to inspire her ideas.

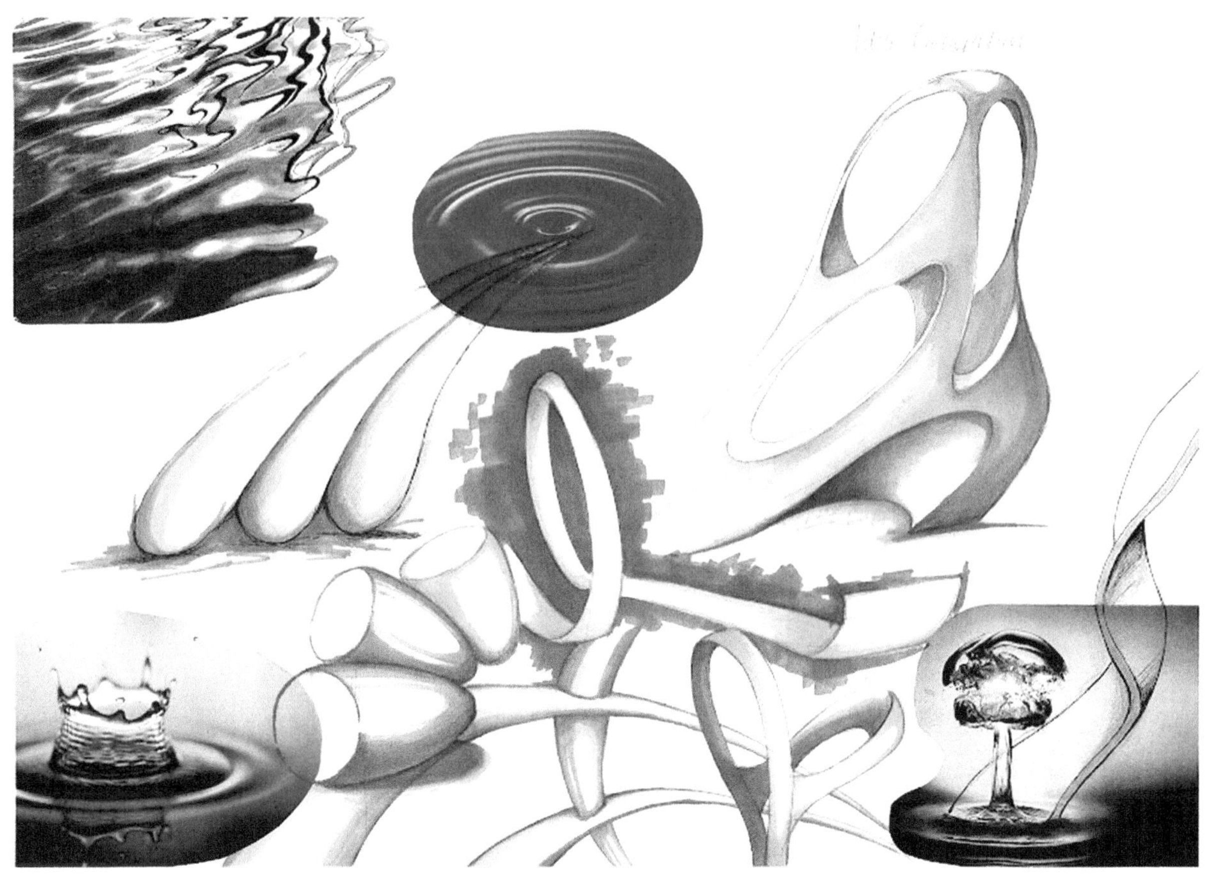

ISBN: 9780170233279

Emma Rouse — Epsom Girls Grammar School

The following work produced by Emma Rouse demonstrates the use of a human skeleton to inform her ideas for a lamp. She has sketched an in-depth analysis of this form before producing initial ideas for the product.

ISBN: 9780170233279

2 freehand sketching

Freehand sketching is used to communicate your design ideas. Design ideas are your responses to a design brief. Freehand sketches are used to explore and explain ideas and must be produced without the aid of instruments or any electronic technologies. Communicating your design ideas involves describing the design features such as showing form, shape and function.

To effectively communicate design ideas involves showing in-depth information about the intent of the design features. In-depth information is a body of work, related sketches that include exploded, assembled, sectional and sequential views that explain the design features. These types of sketches could be used to explain how your design functions or how the components of your design fit together.

Two dimensional and three dimensional freehand sketching techniques must be used to communicate your design ideas. Two dimensional sketches can be orthographic, elevations or sectional. Three dimensional sketches can be isometric, oblique, planometric or perspective.

Techniques that can be used to enhance or aid your sketches are quick rendering, crating and the use of line construction. To explain an idea thoroughly, sketch the design idea from many different directions. For example, turn the object around on the page.

ISBN: 9780170233279

When sketching, lightly construct the object before outlining the image. Hold the pencil loosely, use your whole arm and do not worry about sketching past an intended point. If you keep the sketching really tight it will be harder to represent good proportion in your sketches. DO NOT be afraid to go over the area several times with construction lines. Move your page around as you are sketching, this will also give you more freedom.

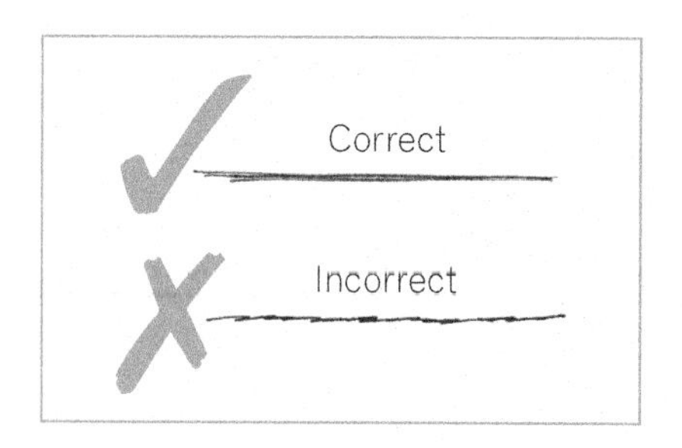

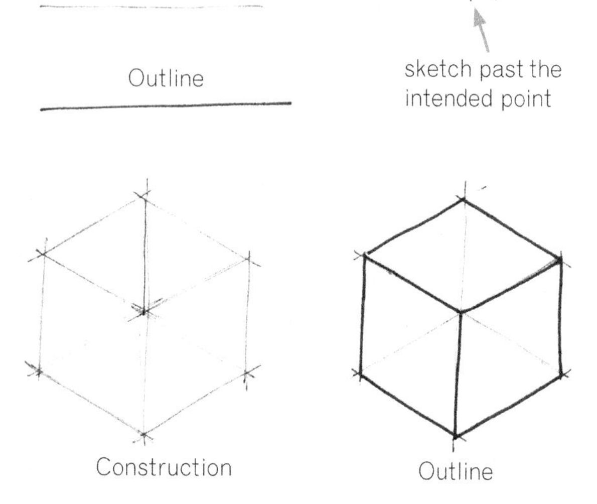

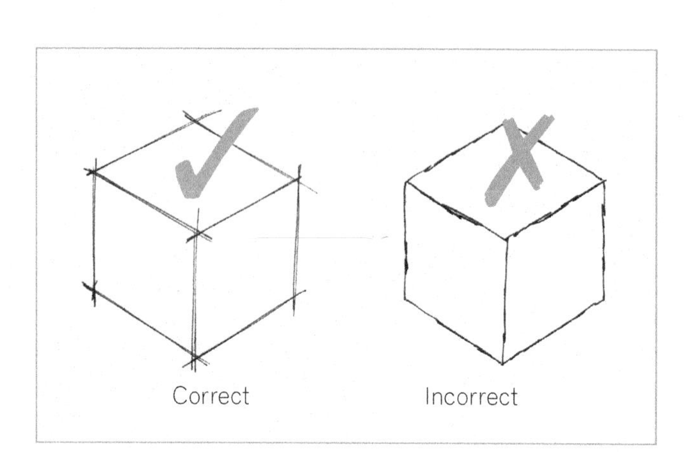

Basic sketching

The following sketches are basic geometric solids using the crating method (page 32). The crate or box which is sketched in blue is a starting point to construct any shape. Crating allows you to sketch using a certain method, for example oblique, isometric, planometric or perspective. Sketch the initial crated box in the style of sketching you have chosen to represent a design idea. Crating also allows you to keep the object or design idea that you are sketching in proportion. The following four basic solids, the cylinder, hexagonal prism, octagonal prism and the triangular prism have all been sketched in proportion using the technique isometric.

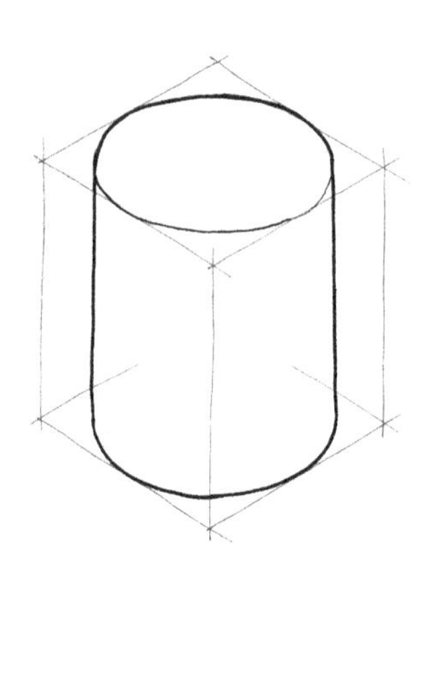

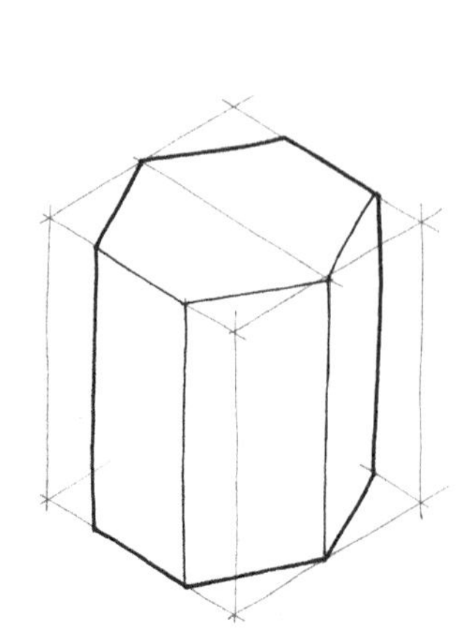

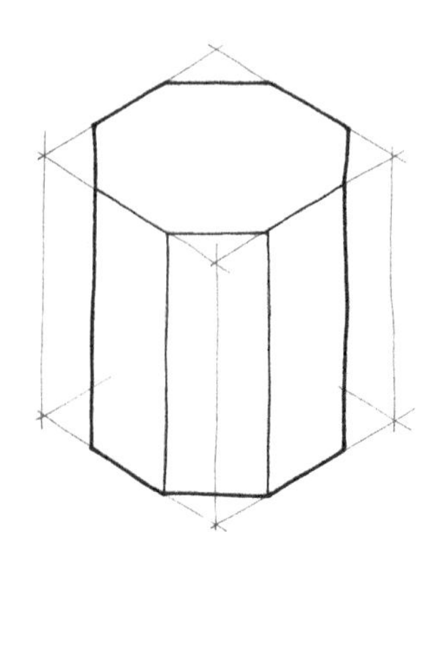

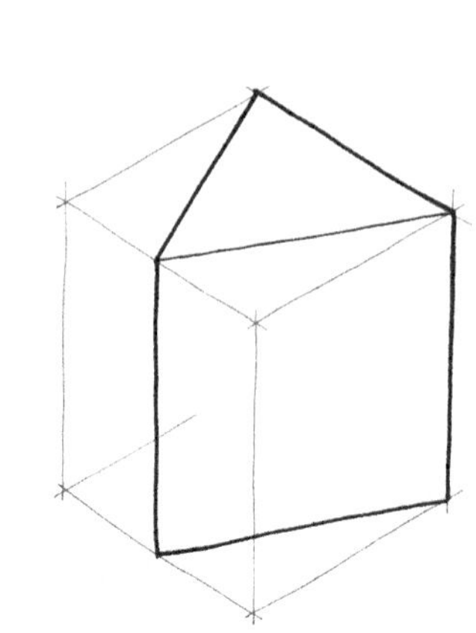

ISBN: 9780170233279

Sketching basic geometric shapes (2D)

To help construct basic geometric shapes, firstly sketch a box where the intended object is to be placed. Divide the box up to gain the required points.

Triangle

Step 1 • Divide the box in half.

Step 2 • Join the points on the base and the top middle point. Outline the triangle.

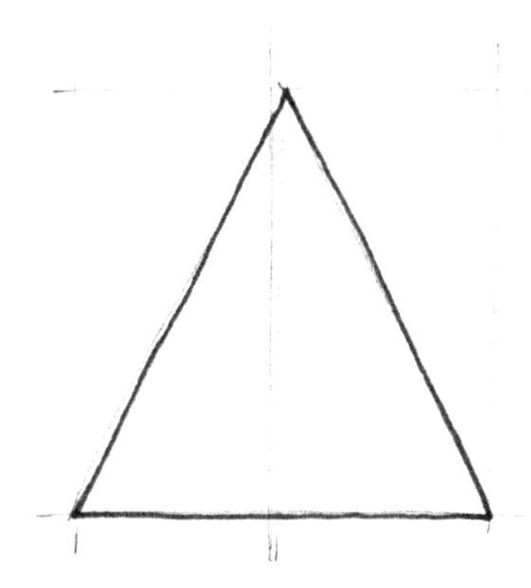

Circle

Step 1 • Divide the box through the corner points to find the centre.

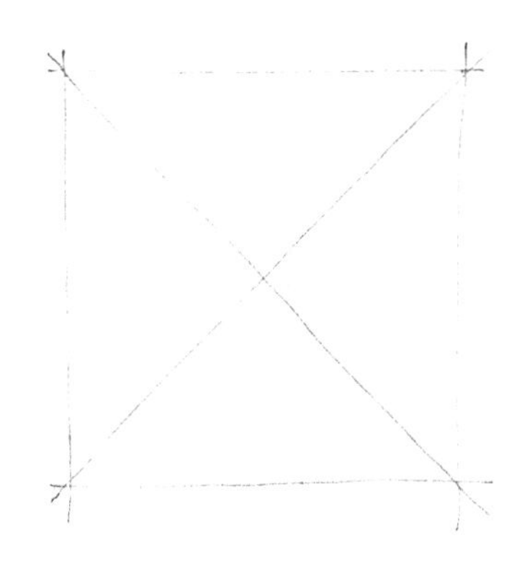

Step 2 • Sketch lines vertically and horizontally through the centre.

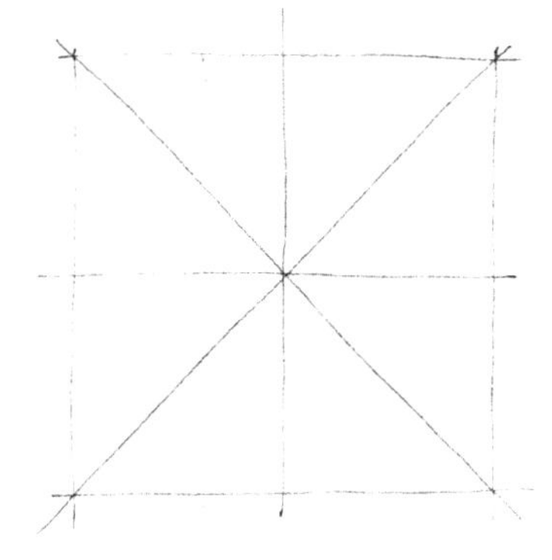

Step 3 • Use these points as a guide when sketching the circle. Outline the circle.

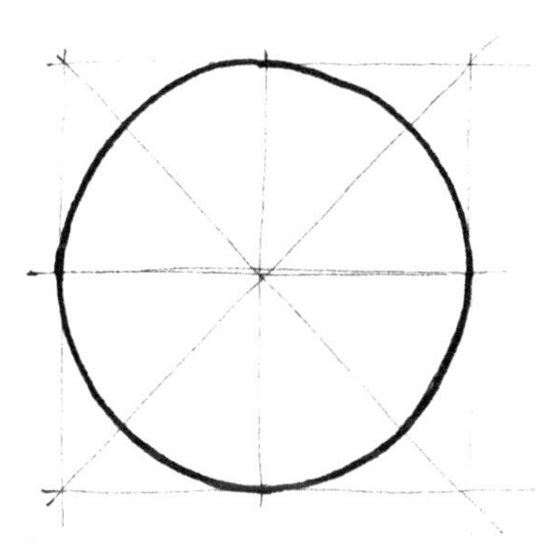

Hexagon

Step 1 • Divide the box vertically into three sections.

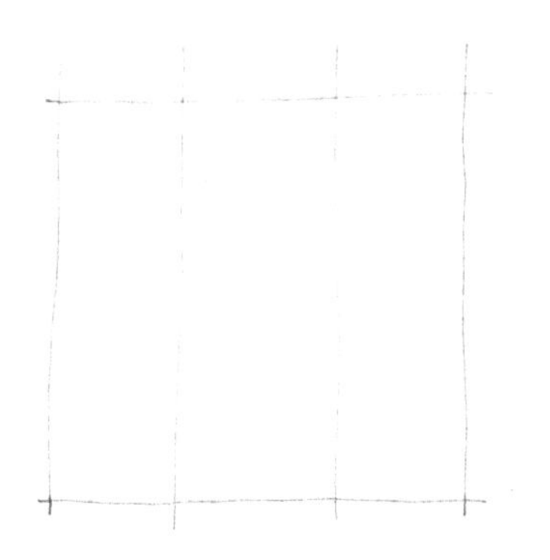

Step 2 • Divide the box through the centre horizontally.

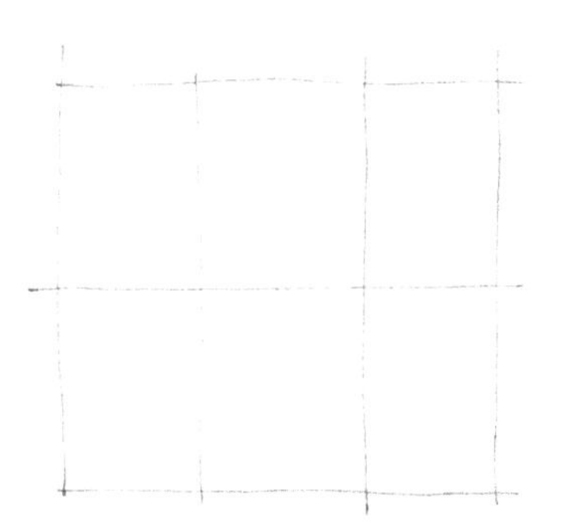

Step 3 • Join the six points you have just created on the outside edge of the box. Outline the hexagon.

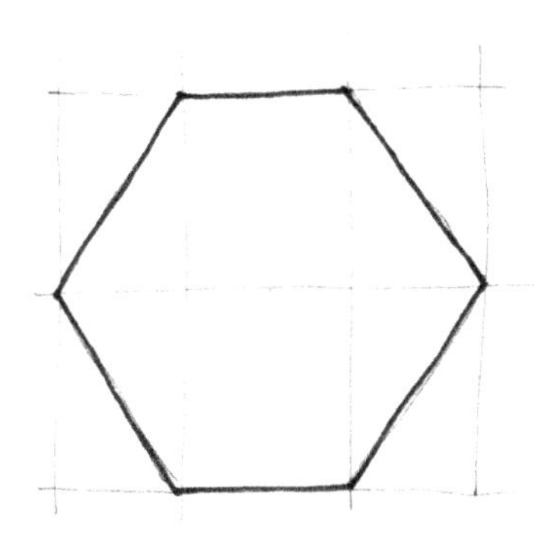

ISBN: 9780170233279

Sketching basic geometric solids (3D)

Once you have learnt to sketch basic geometric solids you will then be able to apply these techniques to more complex objects.

Crating is a technique to make the construction of objects easier. Crating allows you to break down a complicated object into simple geometric solids. The object is sketched in a box which encases the entire object.

The following geometric solids have been sketched using the pictorial drawing technique isometric. You can apply the crating method when using the following techniques – isometric, planometric or perspective.

Sketching a cube

Step 1 • Start by sketching a Y. Try to keep all the angles of the Y as even as possible e.g. 120˚ each.

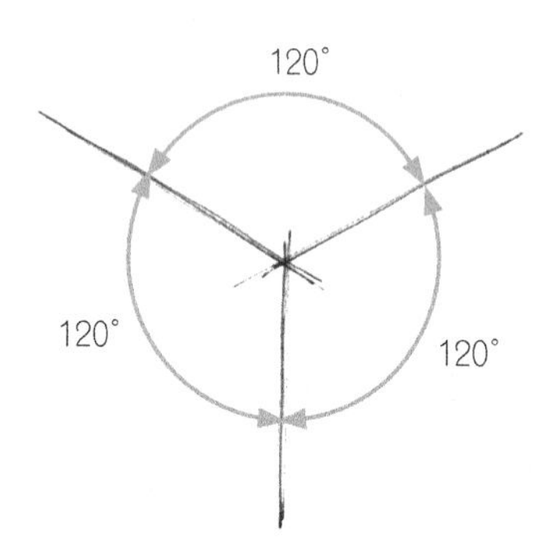

Common mistake: If the initial Y is not drawn with equal angles it will distort the final sketch.

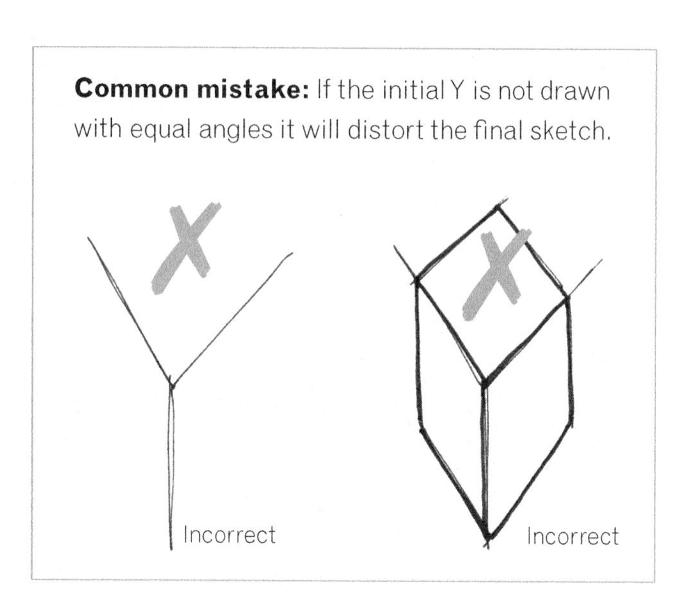

Step 2 • Sketch the background outline parallel to the Y.

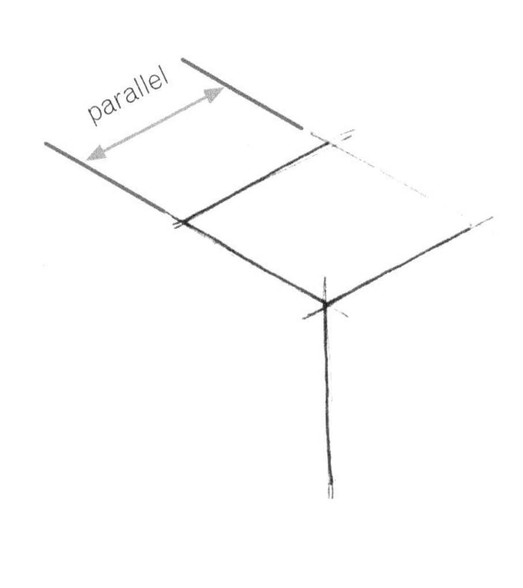

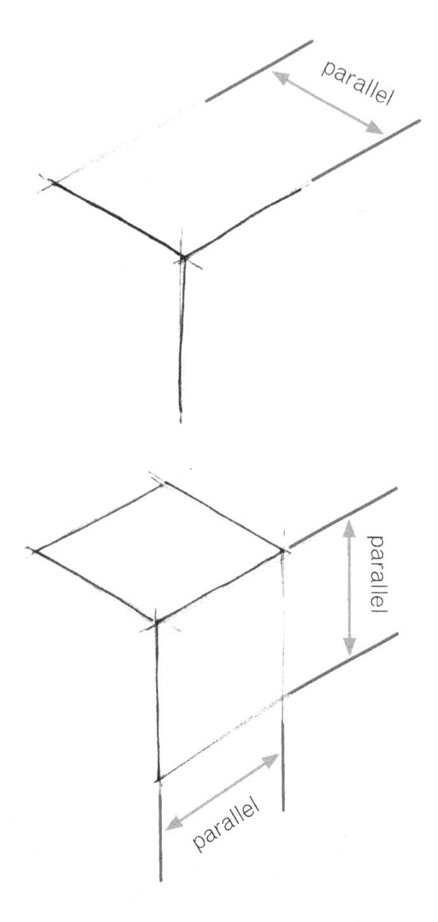

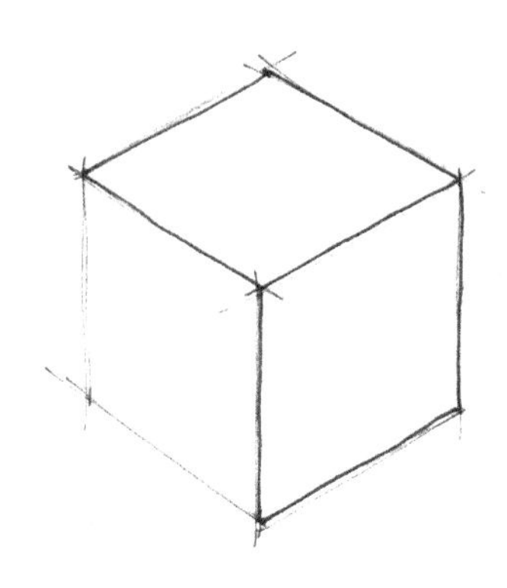

ISBN: 9780170233279

Sketching a cylinder

Step 1 • Sketch a box.

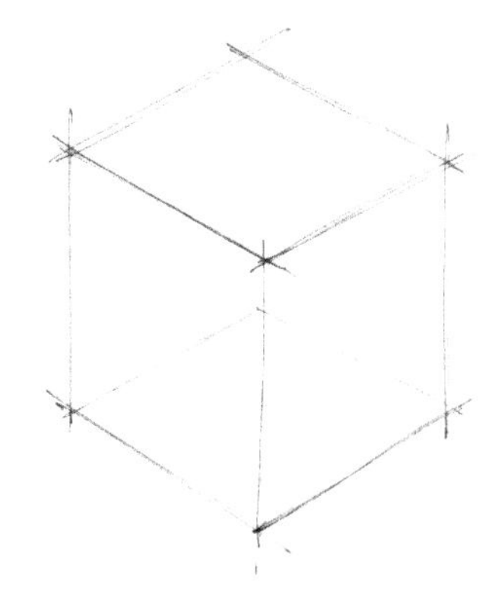

Step 2 • Sketch lines between the corners on the top and bottom surfaces to find the centre.

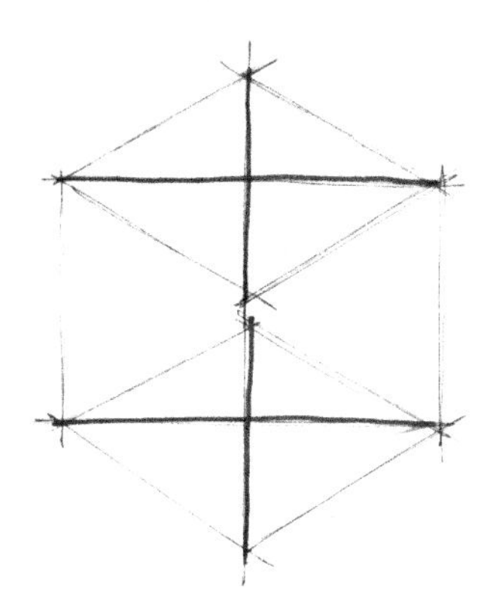

Step 3 • Sketch lines through the centre point. These must be parallel to the outside edge.

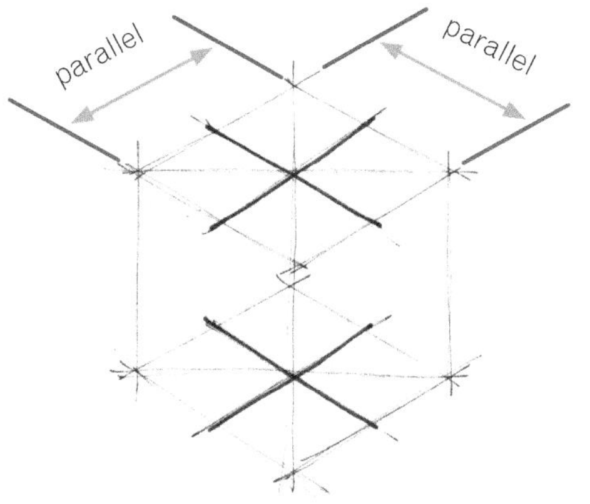

Step 4 • Sketch two arcs from the points made by the lines running through the centre. Think of these as long arcs.

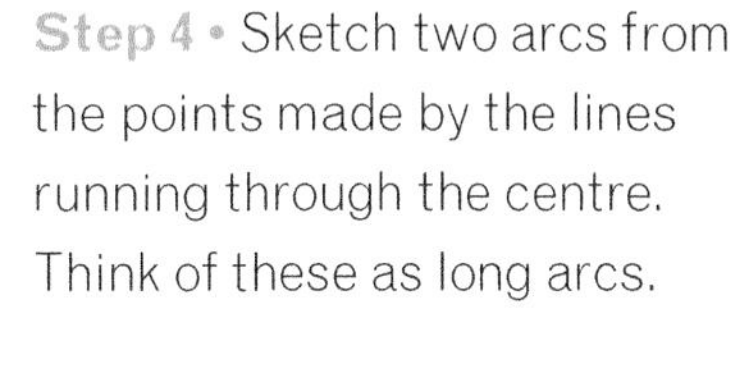

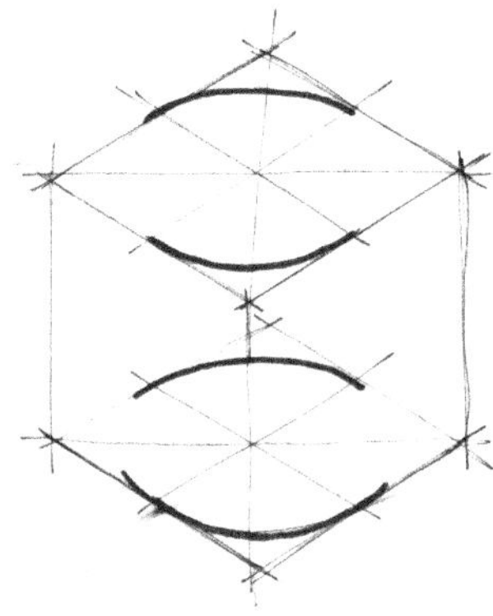

Step 5 • Sketch two arcs into the corner areas. The finished circles on the top surfaces should be elliptical in shape.

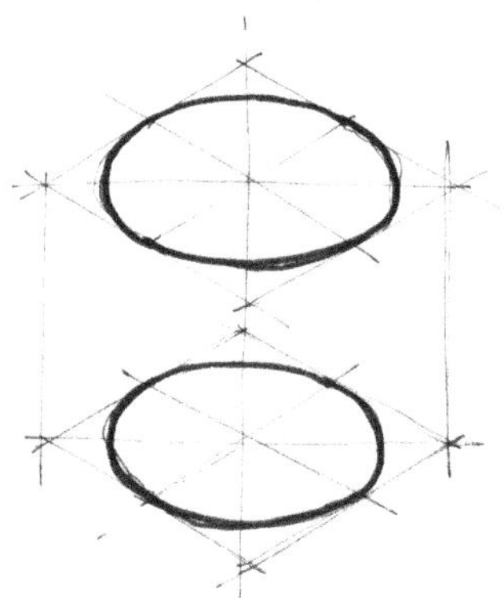

Step 6 • Join the outside edge of each of the circles with a straight line.

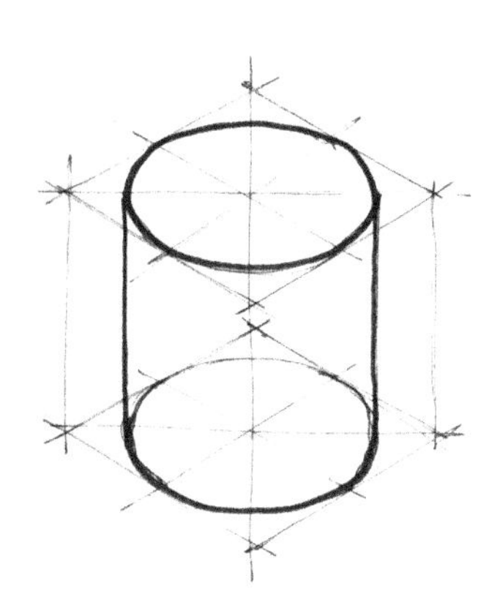

Common mistake: Often students will not draw these centre lines parallel to the outside edge. The final cylinder will then look distorted as in the example given.

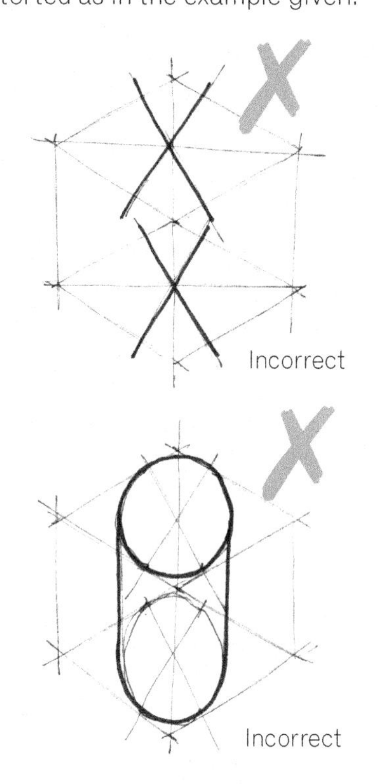

Common mistake: The circles on the top and bottom surfaces are not elliptical in shape because the last two curves have been drawn incorrectly and make the circles look square.

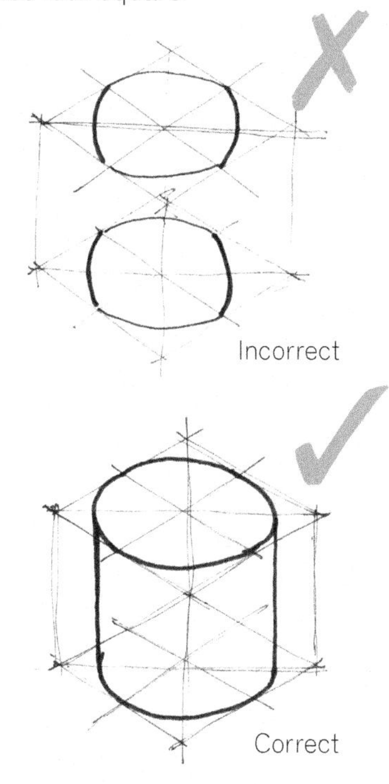

ISBN: 9780170233279

Sketching a hexagonal prism

Step 1 • Sketch a box.

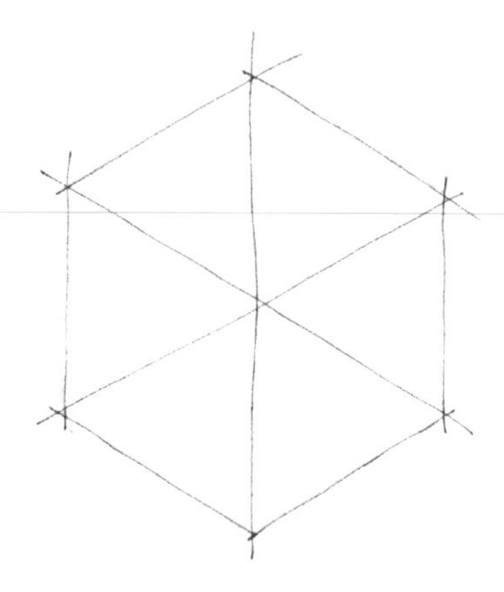

Step 2 • Divide the bottom of the box through the centre and into three in the opposite direction. Make sure the lines are parallel to the outside edges.

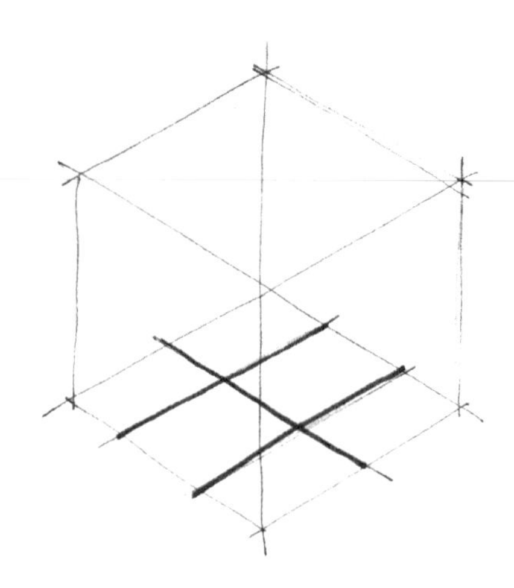

Step 3 • Join the six points that are created from the division.

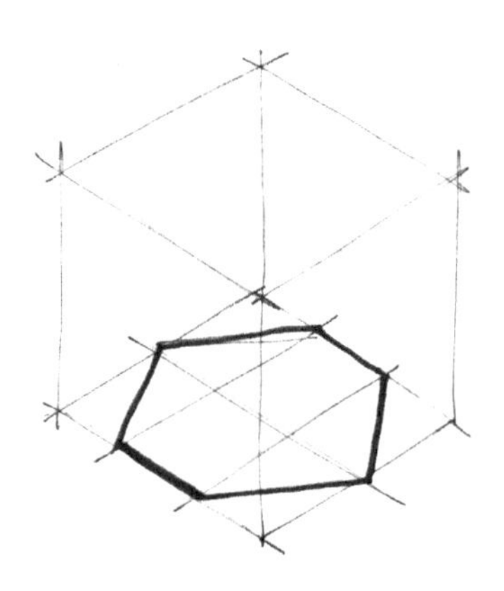

Step 4 • Repeat the above steps on the top surface.

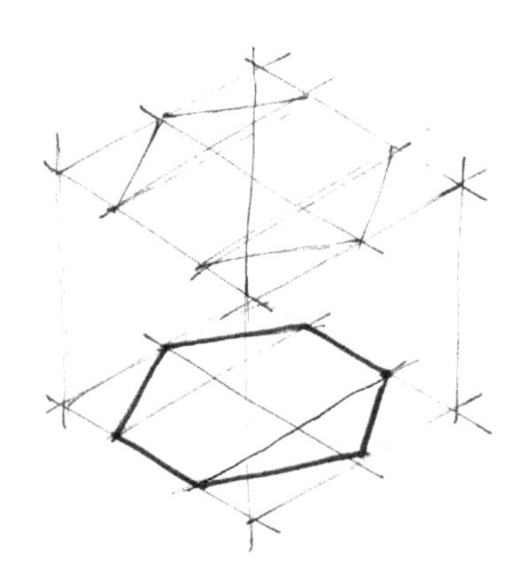

Step 5 • Sketch vertical lines from the points on the top surface to the bottom.

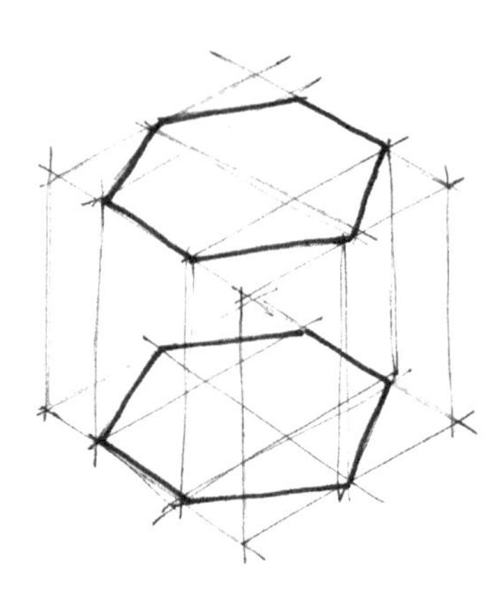

Step 6 • Outline the hexagonal prism.

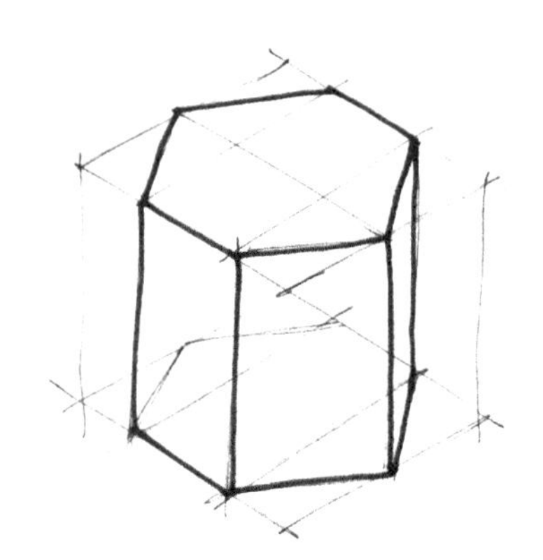

ISBN: 9780170233279

Sketching pyramids

To construct all pyramids you need to follow these steps:

Step 1 • Sketch the box in which the intended pyramid is to be contained.

Step 2 • Sketch the intended type of shape for the pyramid on the base of the box as previously explained. (Square, circle, hexagon.)

Step 3 • Join the corner points on the top surface to find the centre.

Step 4 • Join the bottom shape to the point on the top surface.

The shapes shown are a square-based pyramid, a cone and a hexagonal-based pyramid.

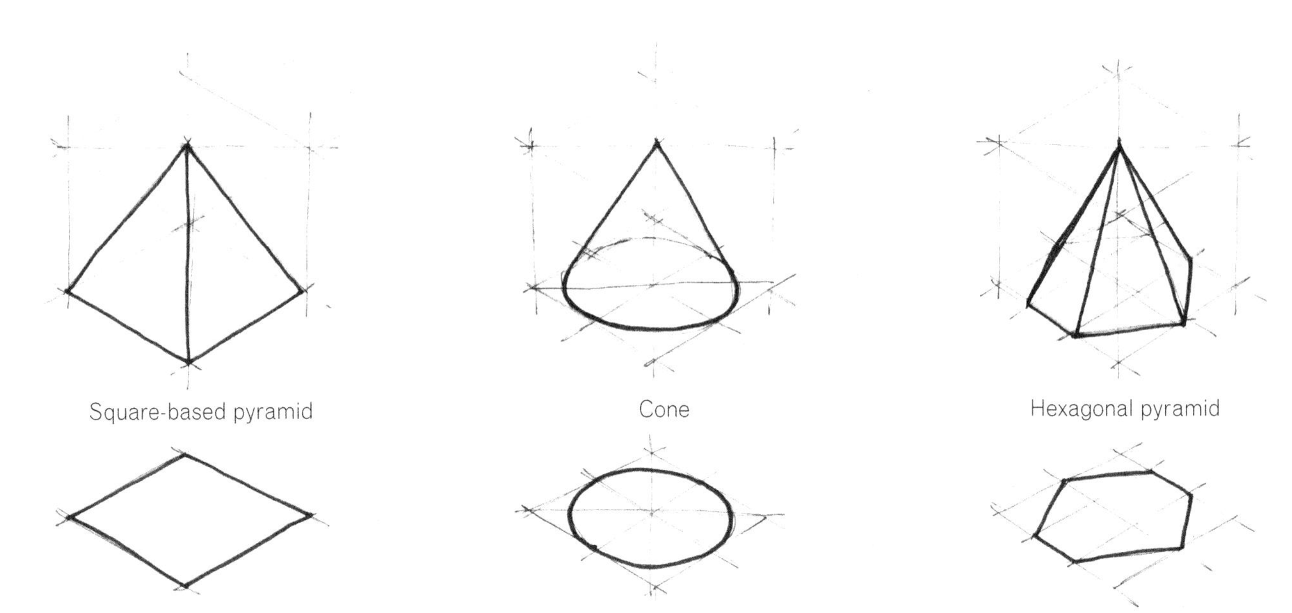

The following sketches add a little more difficulty as they are truncated pyramids. Each of the pyramid shapes has been cut off at the top. You will notice that a square has been constructed on the top surface to guide where the top of the solid should be constructed.

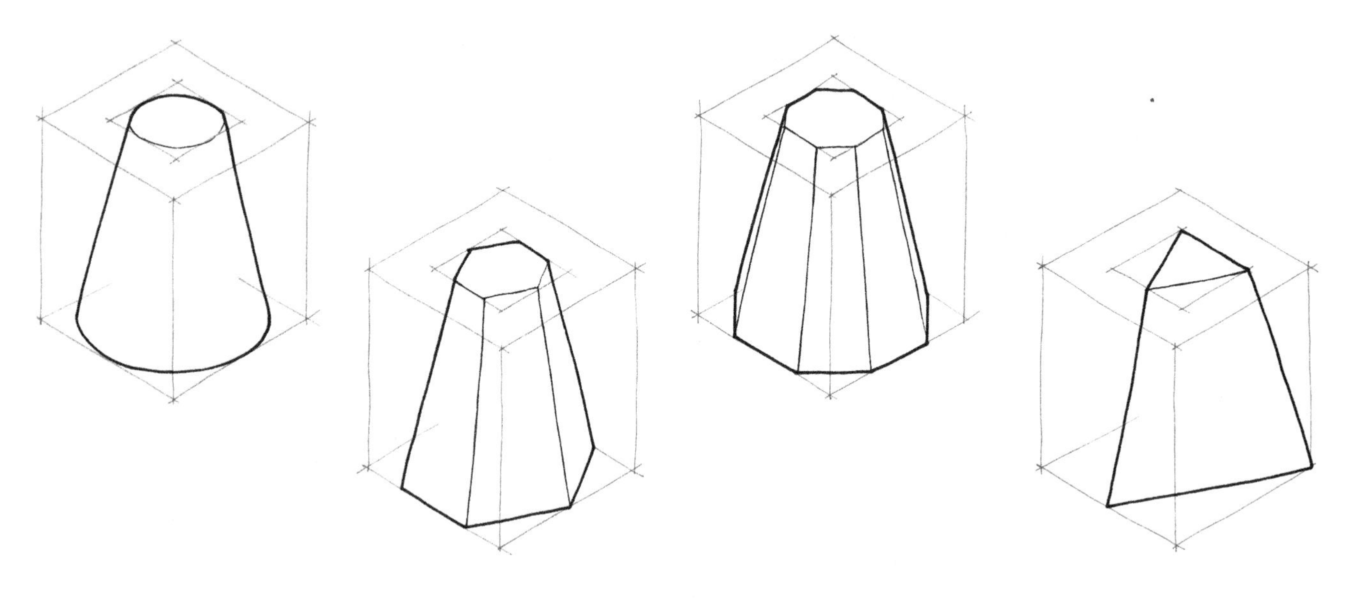

ISBN: 9780170233279

The following are the types of 2D sketching that can be used to communicate design ideas throughout the development.

- Orthographic projcction
- Sectional sketches
- Plans and elevations
- Surface developments.

The mechanics of the above types of drawings are explained in more detail in *Chapter 5 — 2D Instrumental Drawing*. For the purpose of freehand sketching these types of drawings are sketched, instead of drawn using rulers.

Orthographic projection

A sketch using orthographic projection shows different views of an object and what each view looks like if you are only looking straight at that view. Each view is projected through a reference line. The plan view of the camera gives a bird's-eye view. The main elevation is the side with the most information. The right hand elevation shows what the camera looks like from the side view. As this is a sketch the sizes are estimated but still explain the object. The heights and widths are projected from one view to the next, through the reference line. The information on the elevation is projected onto the 45 degree line and straight across to the plan view. The blue lines show the projection onto all views.

You will notice that the orthographic sketch of the camera is dimensioned. Dimensioning is the formal labelling of the measurements. You need to dimension some of your sketches in the development of ideas so that you can communicate what your sizes will be. You will find the standards for dimensioning in *Chapter 5 — 2D Instrumental Drawing* as the final working drawings must also be dimensioned. The main points to follow when dimensioning your sketches are:

1. Leader lines project the start and end point for the length of the arrow line.
2. Arrow lines are a dark line with arrow heads to indicate where the measurement starts and finishes.
3. Position the number in the middle of the arrow line. The number should be positioned on top of or to the left hand side of the arrow line.
4. Measurements should be in mm unless otherwise stated.

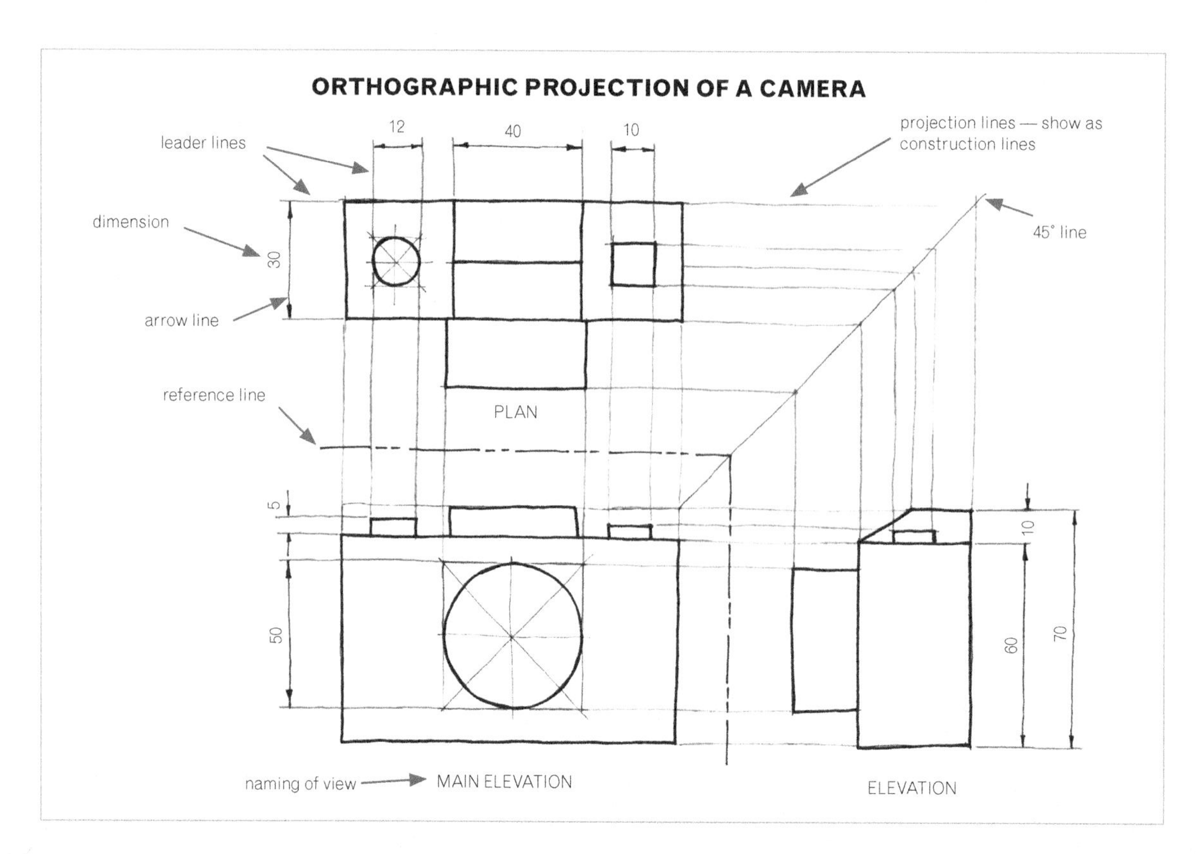

ISBN: 9780170233279

Sectional sketches

A sectional view is also a method in which you can explain the assembly of components or how something functions on the inside. The example shows a cylinder and the plane where it is being cut. The orthographic sketch presents two views, the plan view and the elevation which shows the cylinder cut in half. The two arrows on the plan view indicate where the object is to be cut in the projected view. The material that has been cut through is shown hatched with diagonal lines which are evenly spaced apart.

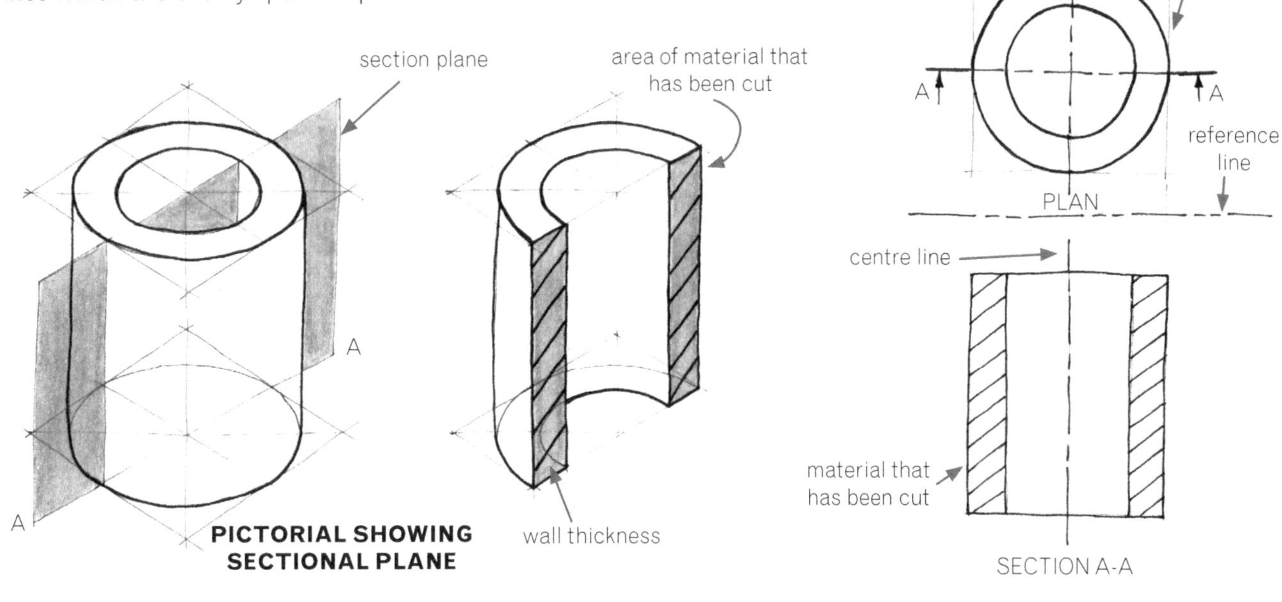

PICTORIAL SHOWING SECTIONAL PLANE

The lamp shows a more complex sectioned sketch. This time the main elevation indicates where the object is being cut. You will notice that parts of the sectional view are hatched in opposite directions. This is to indicate that it is a separate part.

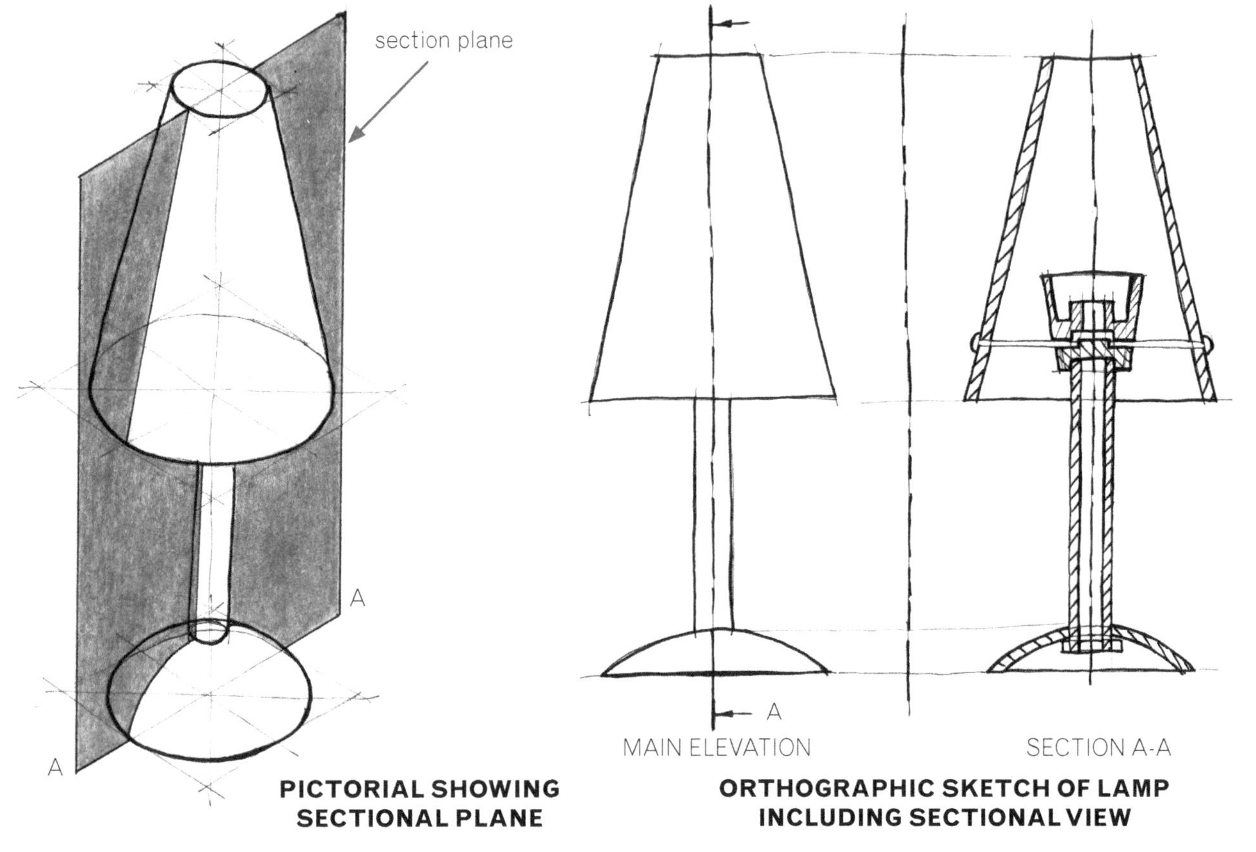

PICTORIAL SHOWING SECTIONAL PLANE

ORTHOGRAPHIC SKETCH OF LAMP INCLUDING SECTIONAL VIEW

ISBN: 9780170233279

Floor plans

A floor plan sketch would be used when exploring ideas for a spatial design brief. A floor plan is a bird's-eye view of the area which shows details such as benches, walls, windows, doors, sinks, cupboards. Floor plan sketches can quickly identify the layout of an area before sketching an elevation or pictorial view.

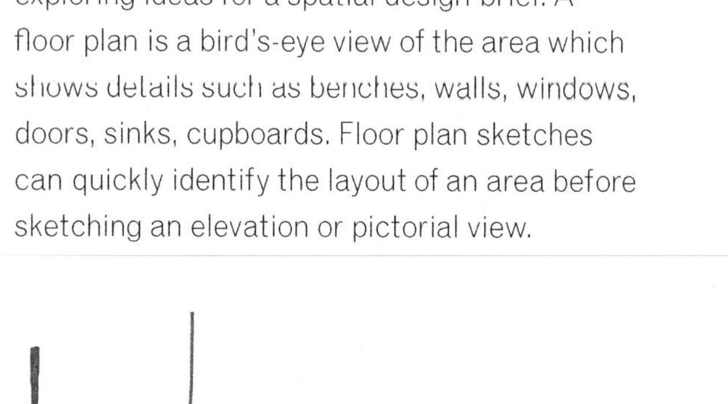

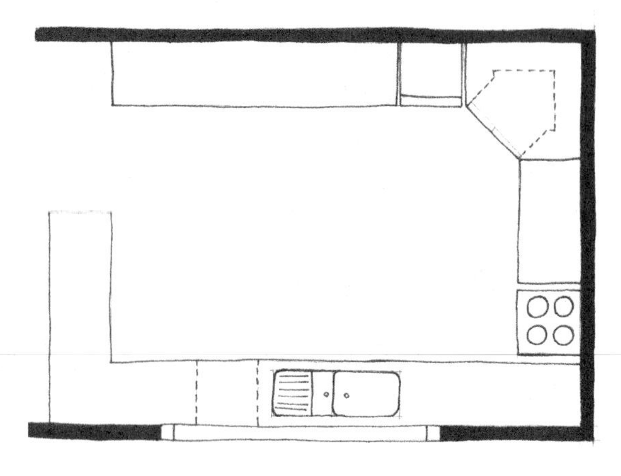

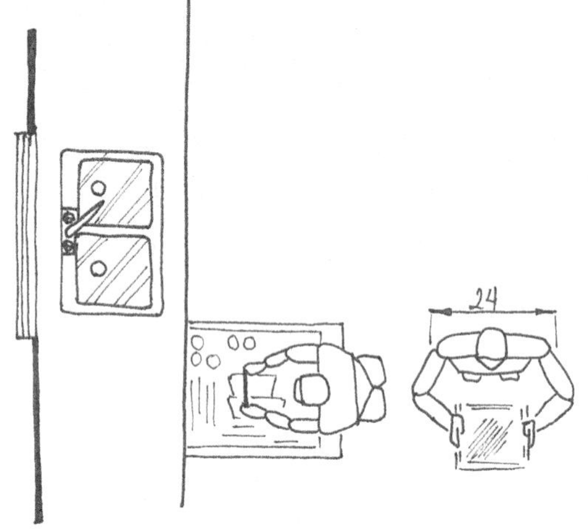

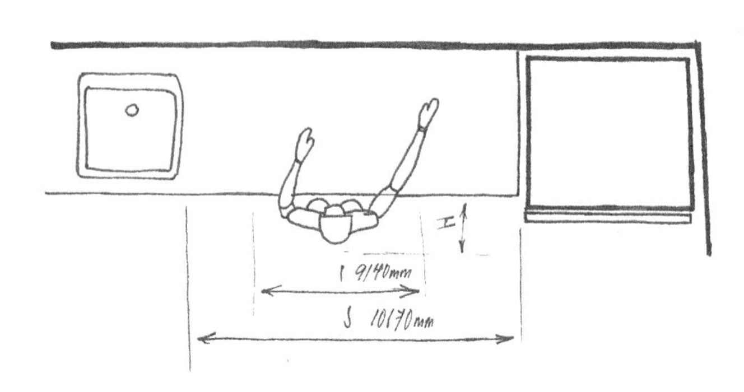

Elevations

An elevation is a side view. You may need to use an elevation to explain one aspect of your design in a product or spatial design brief.

spatial elevation

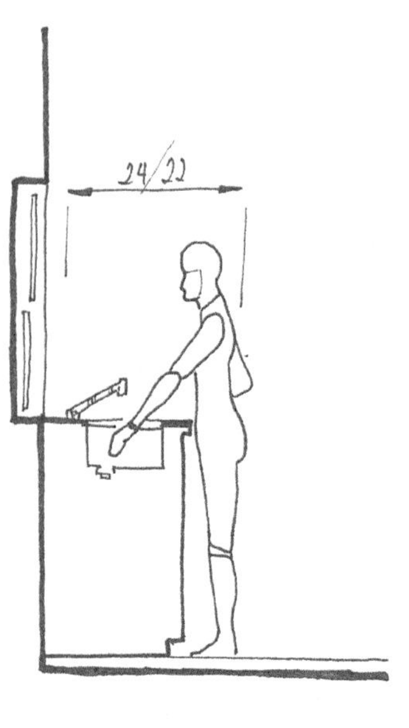

spatial elevation

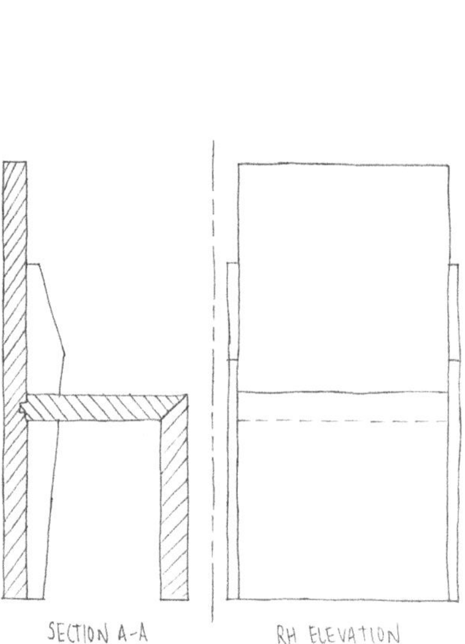

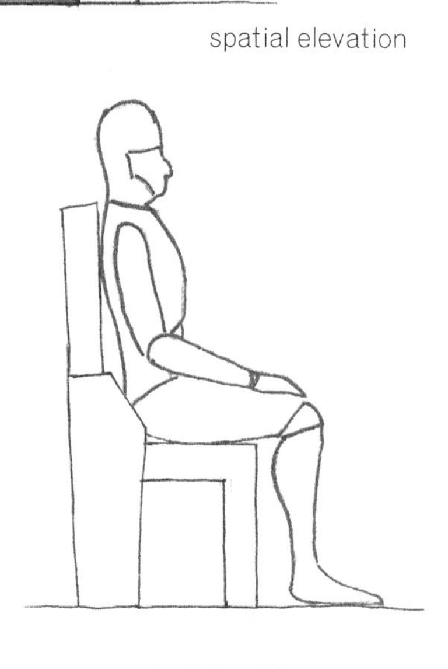

product elevation

ISBN: 9780170233279

The following are the types of 3D sketching that can be used when developing your design ideas.

- Isometric
- Oblique
- Planometric
- 1 point perspective
- 2 point perspective

The mechanics of the above types of drawings are explained in more detail in the section *Chapter 6 — 3D Instrumental Drawing*. For the purpose of freehand sketching these types of drawings are sketched instead of drawn using rulers.

Parallel sketches

Isometric

Isometric sketching gives a realistic view of the object and shows the same amount of information on all three views of the object. Isometric is useful for representing product design. The two rules to remember when sketching in isometric are

1. The object always has a corner facing towards you.
2. The lines are always on a 30° angle or vertical.

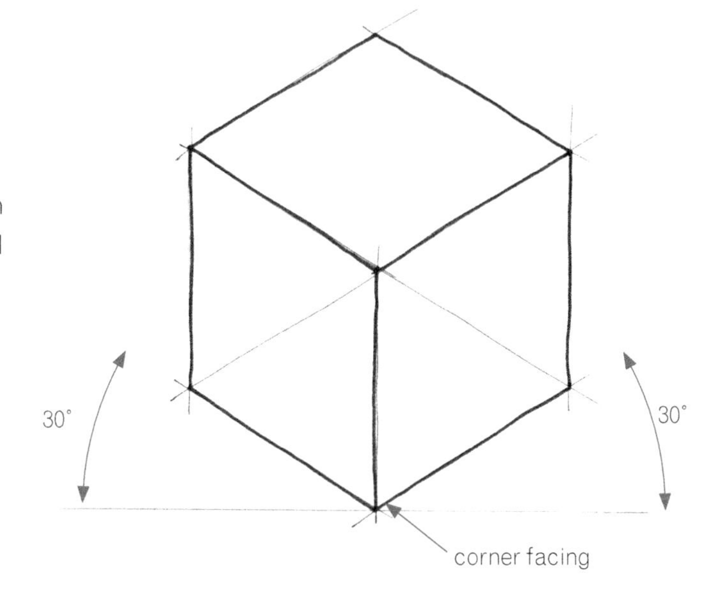

Oblique

Oblique gives a less realistic view than isometric. Oblique is a good sketching method to choose when one side of the object is quite complex. The complex side view can be drawn on the front surface and taken back on a 45° angle to give the 3D effect. The two rules to remember are

1. The main view is front facing.
2. All lines are sketched on a 45° angle, horizontal or vertical.

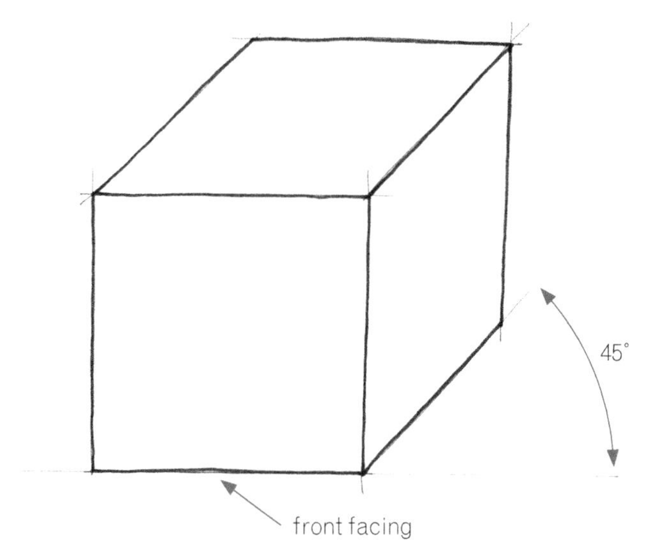

ISBN: 9780170233279

Crating

The camera is constructed using the crating method in both isometric and oblique. The entire object is constructed using basic geometric shapes. The main shape on top of the camera is a little more complex. The construction of this part has been demonstrated in both isometric and oblique. You will notice that the crating method is easily applied to the different sketching techniques. It is just a matter of sketching a box which encases the object and slicing away the unwanted parts. You can see from both these examples that isometric does give a more realistic view.

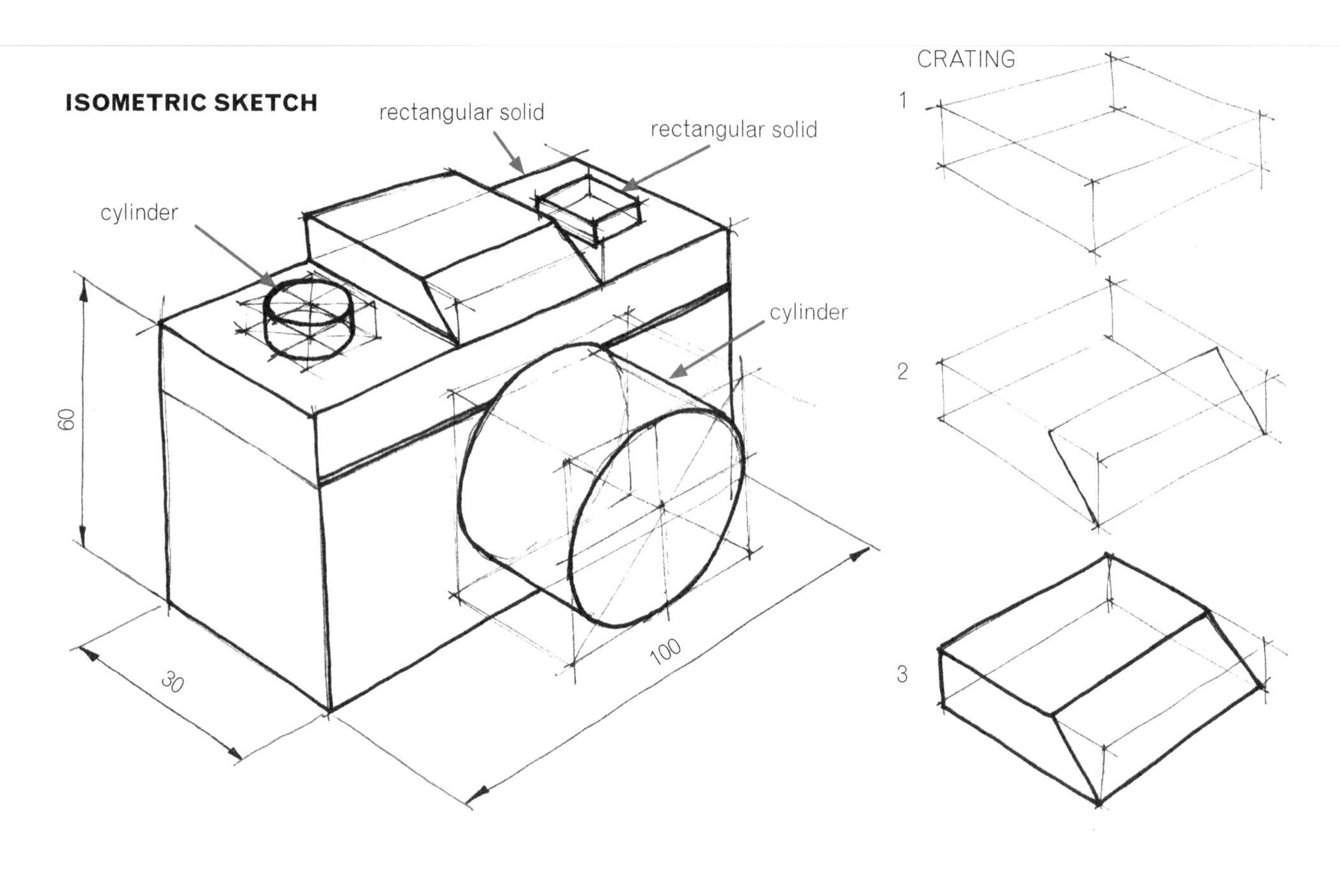

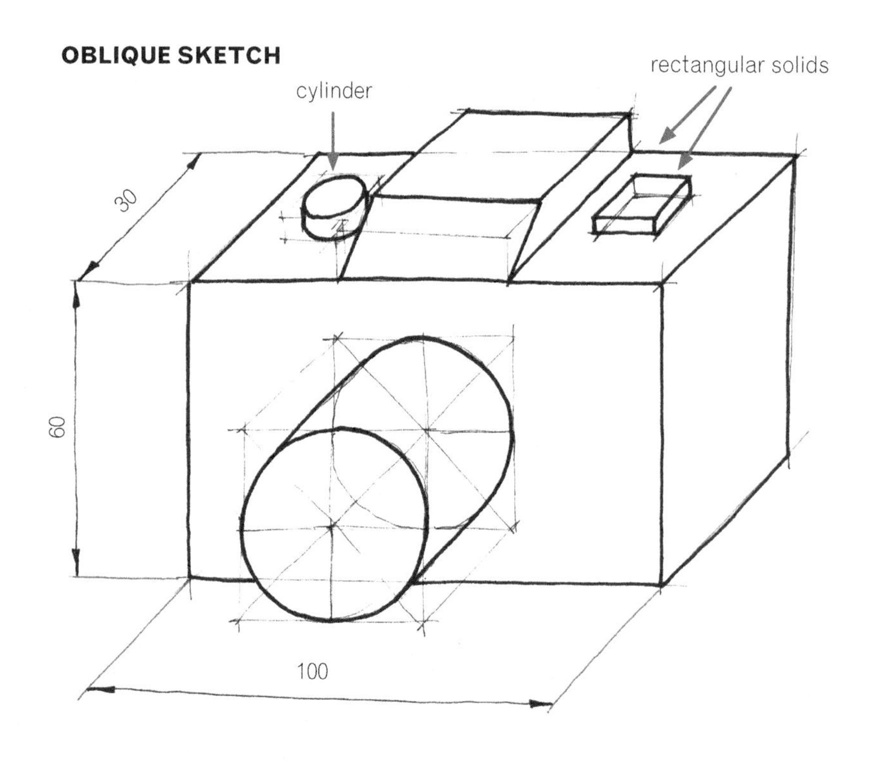

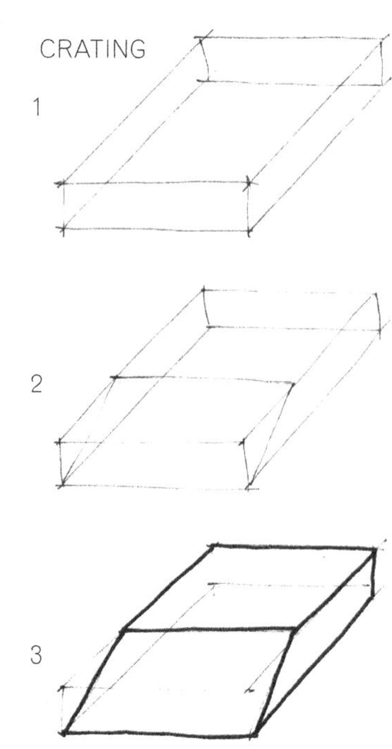

ISBN: 9780170233279

The oblique and isometric view of the camera has some measurements dimensioned. The same rules apply as with dimensioning an object sketched in two dimensions. The angles of the leader lines and arrow lines continue with the context of the sketching. For example in the isometric view all arrow lines and leader lines are on a 30 degree or are vertical. For the oblique view all the arrow lines and leader lines are on a 45 degree angle, horizontal or vertical.

Rounding

Singular rounding

Singular rounding is rounding in one direction only. Curved edges can be broken into sections when constructing.

A shows the construction of an isometric circle divided into four sections. The examples **B** and **C** demonstrate the use of each quarter of the isometric circle to construct the curved edges.

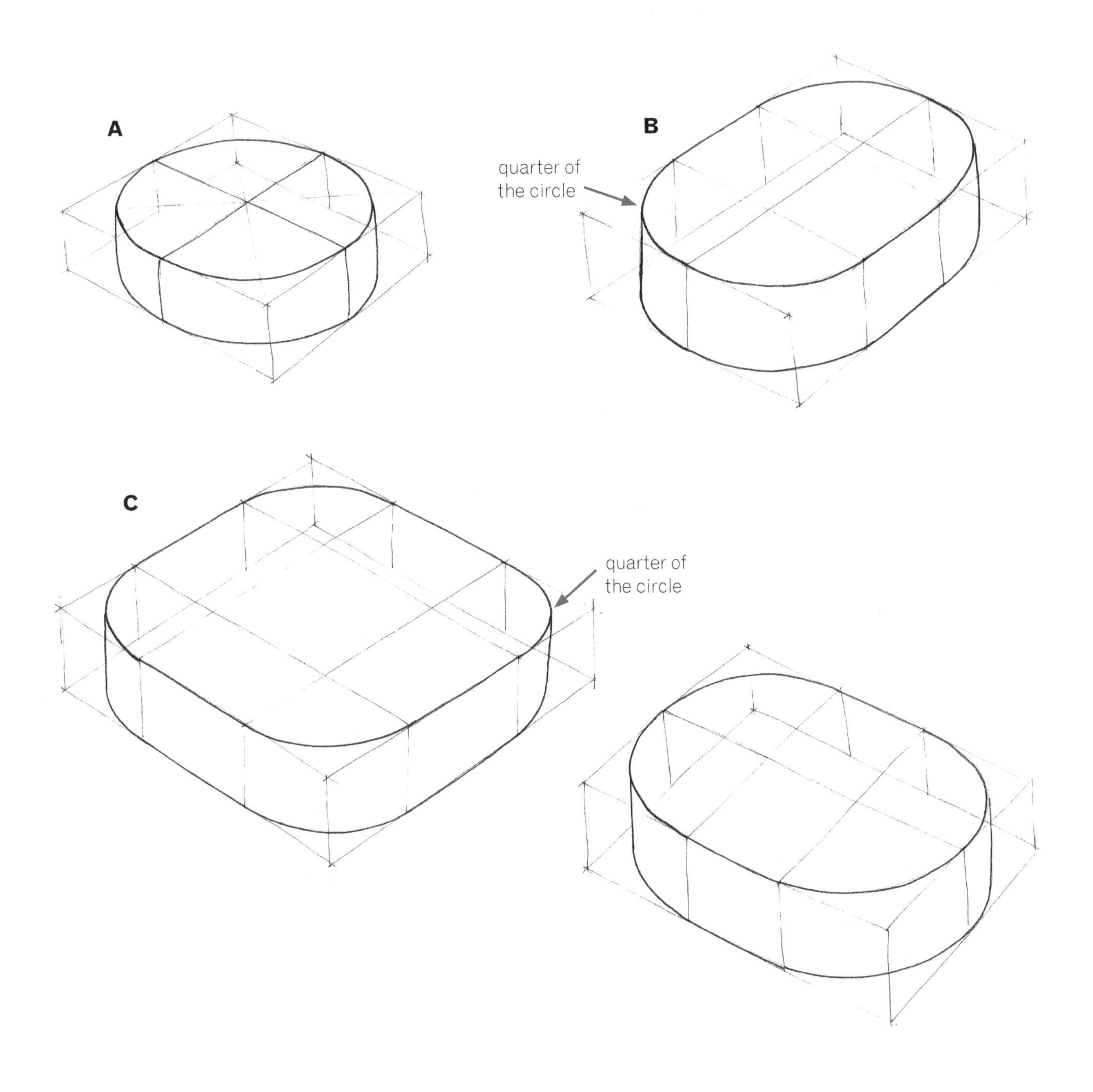

ISBN: 9780170233279

Sketching a watch

The following will give you a good understanding of the crating method using isometric. The watch will be reasonably difficult to construct as it contains quite a few circles. Once you have mastered the construction you will be able to apply these skills to your own design work.

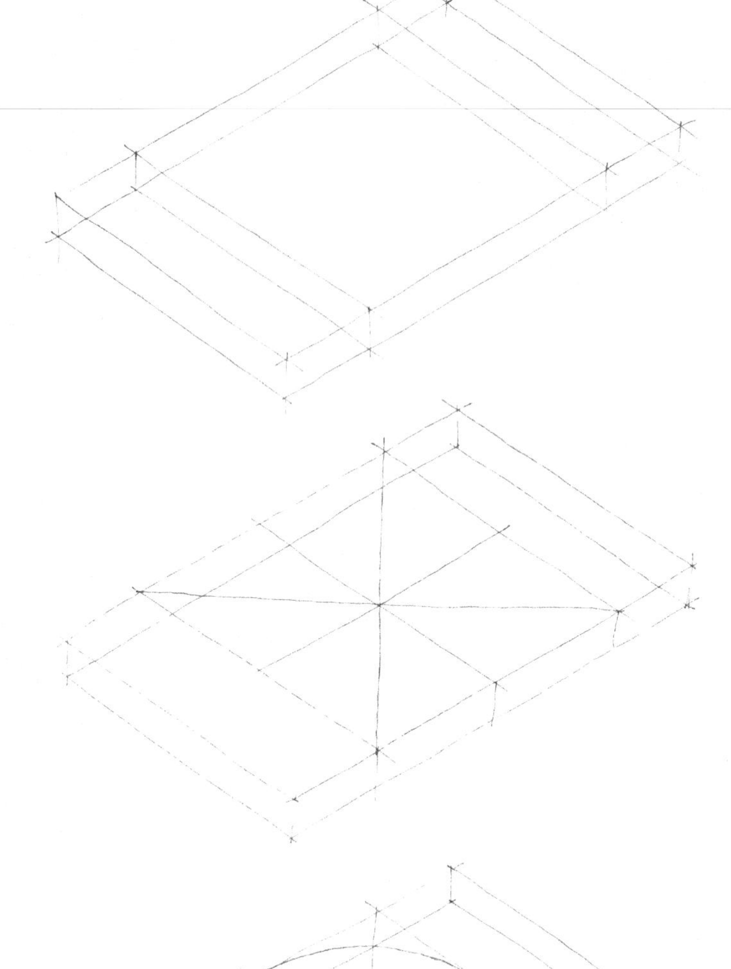

Step 1 • Sketch the outside box for a watch and divide the section into three segments as shown.

Step 2 • Divide the middle section by drawing lines from corner to corner to find the centre

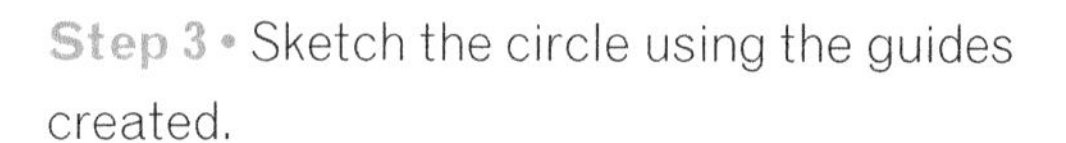

Step 3 • Sketch the circle using the guides created.

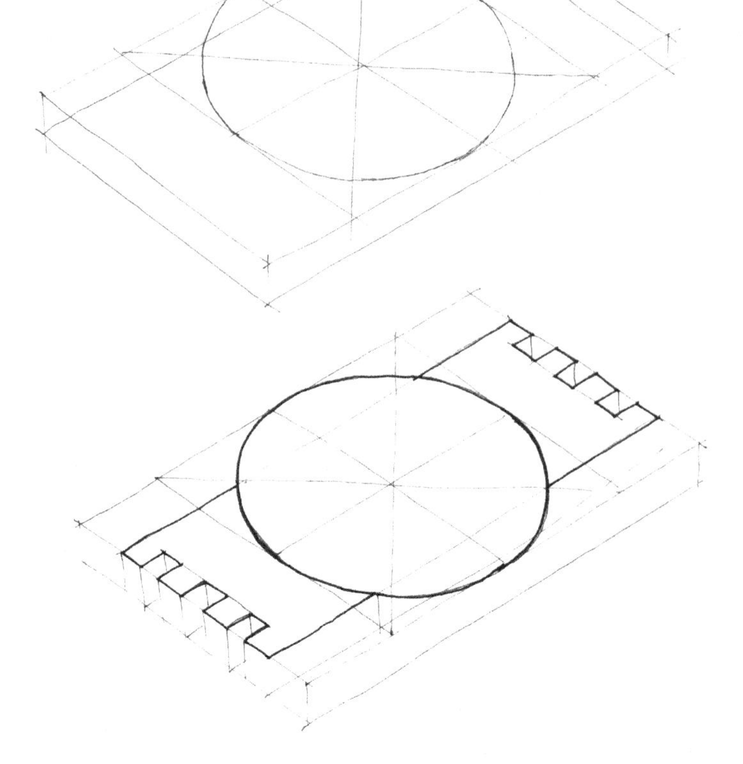

Step 4 • On the top surface draw the part of the watch that would attach to the strap on both sides of the circle. Make sure you keep all lines parallel.

ISBN: 9780170233279

Step 5 • Sketch the depth of the shape that you have constructed on the top surface.

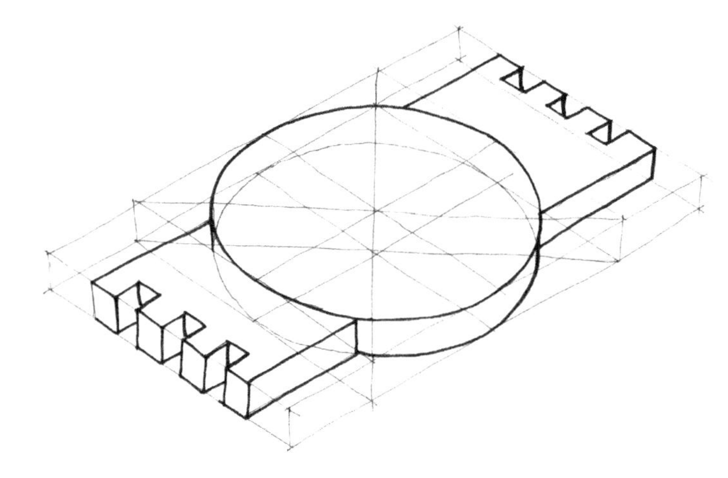

Step 6 • Sketch curves joining the parts.

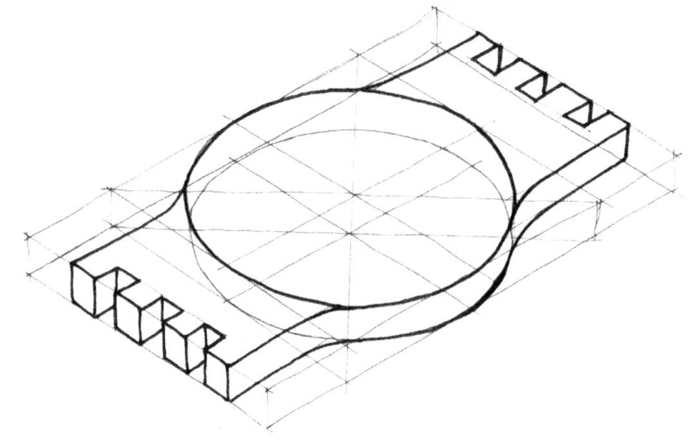

Step 7 • Sketch a circle which is raised up from the top surface.

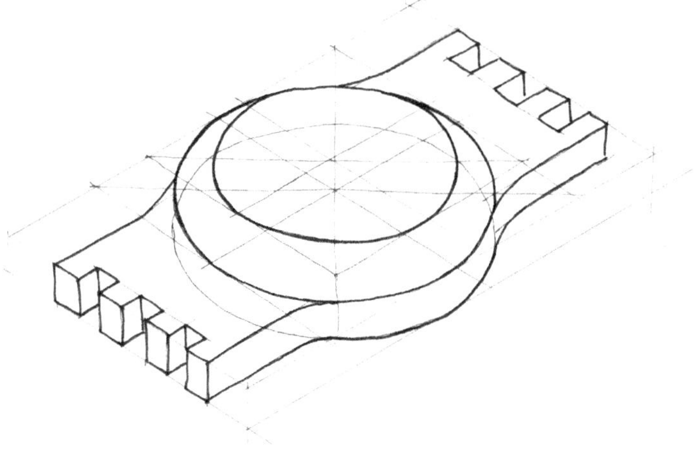

Step 8 • Add the details of the watch. The hand, numbers, button. Outline the product.

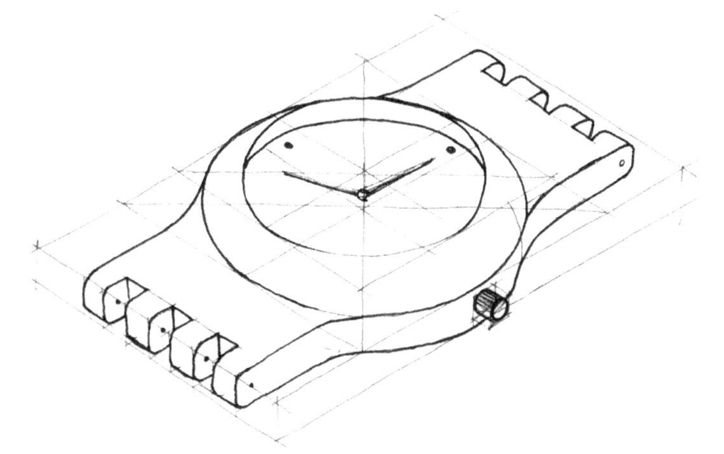

ISBN: 9780170233279

Multiple rounding

Constructing a sphere

To round an object in multiple directions you need to firstly understand how to construct a sphere. Following the steps below you will be able to construct a sphere and understand how it can be divided into sections.

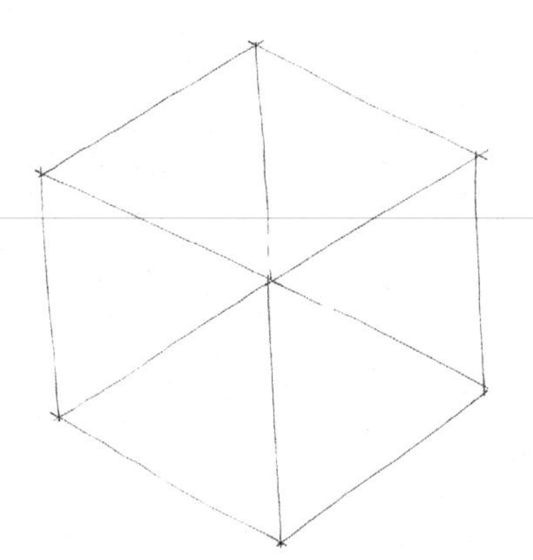

Step 1 • Construct a box in isometric.

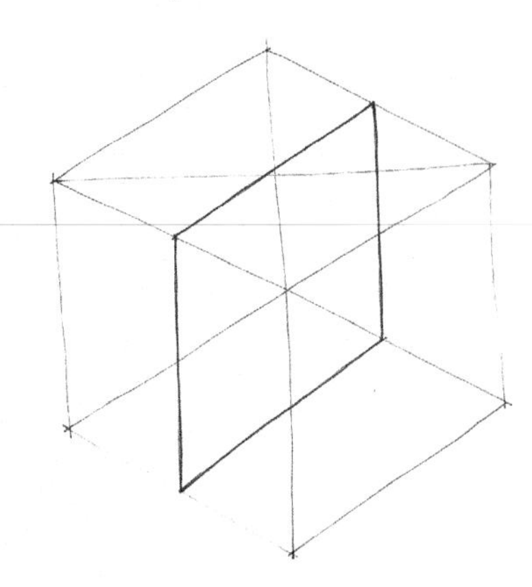

Step 2 • Divide the box in half in one direction.

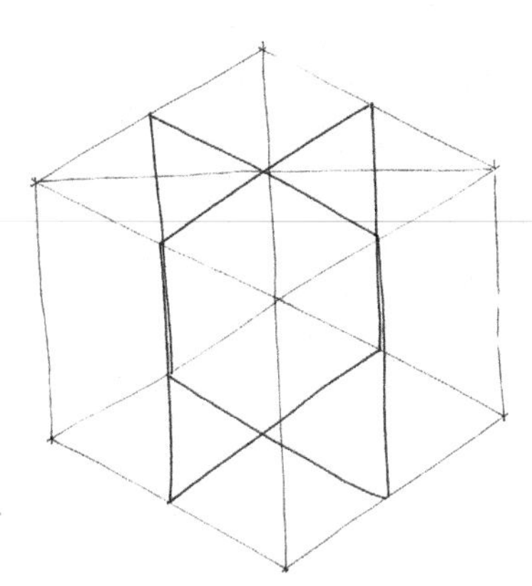

Step 3 • Divide the box in half in the other direction.

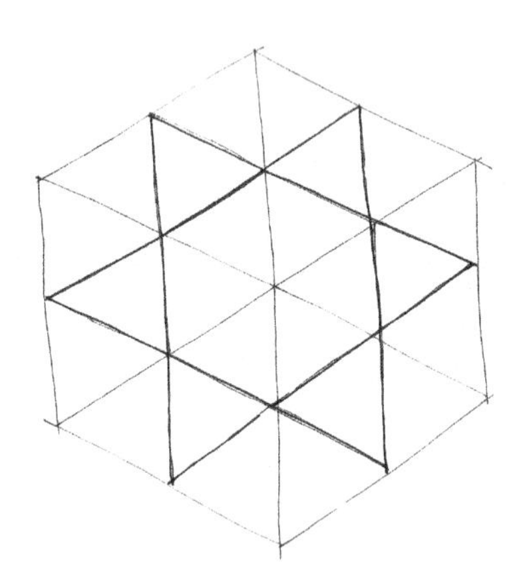

Step 4 • Divide the box in half vertically.

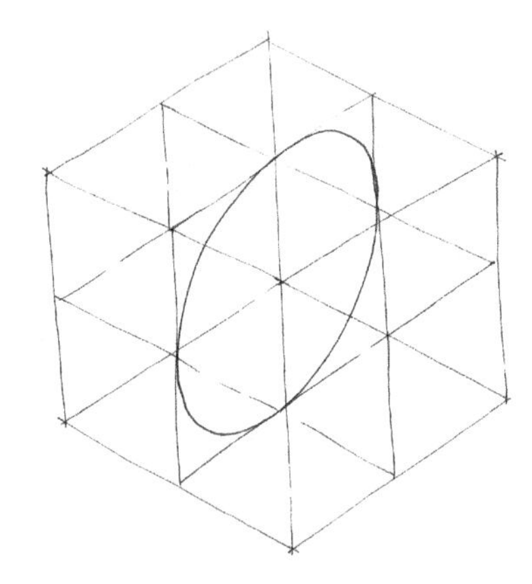

Step 5 • Construct an isometric circle in the created square.

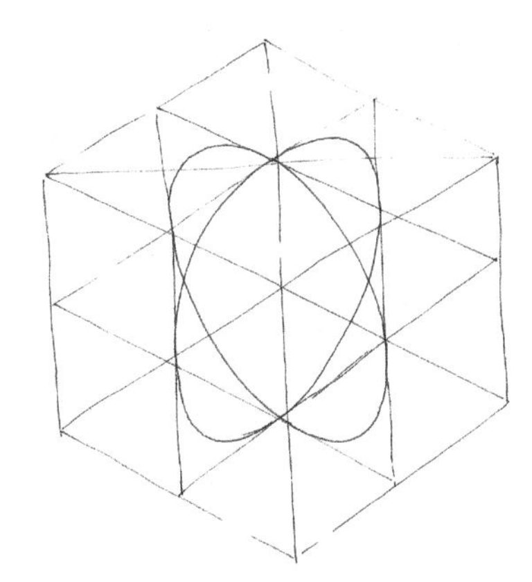

Step 6 • Constrict an isometric circle in the opposite direction.

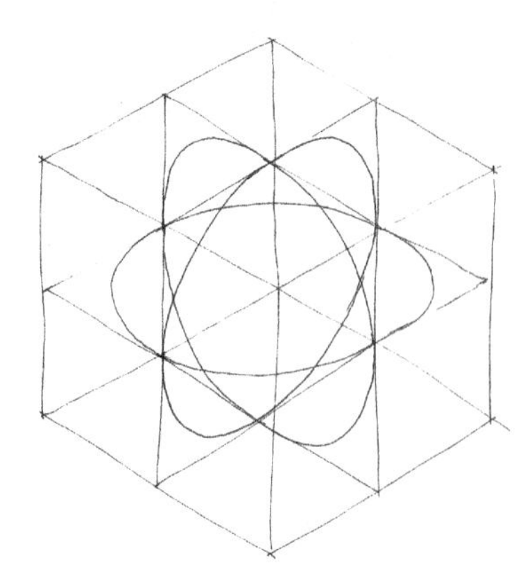

Step 7 • Constrict the isometric circle in the horizontal square.

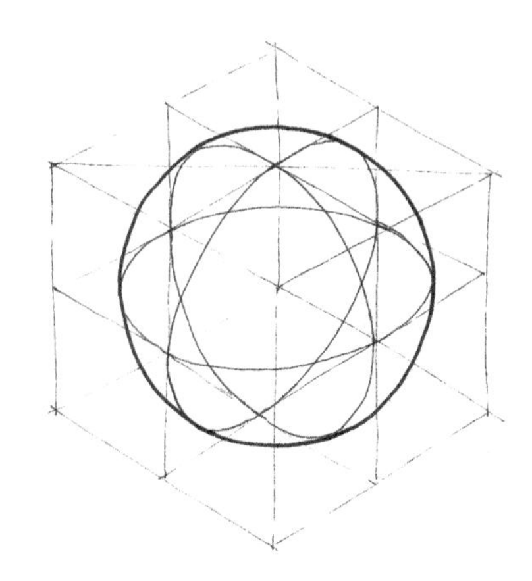

Step 8 • Sketch a circle around the edges of the isometric circles.

ISBN: 9780170233279

Half and quarter spheres

Example **A** demonstrates the sphere divided into halves and quarters.

If you look around your home at products, most have rounding, which is not only aesthetically pleasing but also is a requirement for manufacturing.

A

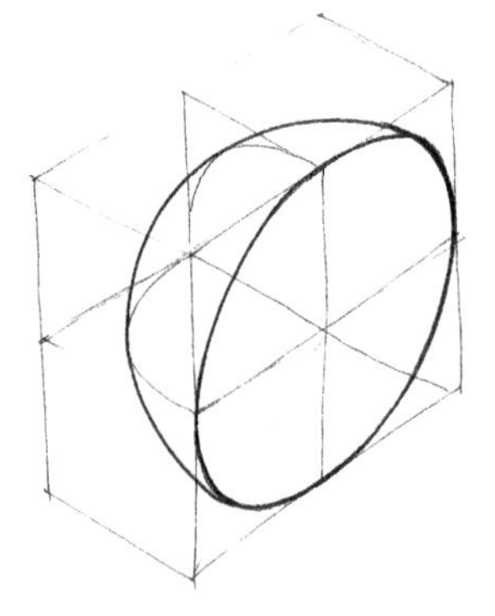

Half a sphere cut horizontally.

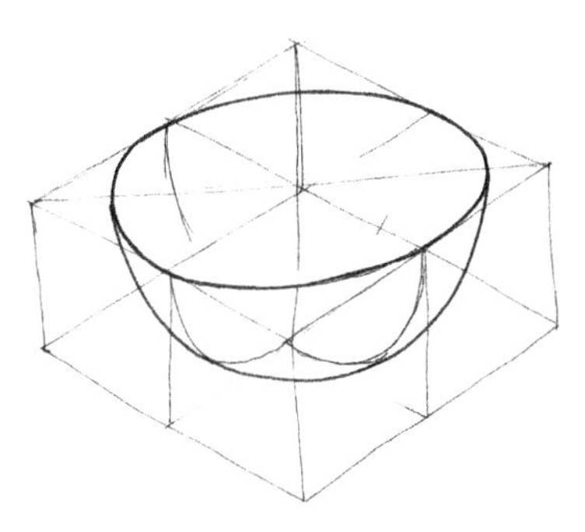

Half a sphere cut vertically.

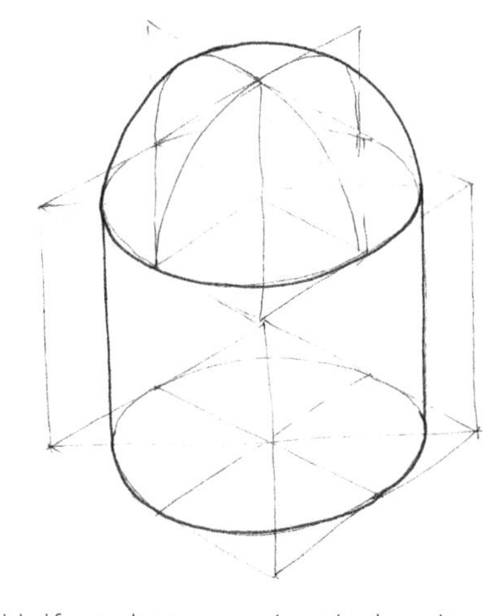

Half a sphere constructed on top of a cylinder.

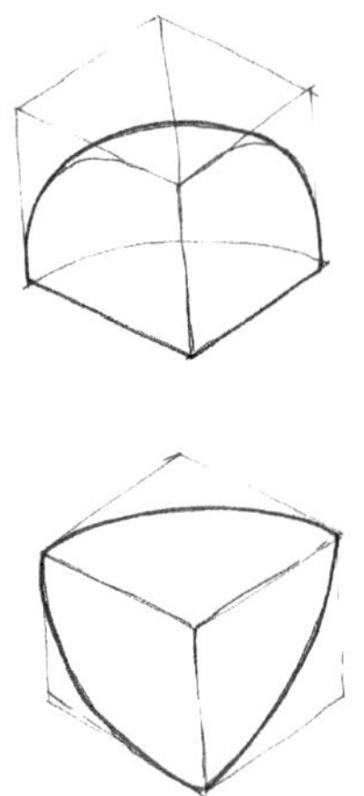

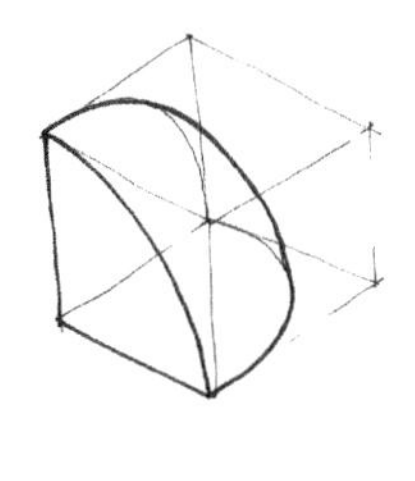

Quarters of a sphere.

Example **B** demonstrates the step-by-step example of constructing a cube with multiple rounding. You will notice that each corner is actually a quarter of the sphere.

B

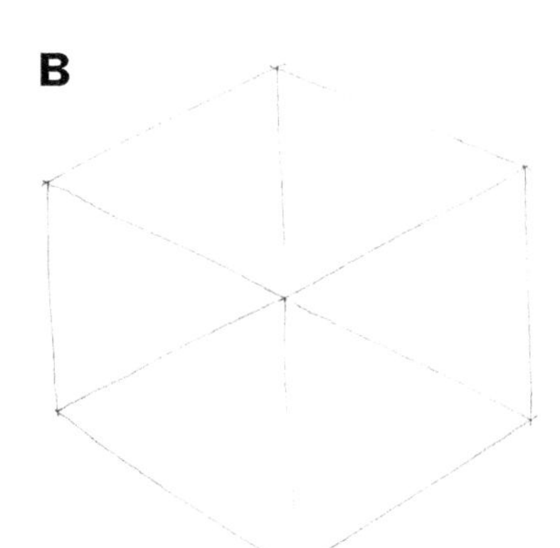

Step 1 • Construct a box.

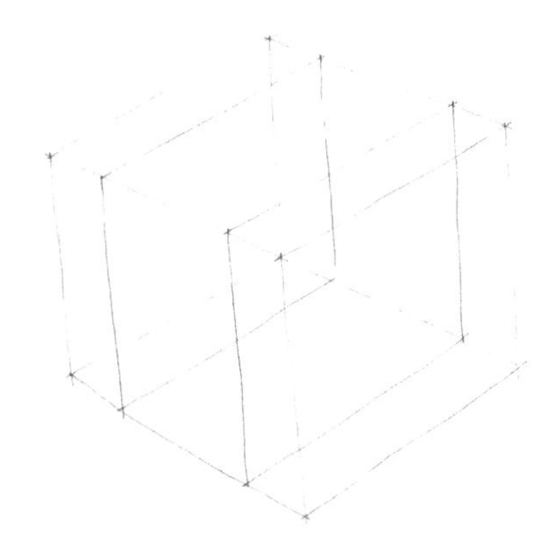

Step 2 • Construct sections vertically.

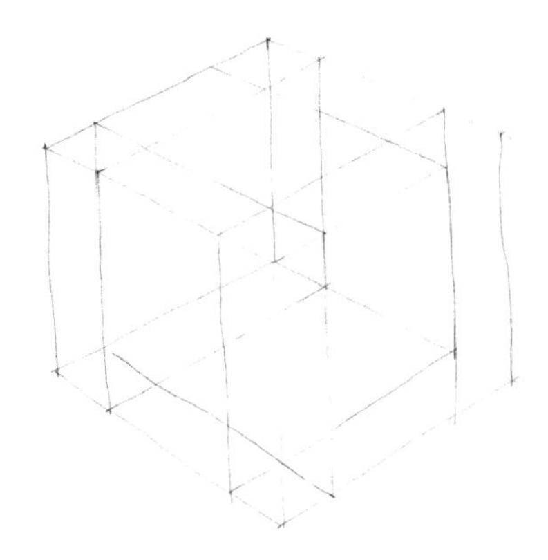

Step 3 • Construct sections in the opposite direction vertically.

ISBN: 9780170233279

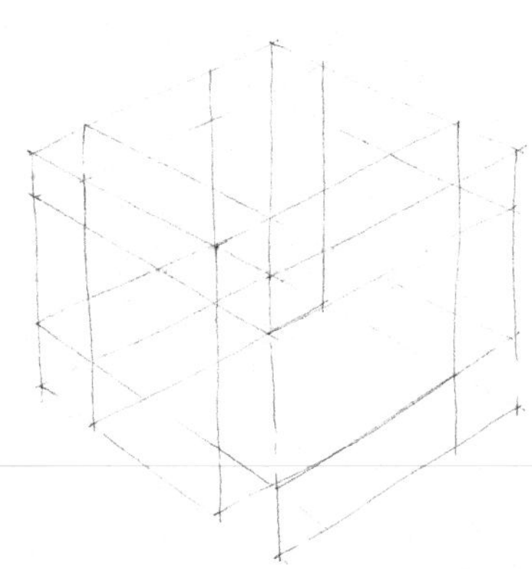

Step 4 • Construct sections horizontally.

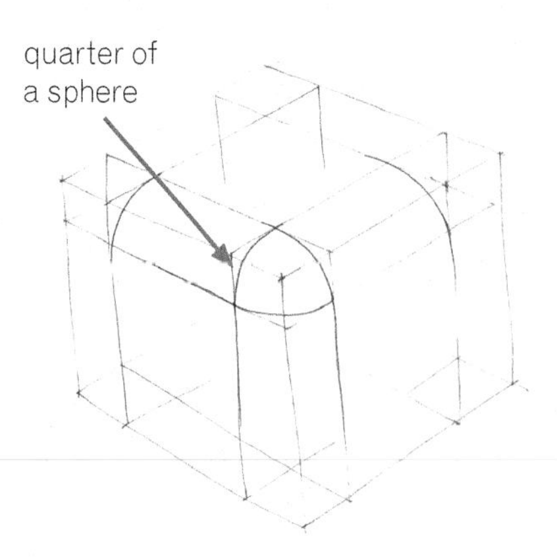

Step 5

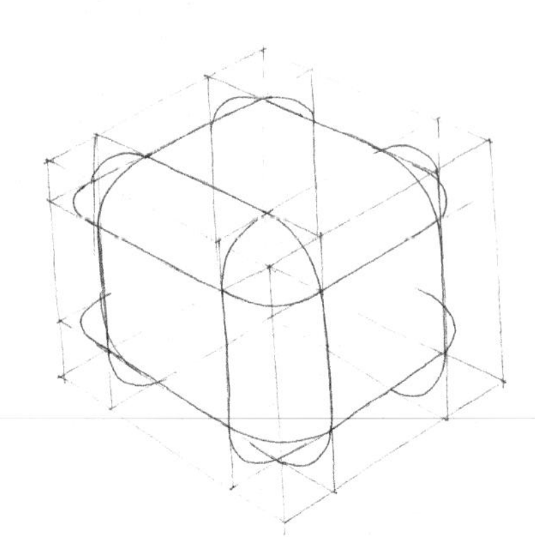

Step 6 • Construct quarter spheres in each corner.

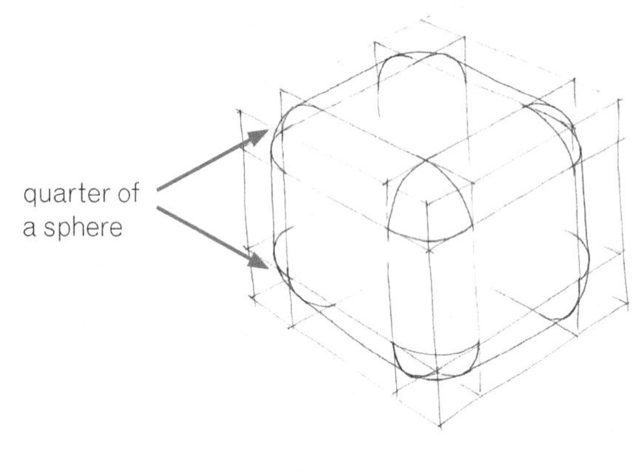

Step 7

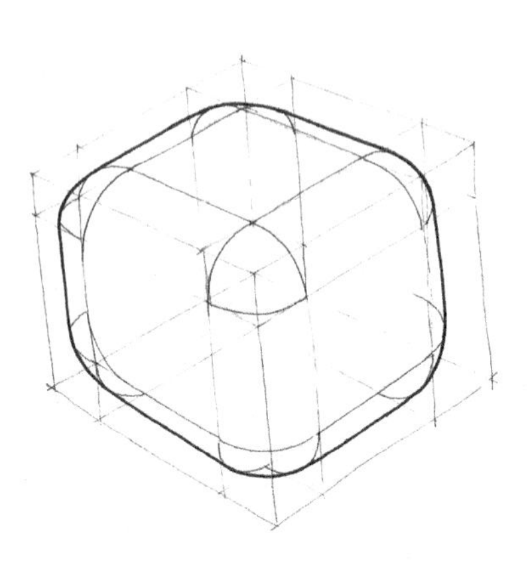

Step 8 • Outline cube with multiple rounding.

Application of multiple rounding

A back-pack clasp has many curved edges and complex parts which are difficult to construct. The step-by-step construction using the crating method is shown here.

ISBN: 9780170233279

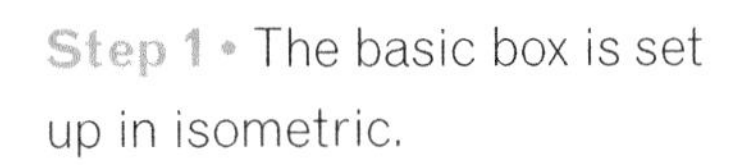

Step 1 • The basic box is set up in isometric.

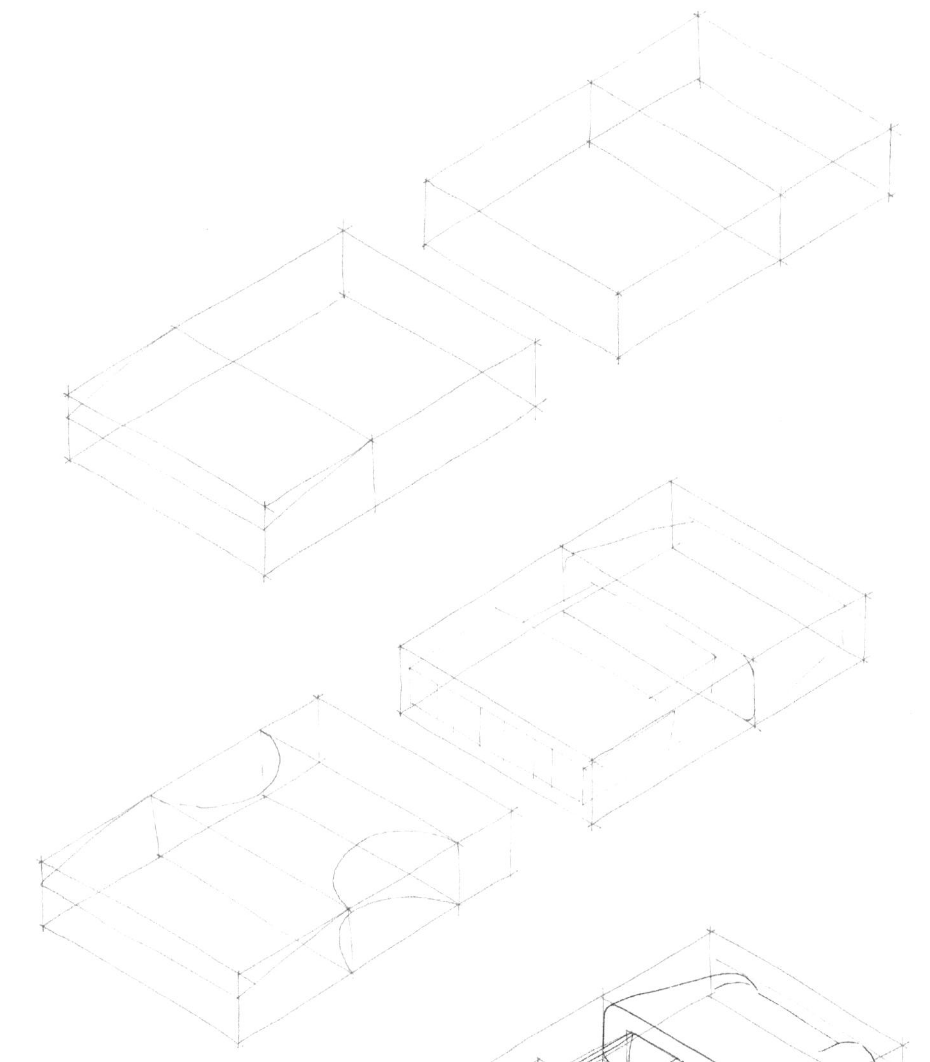

Step 2 • Details of the shape are added.

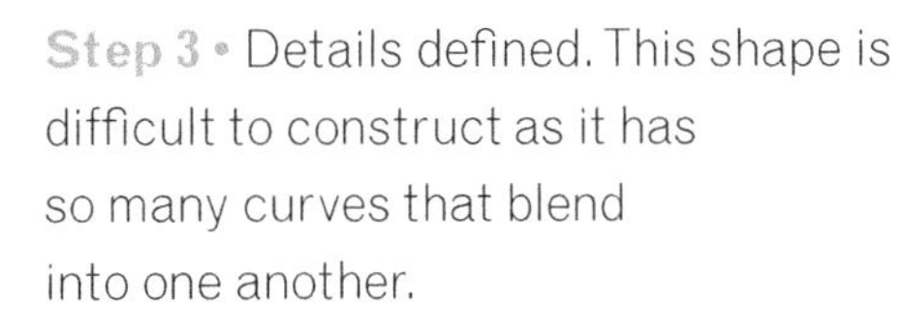

Step 3 • Details defined. This shape is difficult to construct as it has so many curves that blend into one another.

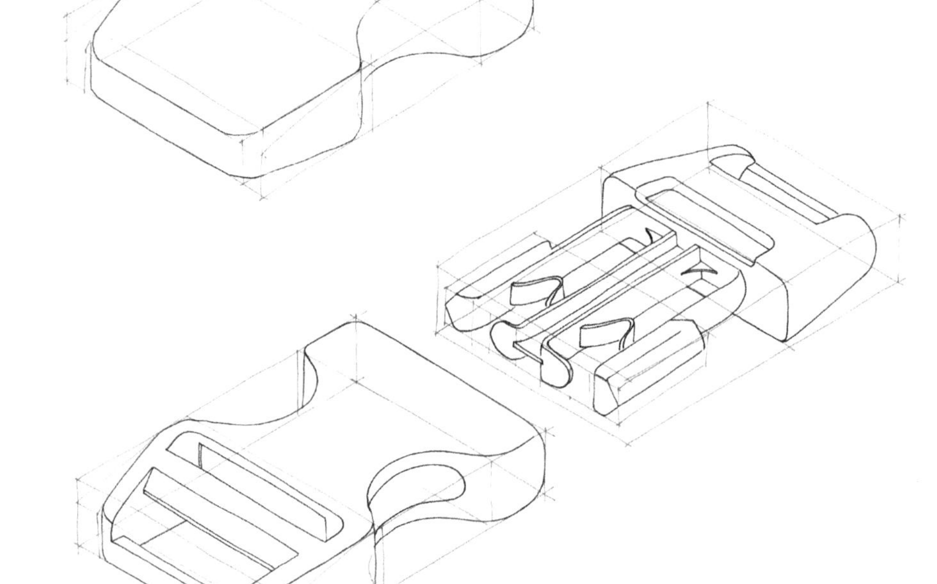

Step 4 • The shape is outlined.

ISBN: 9780170233279

Tubes

Connecting and bending tubes

The detailed sketch of the bike below not only demonstrates the construction of cylinders, rounding and spheres but also the joining and bending of tubes. The enlarged sketches demonstrate the construction of joining one tube to another.

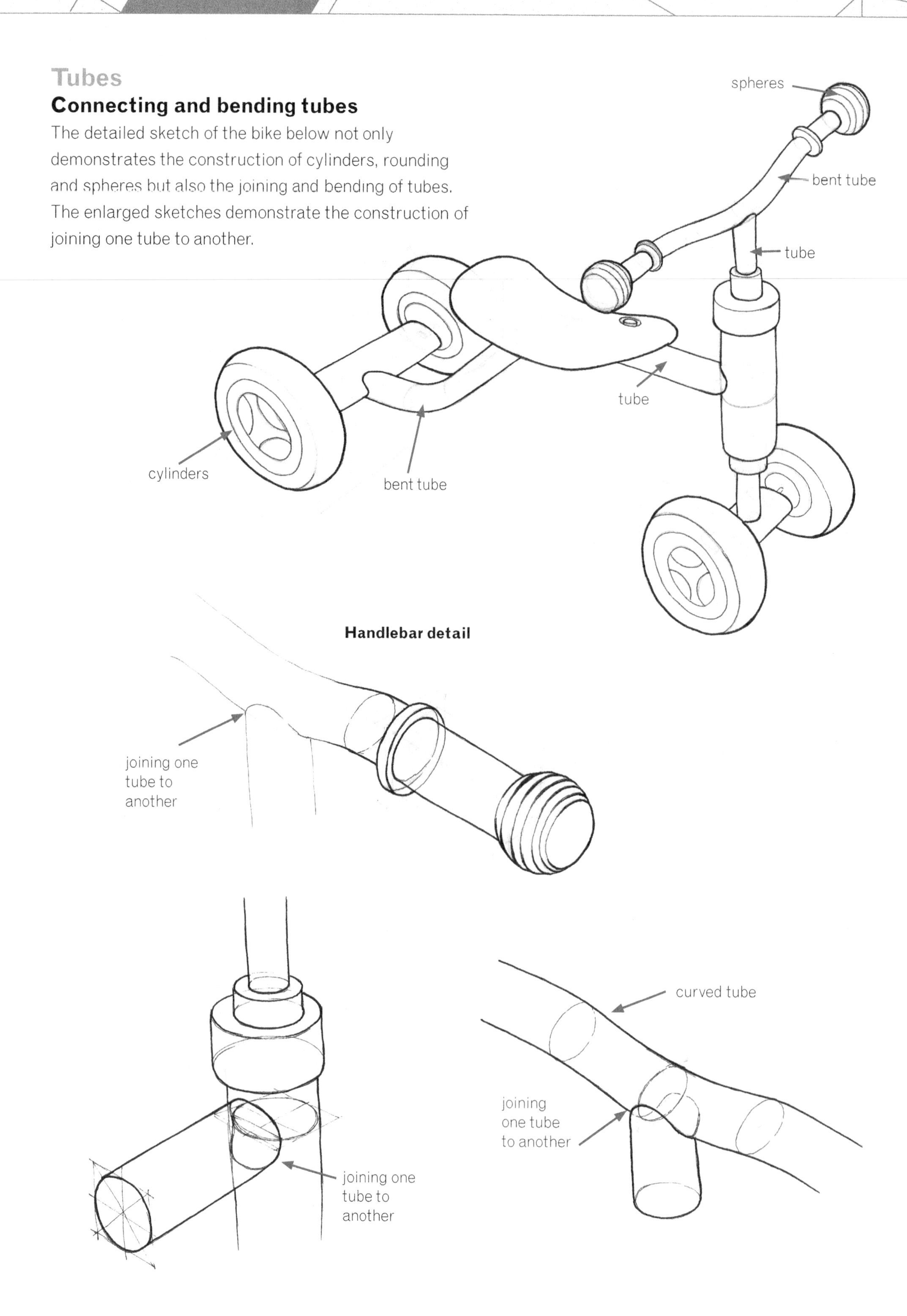

ISBN: 9780170233279

Exploded sketches

Exploded pictorials explain a product with the parts or components pulled apart. Isometric, Oblique and Perspective can all be used to sketch exploded pictorials. The key thing to remember is to project all parts in line with one another, it is easiest to do this from the centre point. You will notice that for the sketch of the watch each part is still sketched encased in a box. Apply the same rules to any type of pictorial sketching.

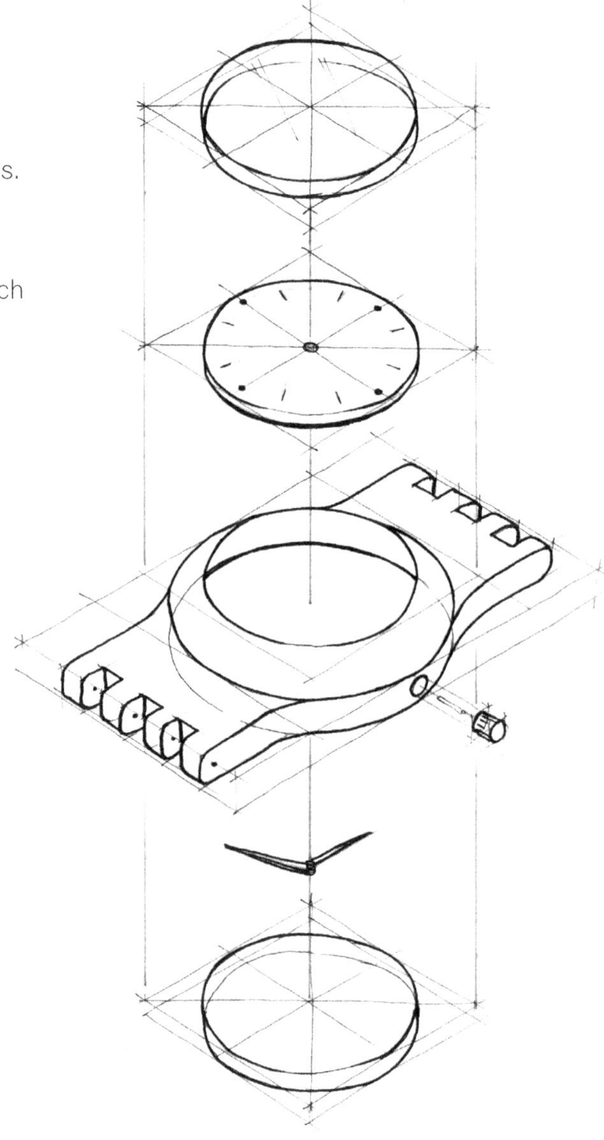

The exemplar of the torch is sketeched in isometric. The cylinders are kept in line with the 30° centre line. Each cylinder is made up of circles. Each of these circles is constructed within a box which has been sketched on a 30° angle. The construction for an object that contains a lot of cylinders or circles can look fairly complex but this method will make sketching in proportion easier.

ISBN: 9780170233279

Paihere Tims — St Peter's College

The following work produced by Paihere Tims and Ambrose Ferrick demonstrates 2D and 3D sketching. They show how the components fit together and also the relationship of the human body to the object.

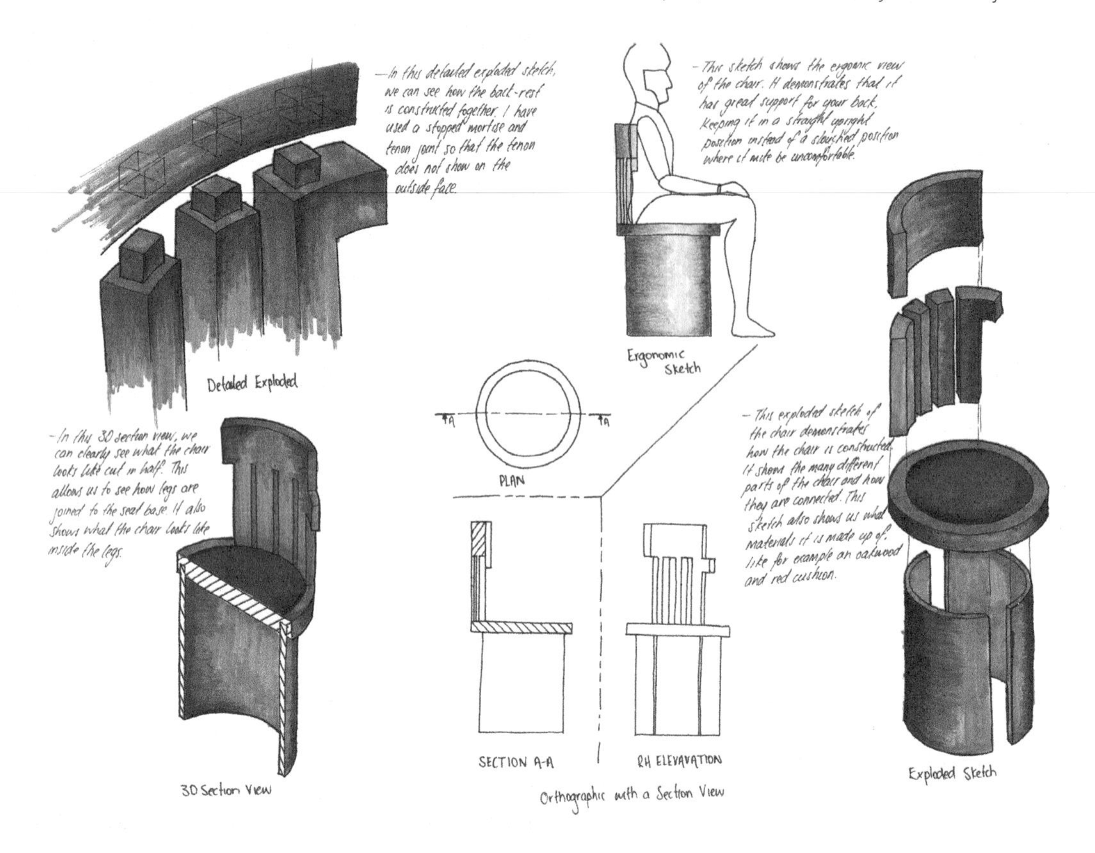

Ambrose Ferrick — St Peter's College

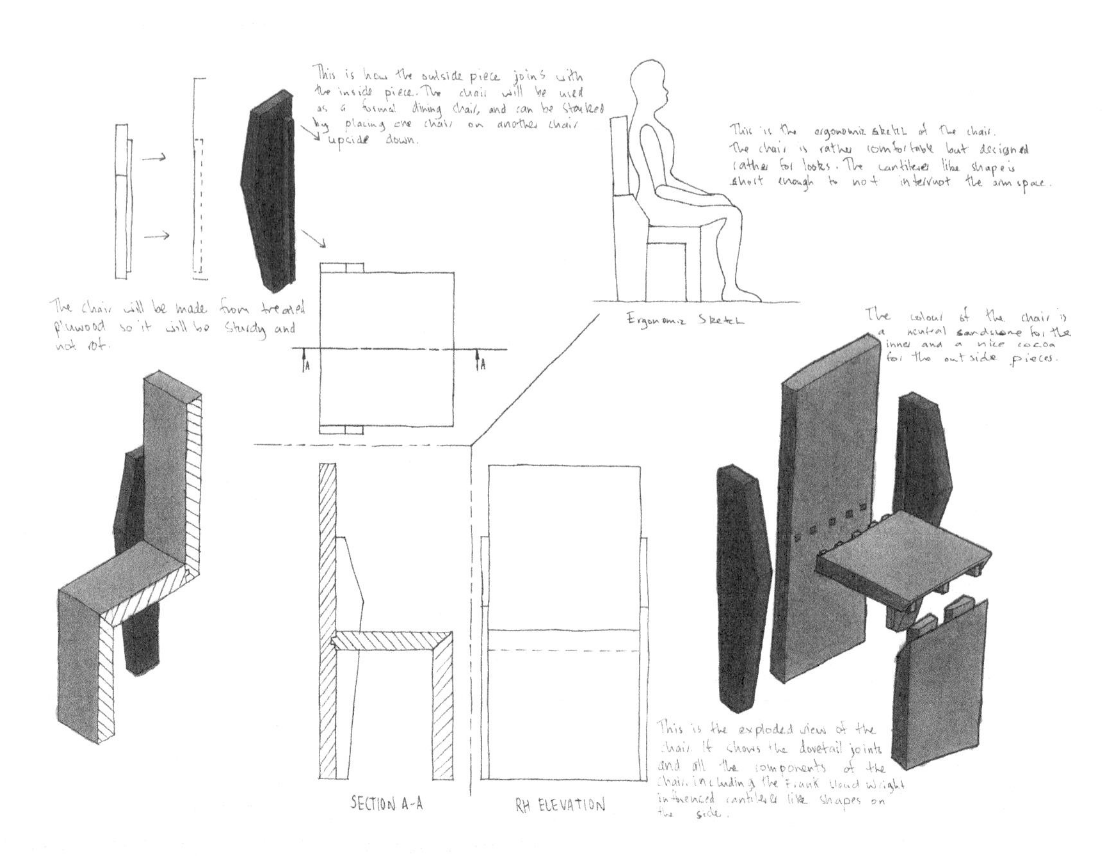

ISBN: 9780170233279

Thanya Chansouk — St Peter's College

The following work produced by Thanya Chansouk demonstrates detailed 2D and 3D sketching to explain his design. He has shown how the components fit together with the use of 2D and 3D exploded and sectional sketches.

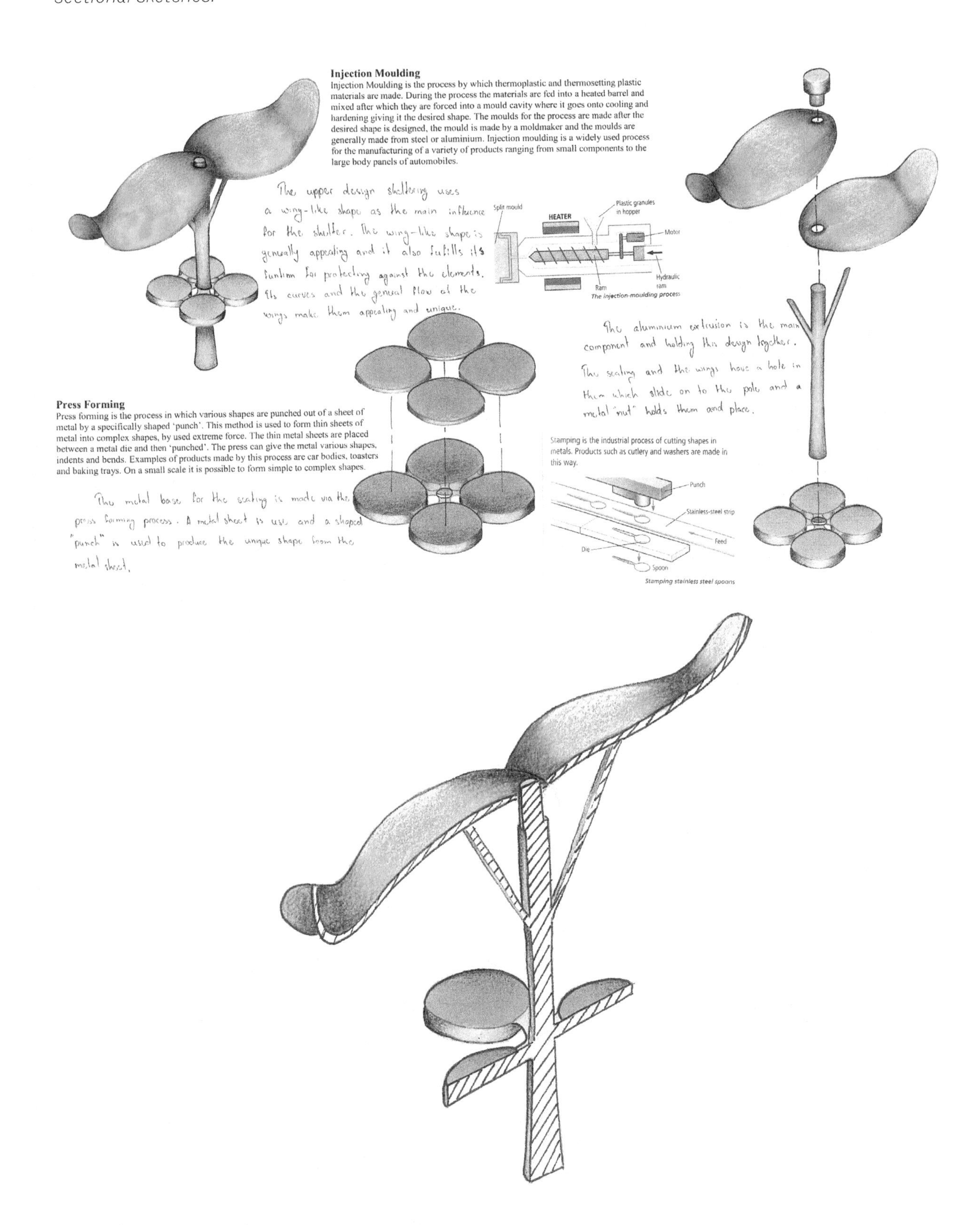

ISBN: 9780170233279

Davin Lim — Wellington College

The following work produced by Daven Lim demonstrates detailed 2D and 3D sketching to explain his design. He has shown how the components fit together with the use of 2D and 3D exploded and sectional sketches.

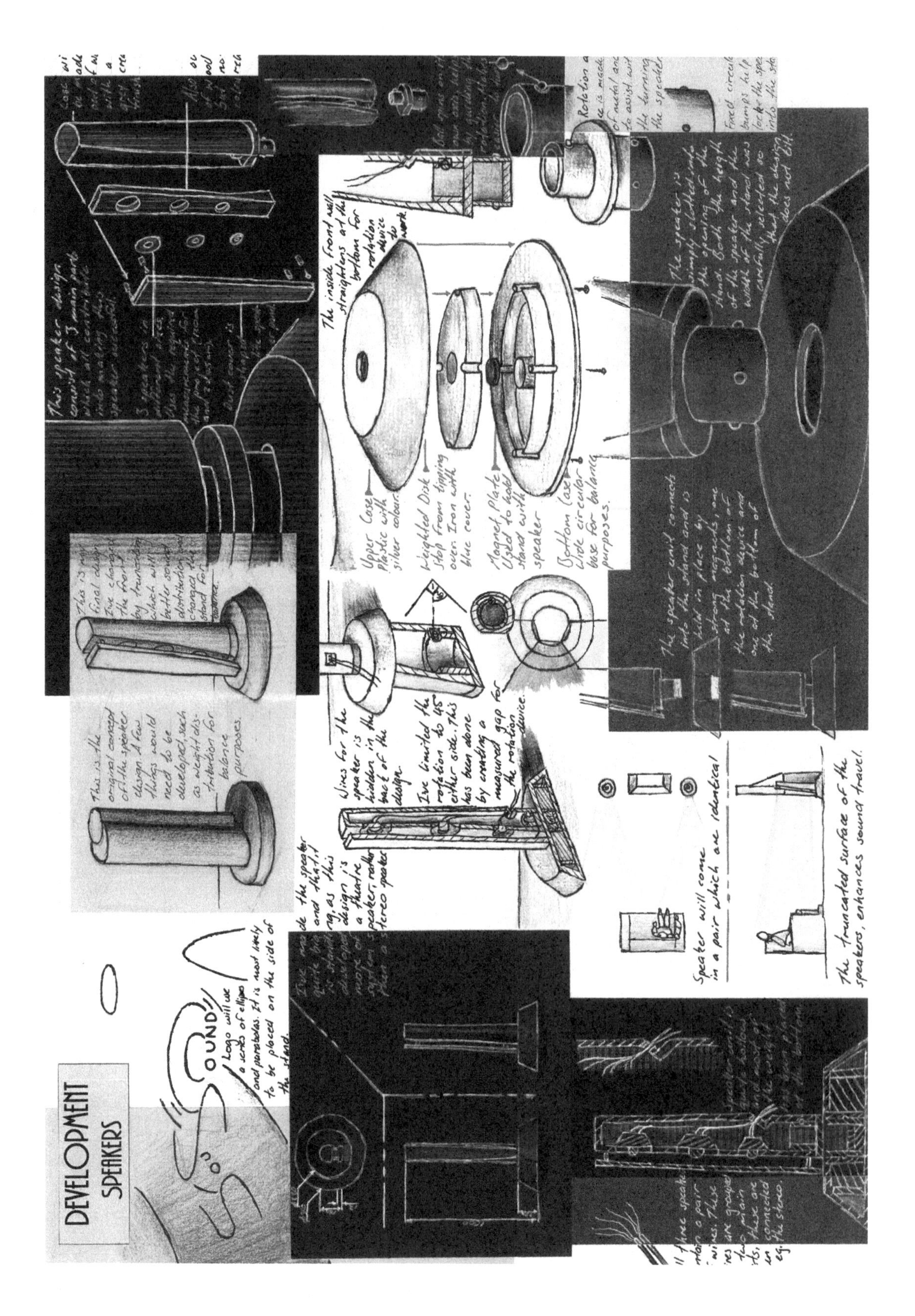

ISBN: 9780170233279

Planometric

Planometric is useful for interior views. The floor plan is drawn on the angles 30° and 60°, or 45° and 45°. Lines are then raised up vertically. The corners of the plan view should be perpendicular to one another.

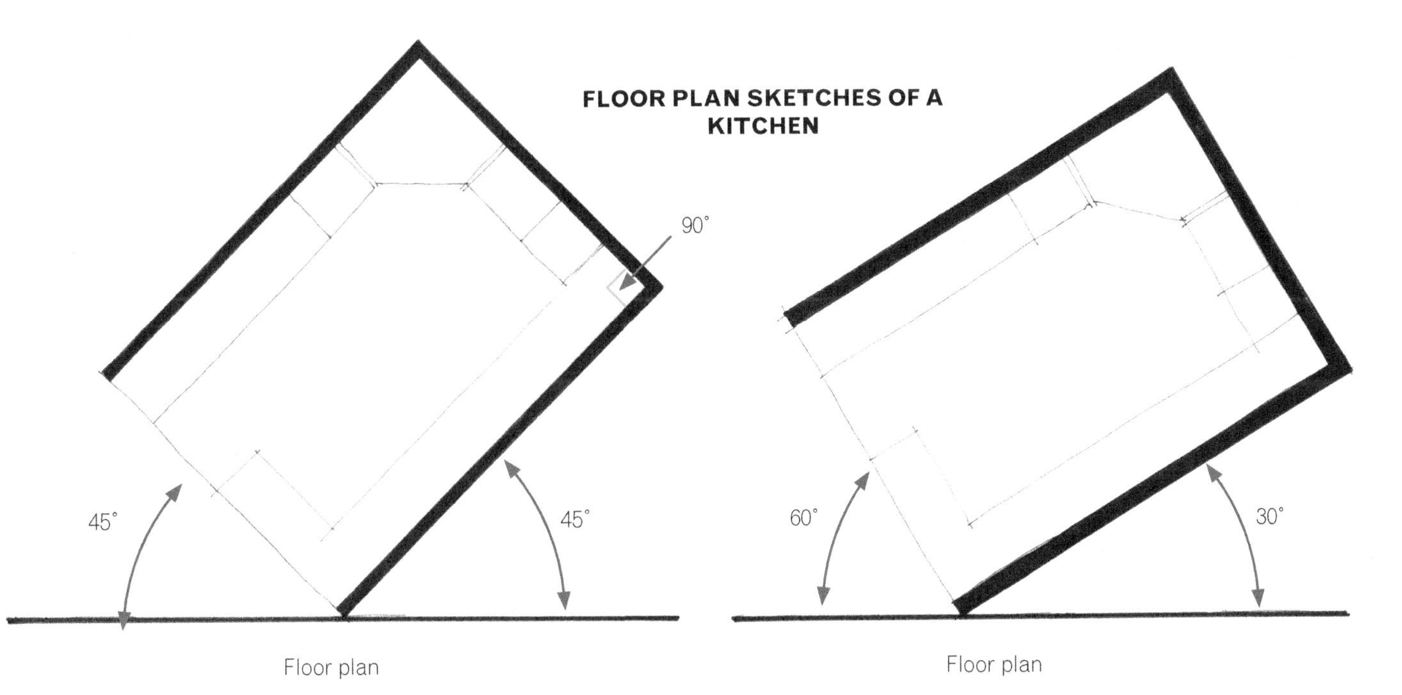

The interior of the kitchen has been sketched on 30° and 60° angles. The planometric sketch does not look as realistic as perspective but is a quick and easy method to present ideas.

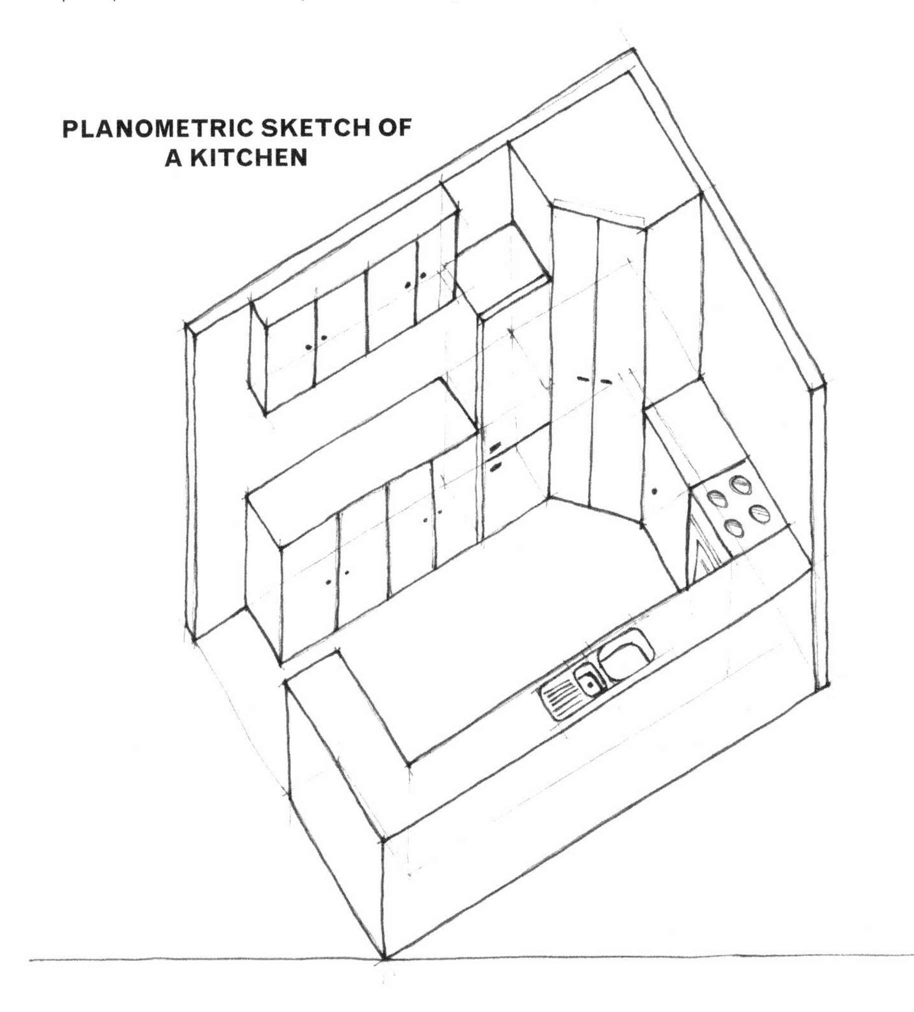

ISBN: 9780170233279

Perspective

Perspective is the most realistic method as it shows objects getting smaller as they go into the distance. It is useful for interior and exterior views but also can be used for drawing products. Objects in perspective can be drawn from different positions giving the advantage of looking up, down or to the side of an object. There are two types of perspective, 1 point and 2 point.

VARIOUS VIEWS IN 1 POINT PERSPECTIVE

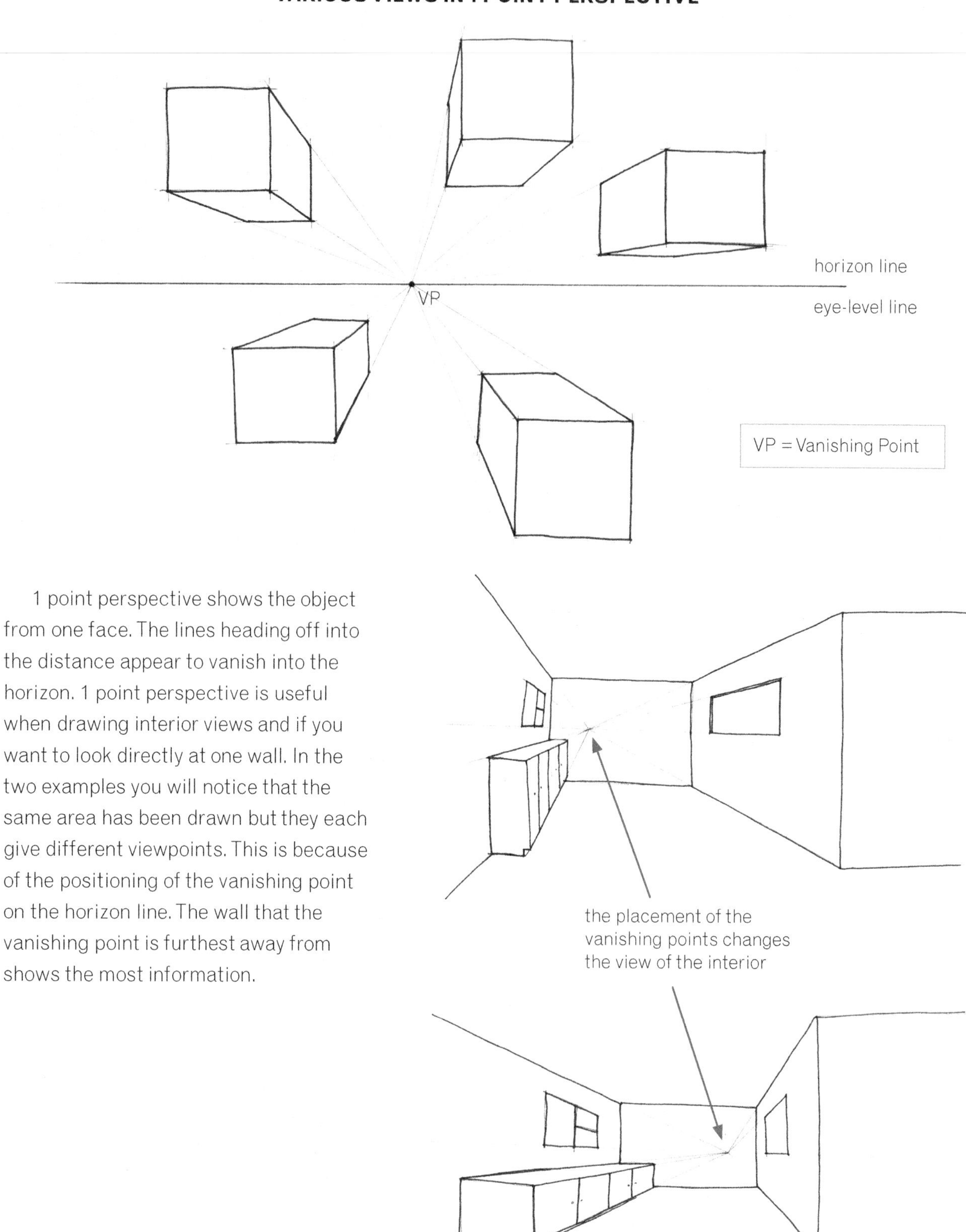

1 point perspective shows the object from one face. The lines heading off into the distance appear to vanish into the horizon. 1 point perspective is useful when drawing interior views and if you want to look directly at one wall. In the two examples you will notice that the same area has been drawn but they each give different viewpoints. This is because of the positioning of the vanishing point on the horizon line. The wall that the vanishing point is furthest away from shows the most information.

ISBN: 9780170233279

2 point perspective always has an edge or corner facing you so is useful for interior and exterior views. The boxes in the example have been sketched in various positions. Some boxes seem closer than others — this depends on the distance from the vanishing point. It appears as if both sides vanish into the horizon. You can get a range of different views dependent on where you position the sketch with reference to the vanishing points. Position the vanishing points as far away as possible from each other. If they are too close they will give you a distorted view. An imaginary cone of vision surrounds the vanishing points. Anything sketched outside the cone of vision will also look distorted as the diagram demonstrates. One important thing to remember is that 2 point perspective requires all lines to be sketched to the vanishing points or vertically unlike 1 point perspective which also uses horizontal lines.

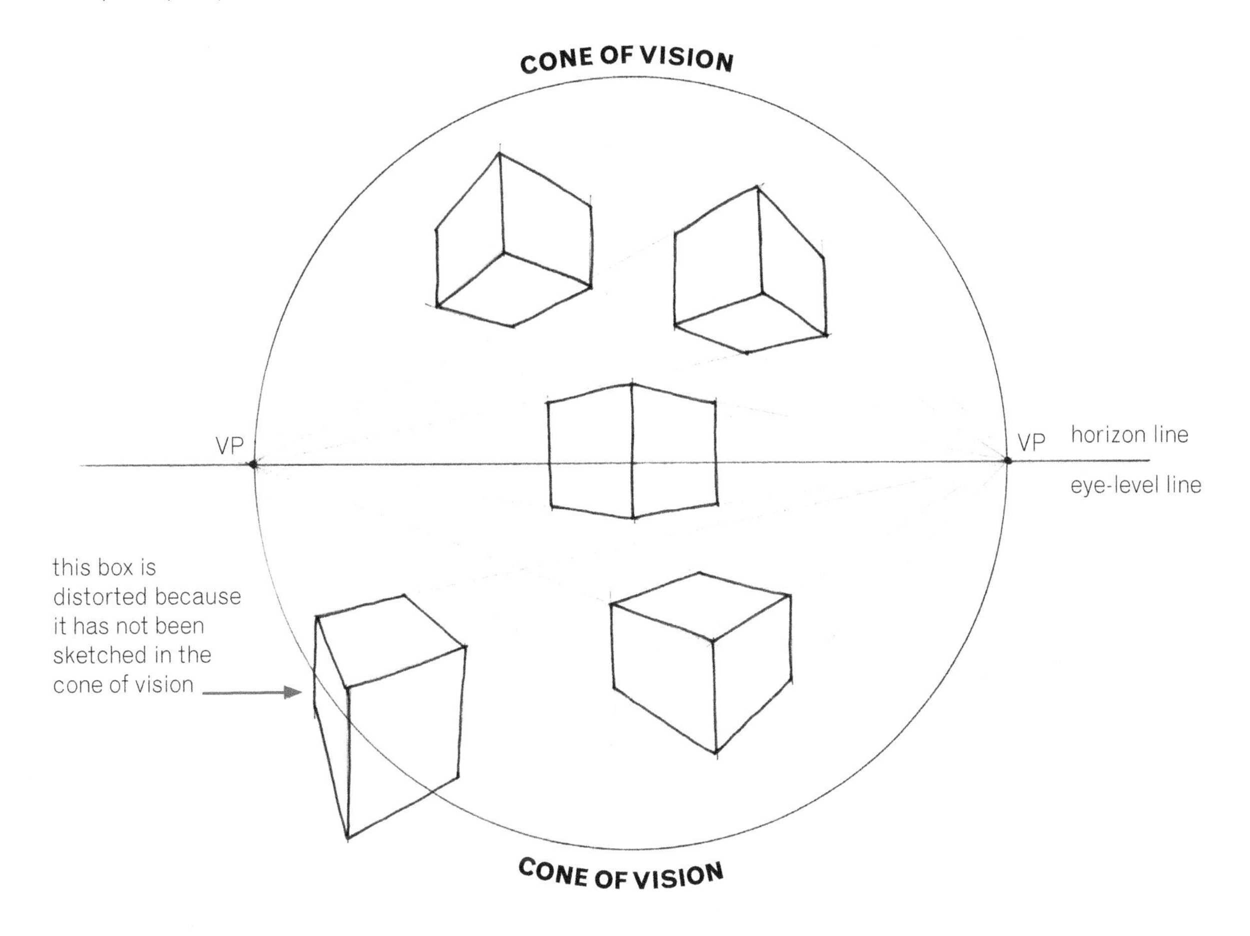

2 point perspective is always seen looking in at a corner wall whereas in the 1 point perspective drawings you always look in at a back wall.

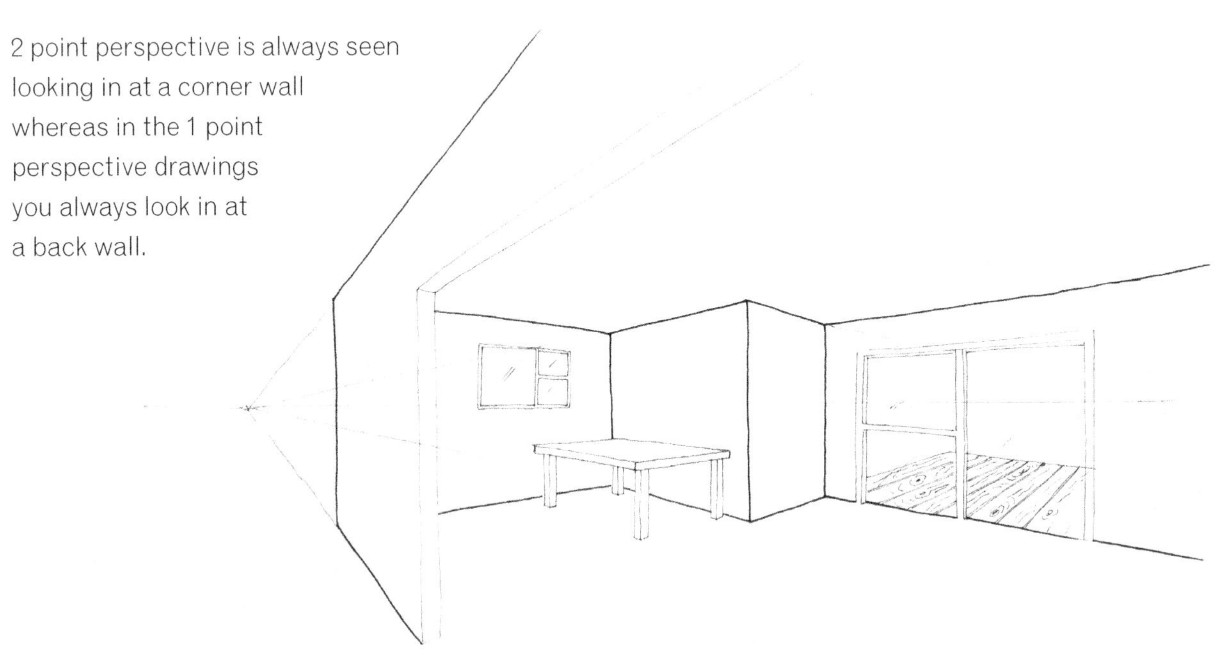

Images that are sketched above and below the horizon line give an indication of scale and height e.g. the horizon line is your eye level line.

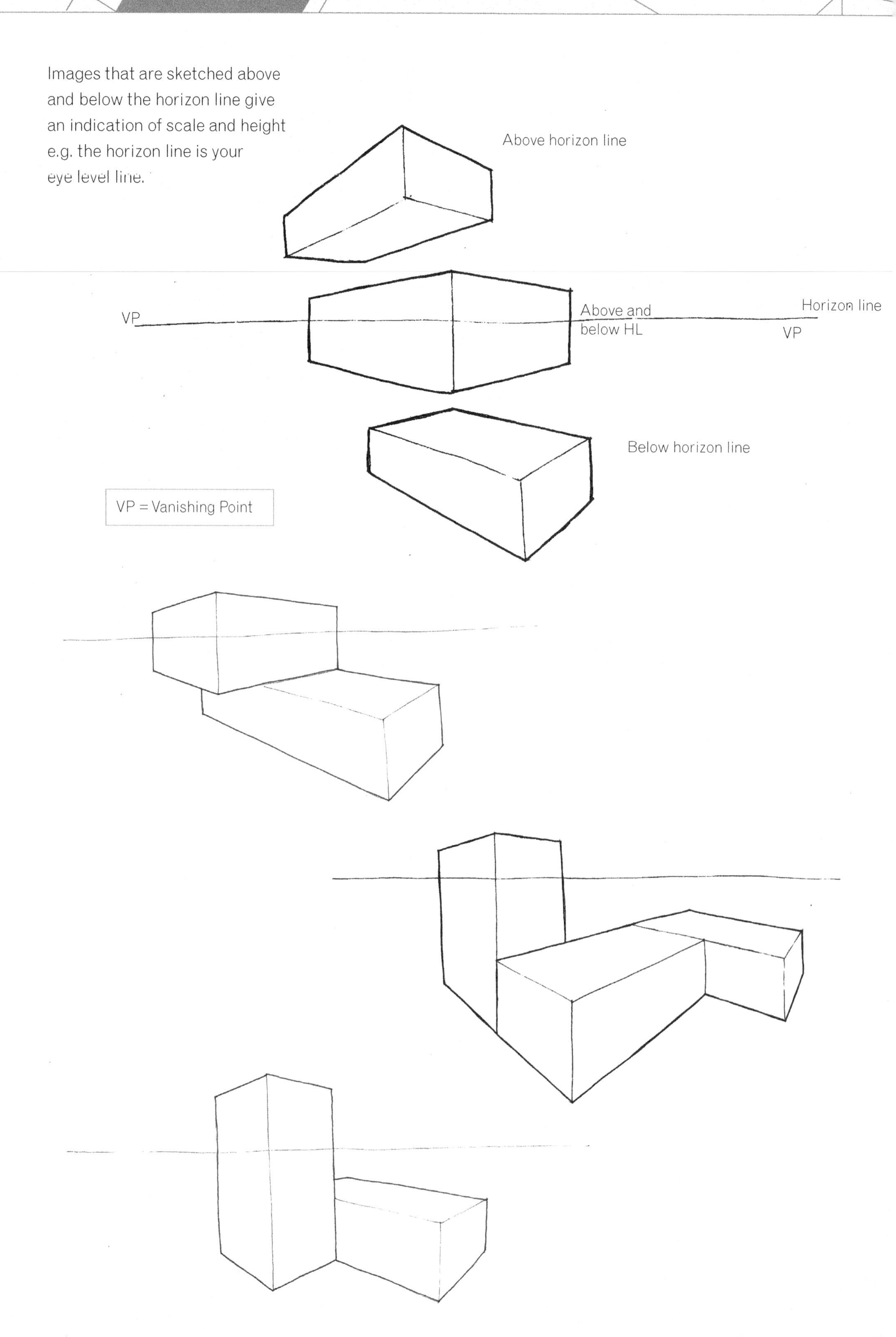

ISBN: 9780170233279

Sketching below the horizon line gives the impression of being above the object and looking down on it.

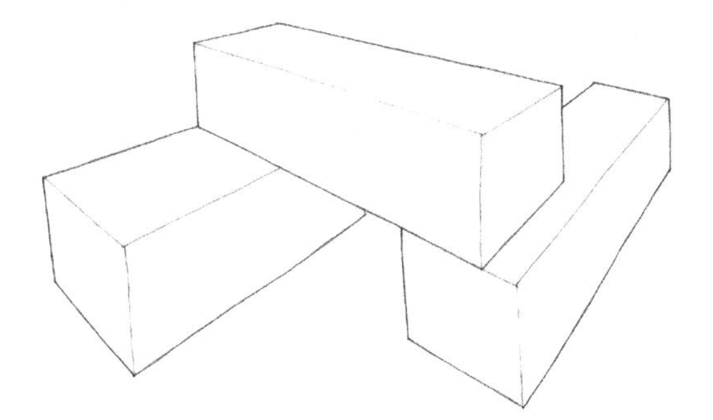

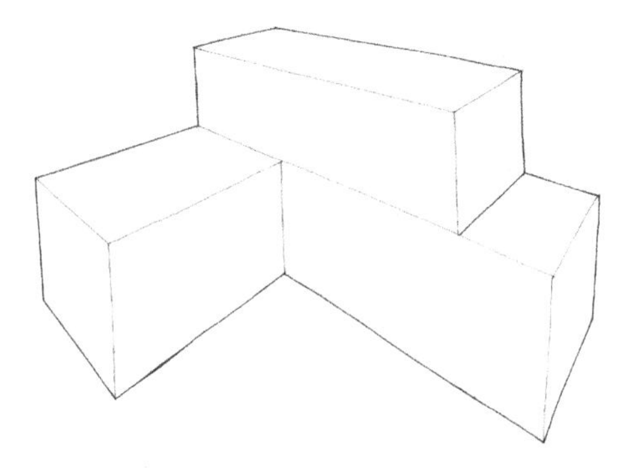

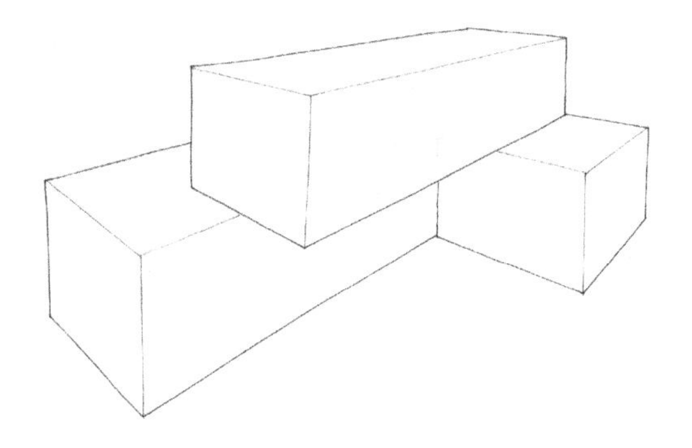

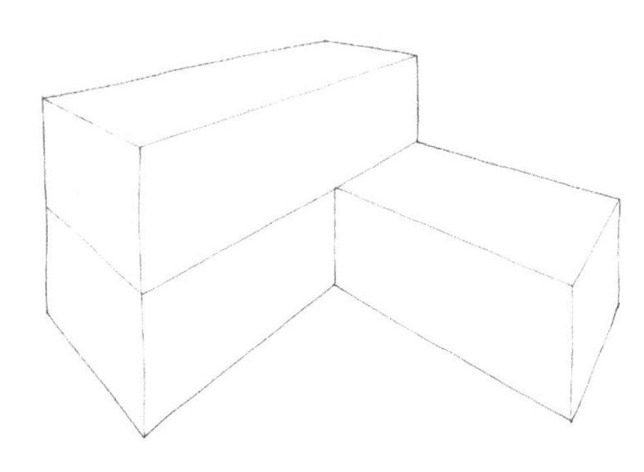

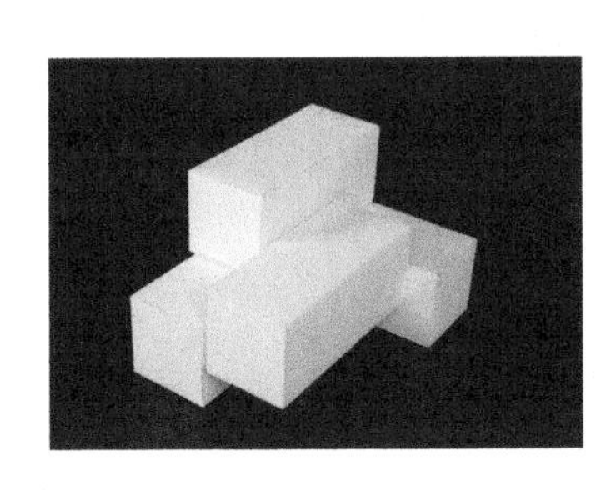
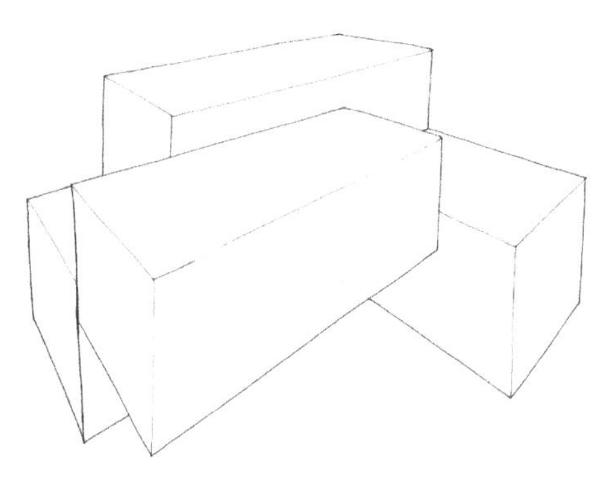

Practice exercise
You could try sketching these objects above and below the horizon line.

ISBN: 9780170233279

Christian Burgos — St Peter's College

The following work produced by Christian Burgos demonstrates 1 point and 2 point perspective sketches to explain the interior and exterior of his design.

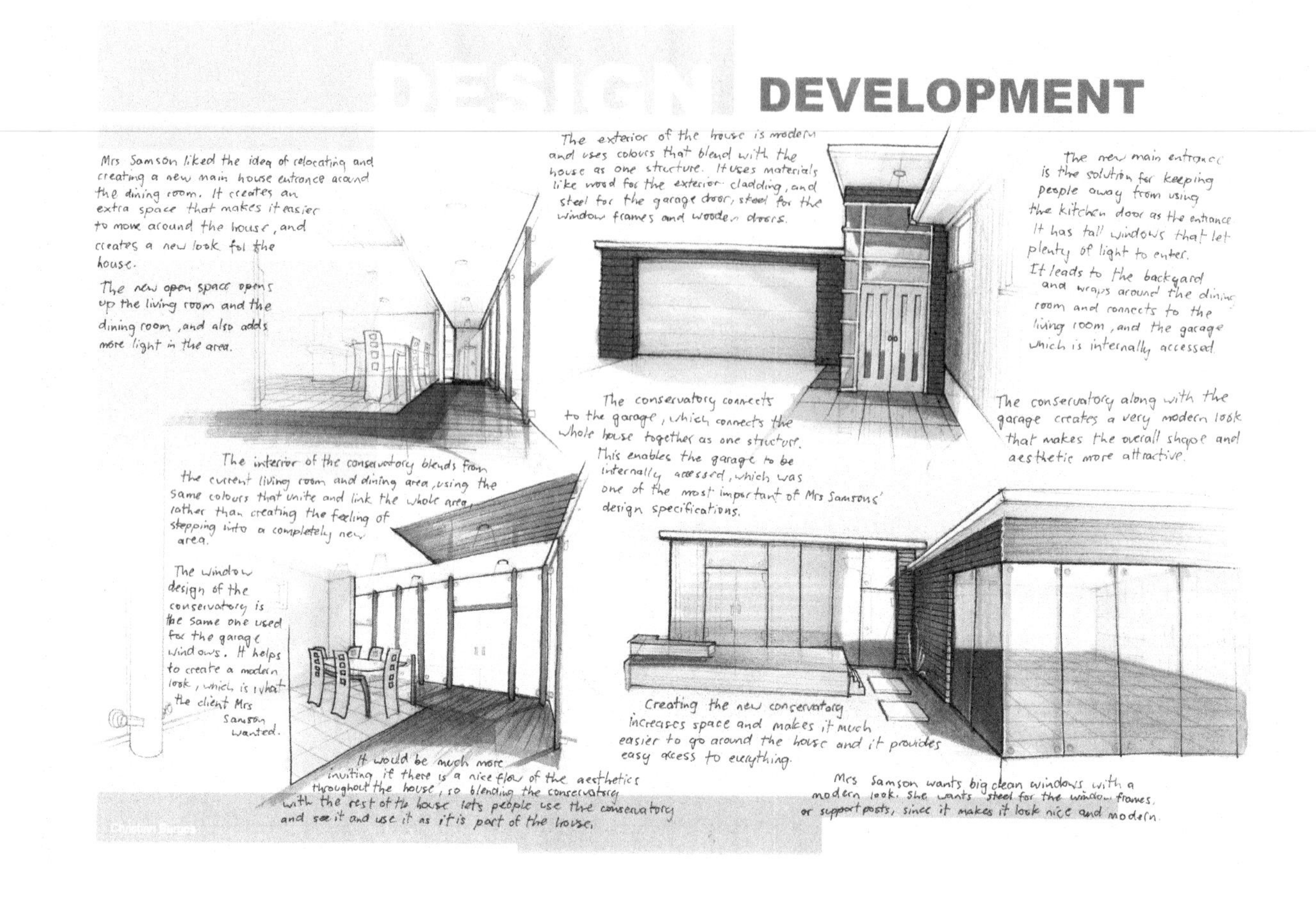

Exterior perspective

The exterior view below gives the impression that you are looking down on the house from above the ground because the entire view has been drawn below the horizon line. You can draw an exterior view with the roof above the horizon line and the floor below the horizon line. This would give you the impression of looking directly at the house from ground level.

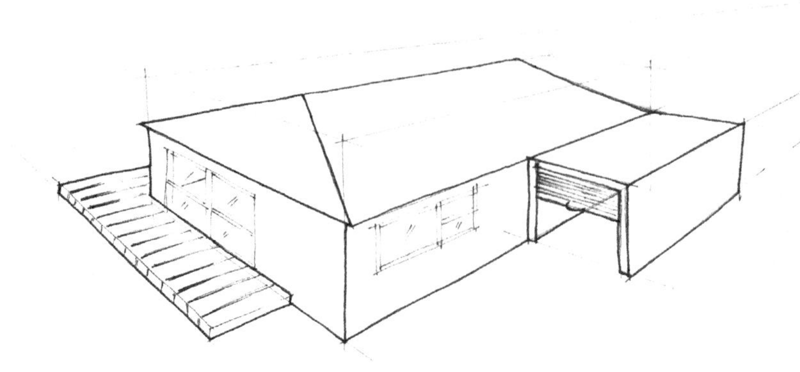

ISBN: 9780170233279

Interior perspective

Perspective is more complex to sketch than the other methods but can look very effective. Try the following exercise, follow each step and you will gain a good understanding of sketching in perspective.

Step 1 • Sketch the horizon line and the back corner line. Join the bottom and top of the corner line to the vanishing points.

Step 2 • Sketch a second corner line. Again join the top and bottom of the line to the vanishing point.

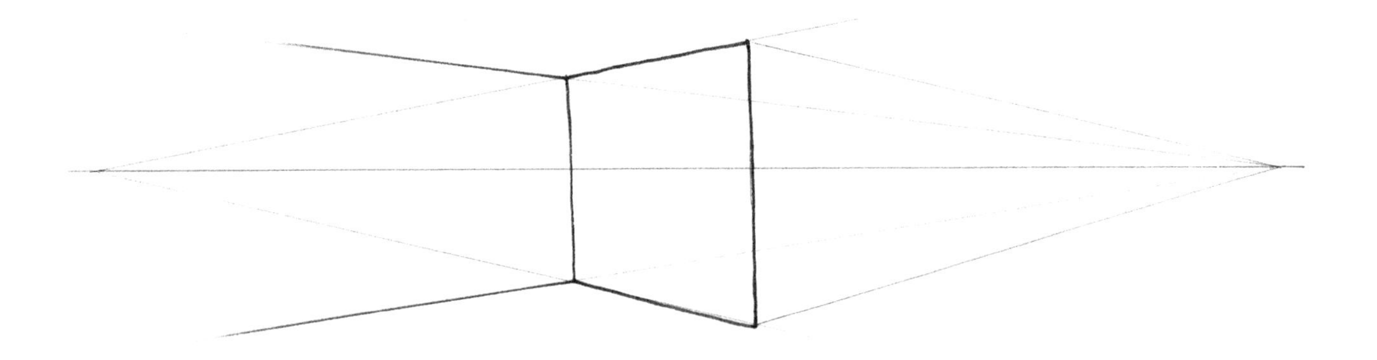

Step 3 • Sketch a third corner line. Join the top and bottom to the opposite vanishing point.

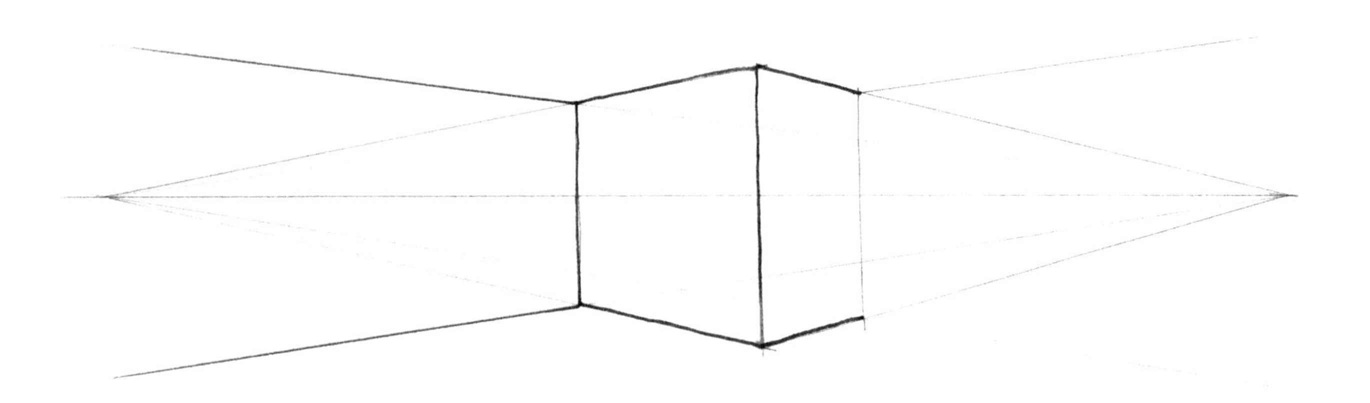

ISBN: 9780170233279

Step 4 • Sketch the fourth corner line.

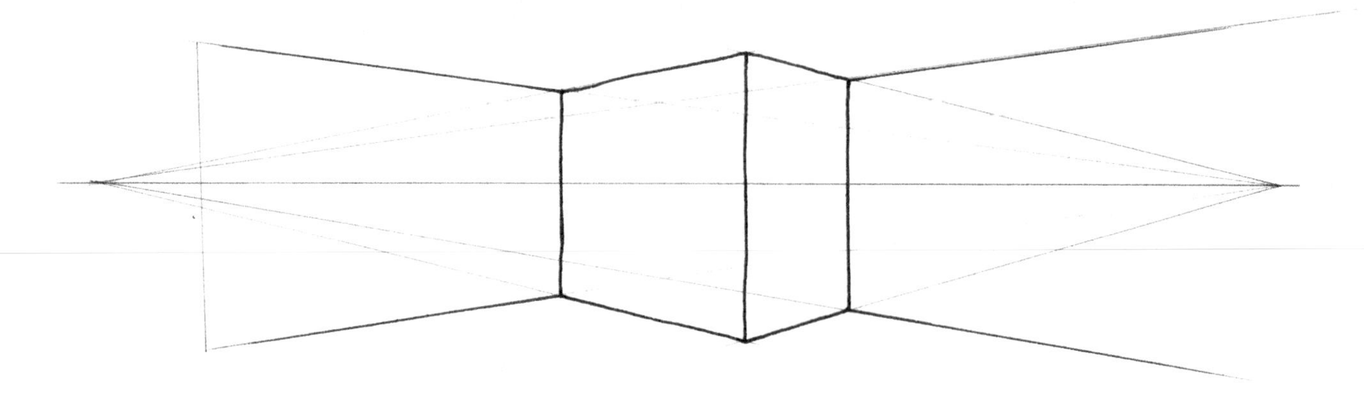

Step 5 • Outline the walls.

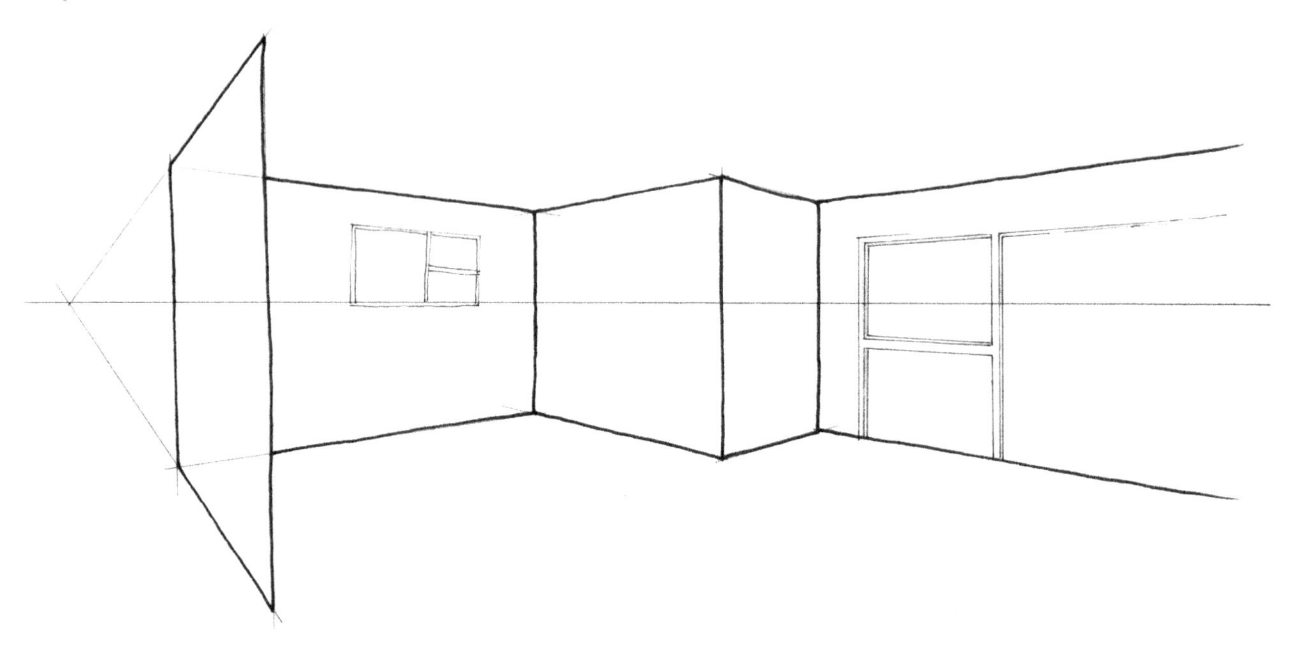

Step 6 • Sketch in the details such as the furniture, window, door and deck.

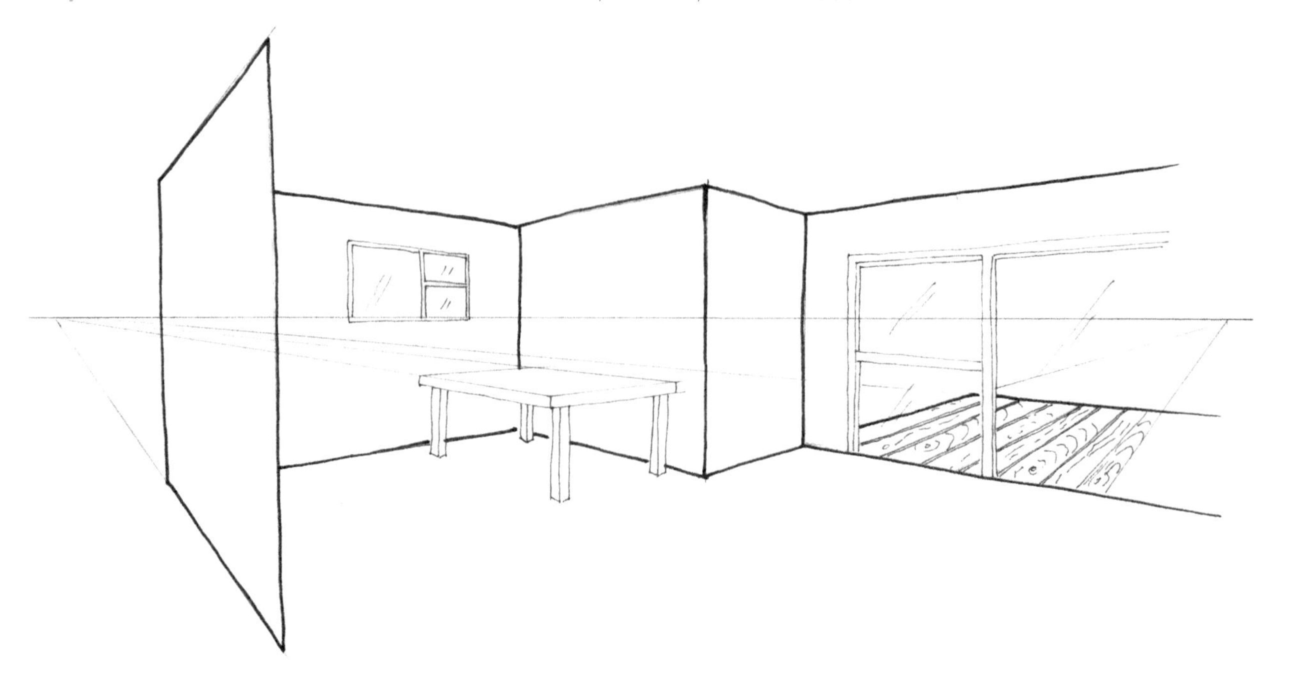

ISBN: 9780170233279

Christian Burgos — St Peter's College

The following work produced by Christian Burgos demonstrates 2D and 3D sketches to explain the interior and exterior. He has used 1 point and 2 point perspective sketches.

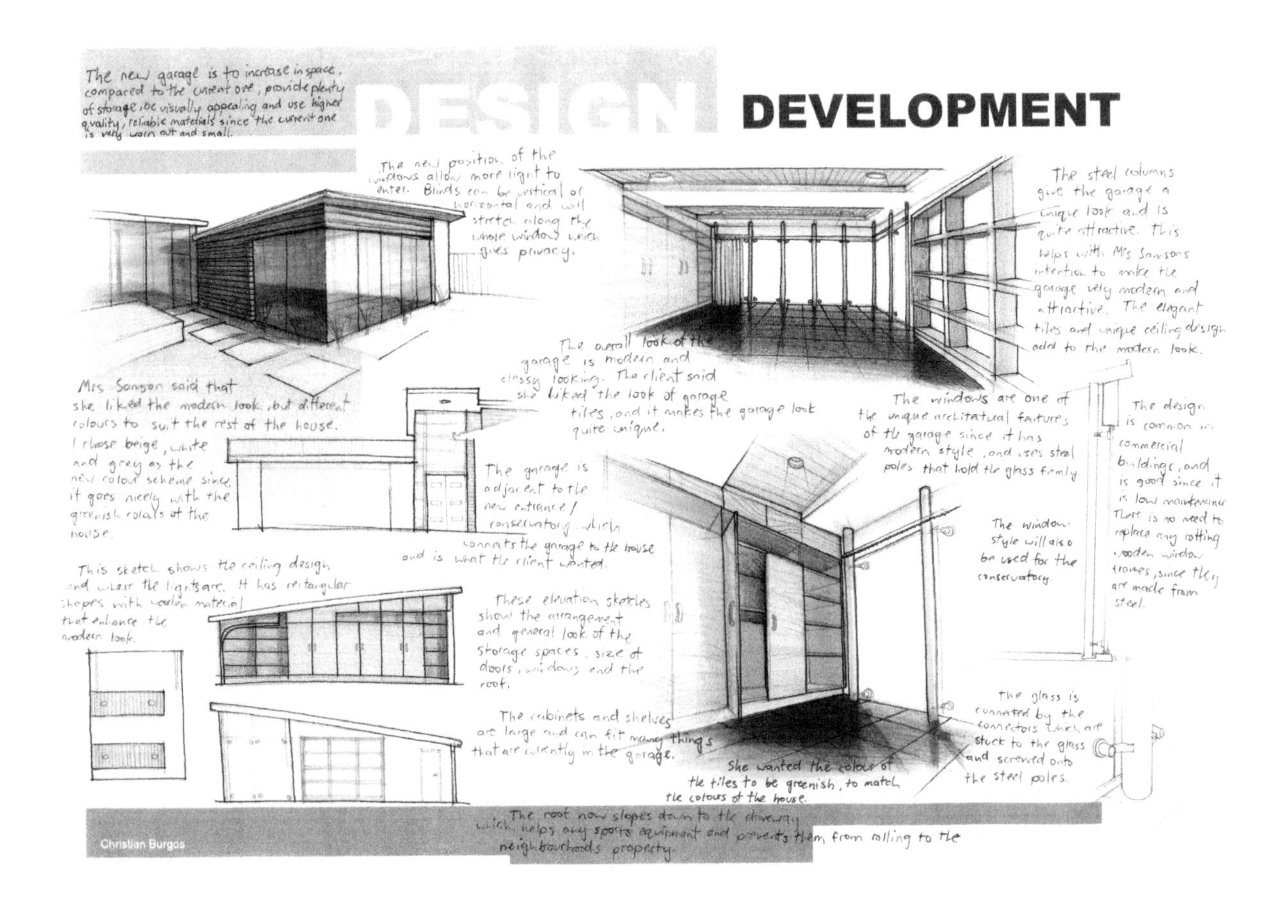

Perspective assembly — crating

The perspective sketch is more difficult to construct using the crating method. The same basic technique is required such as constructing a box to sketch the cylinders inside. The only difference compared to isometric is that the lines for the box need to go back to the vanishing point instead of on a 30° angle.

The step-by-step construction using the crating method is shown here.

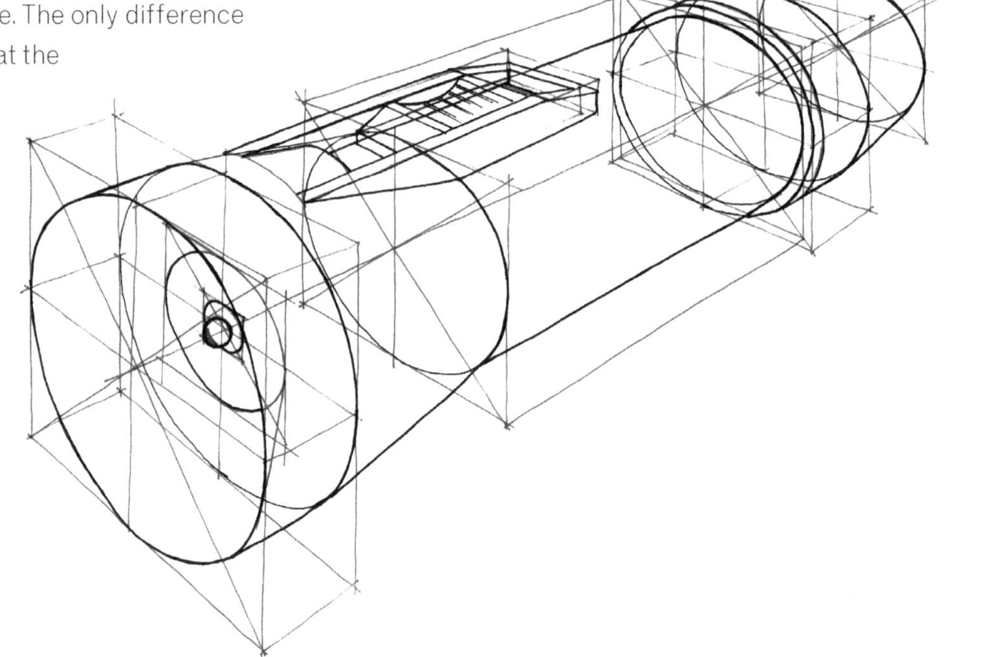

ISBN: 9780170233279

Step 1 • Set your page up correctly for perspective sketching.
VP = Vanishing Point

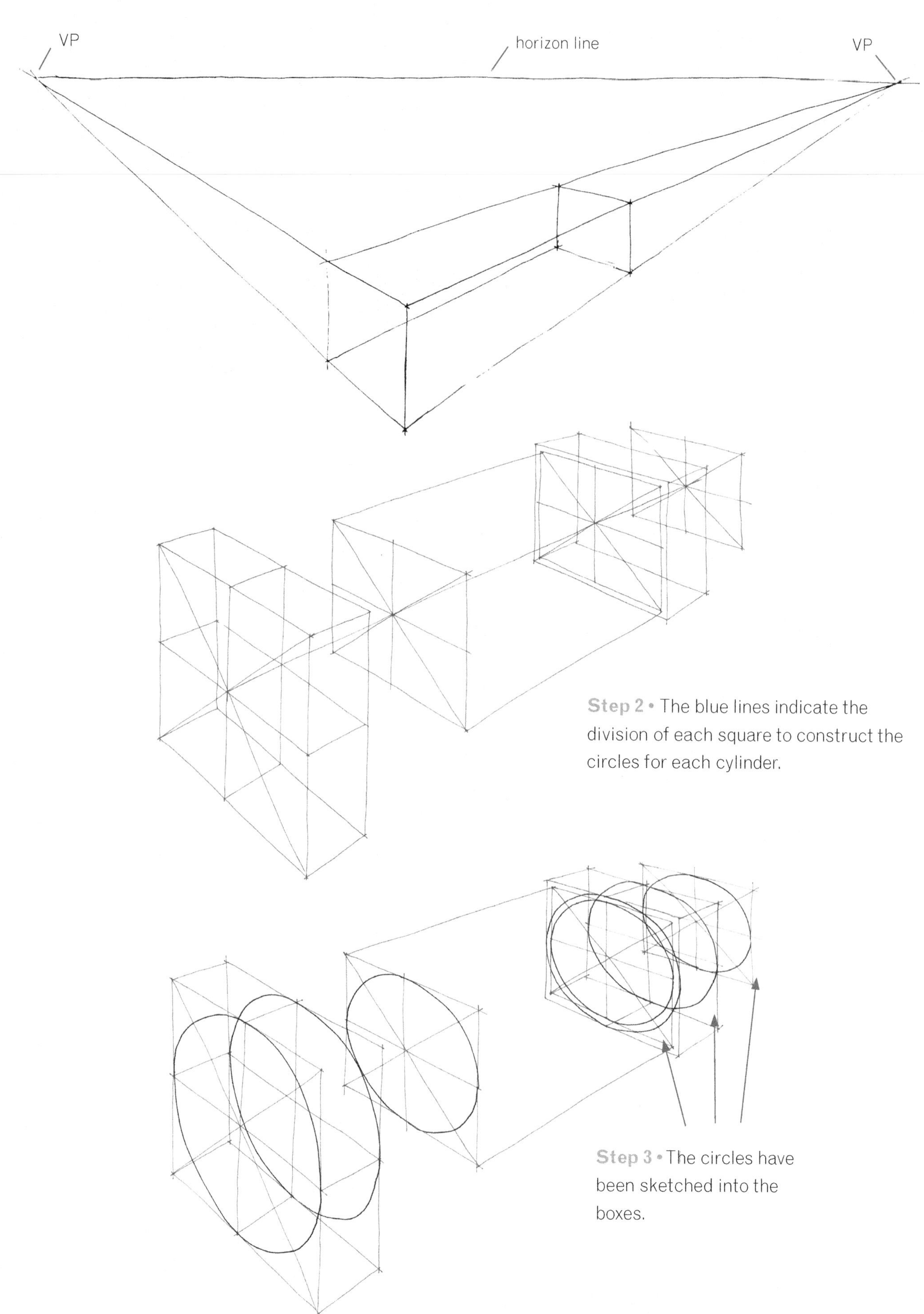

Step 2 • The blue lines indicate the division of each square to construct the circles for each cylinder.

Step 3 • The circles have been sketched into the boxes.

ISBN: 9780170233279

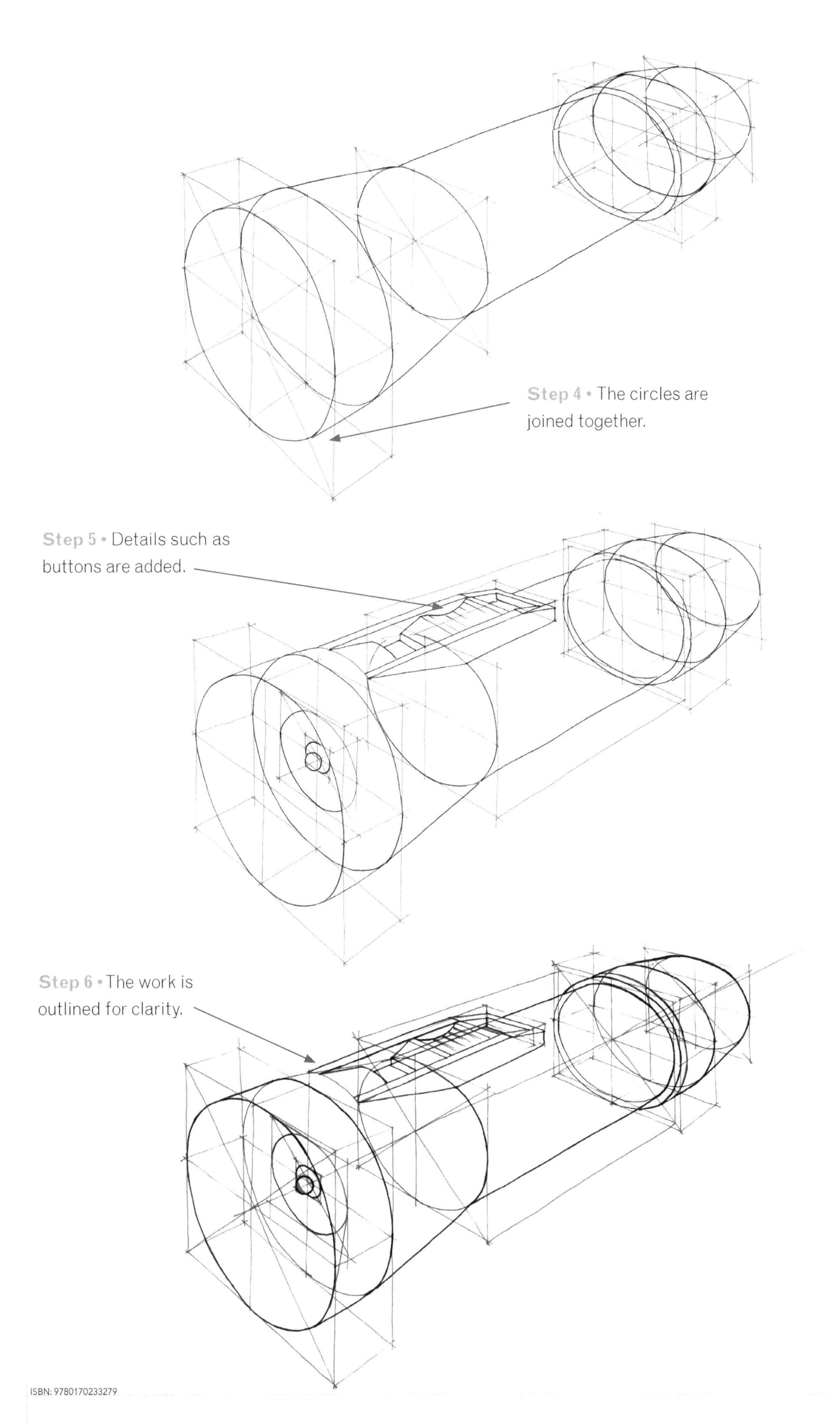

ISBN: 9780170233279

To fully explain an object you need to turn the object around on the page and sketch from many different directions.

H to I

An effective and simple exercise to gain a sound understanding of this is to sketch an H in three dimensions and turn it in different directions looking at the object from many viewpoints until it is an I. You could do the same with the letters N to Z shown on the following page. You will notice that in both examples oblique, isometric and perspective have been used to achieve this. The example N to Z demonstrates another dimension by overlapping each sketch.

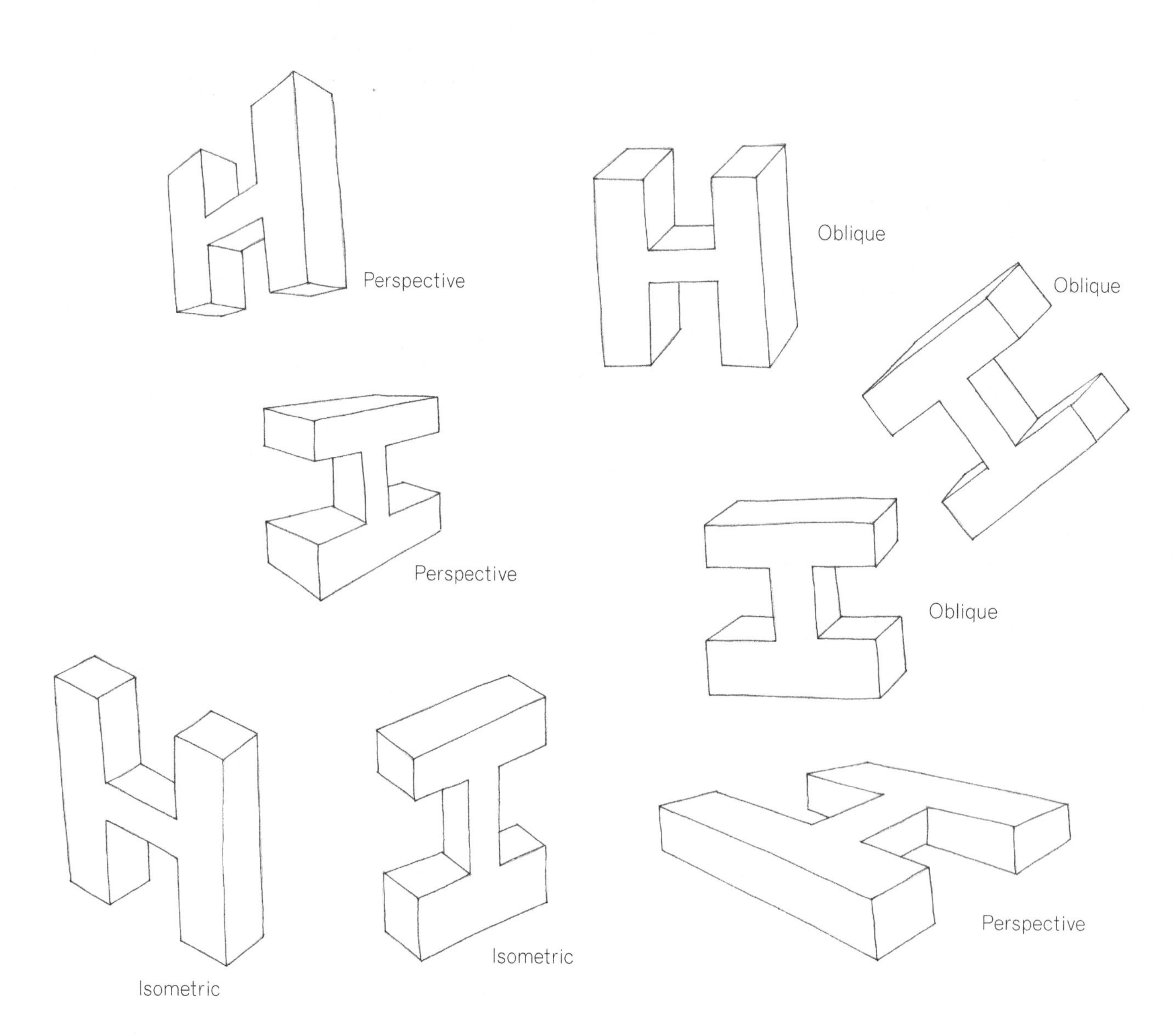

ISBN: 9780170233279

N to Z

Block model sketched from many directions

A complex block model made up of basic geometric shapes has been sketched from many different directions. The detail of the shape and how components join together have been fully explained visually using this method.

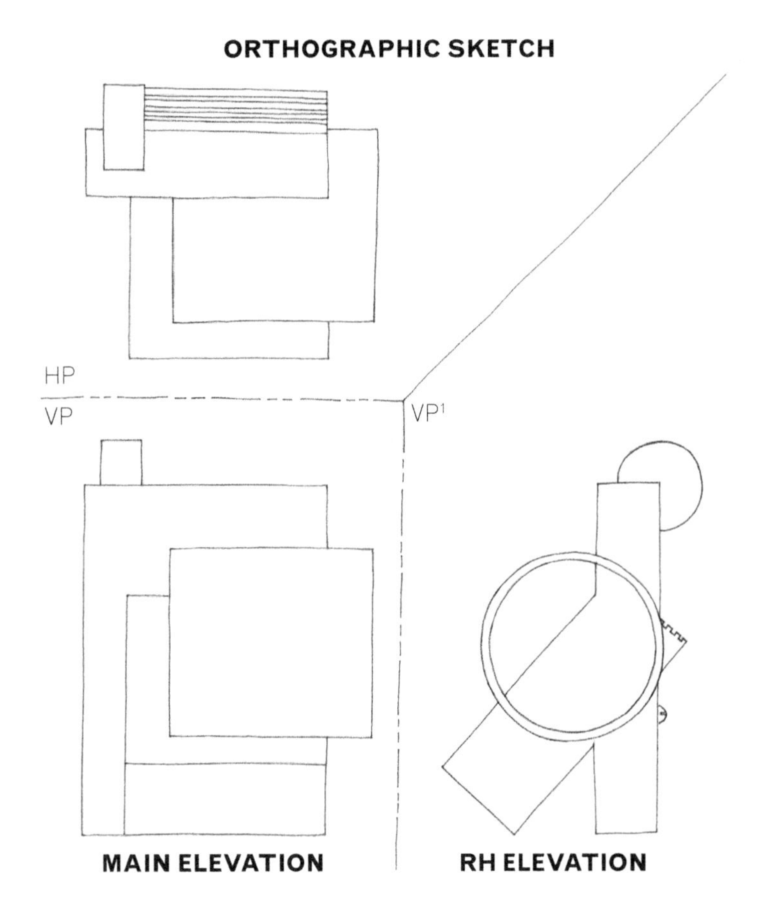

ISBN: 9780170233279

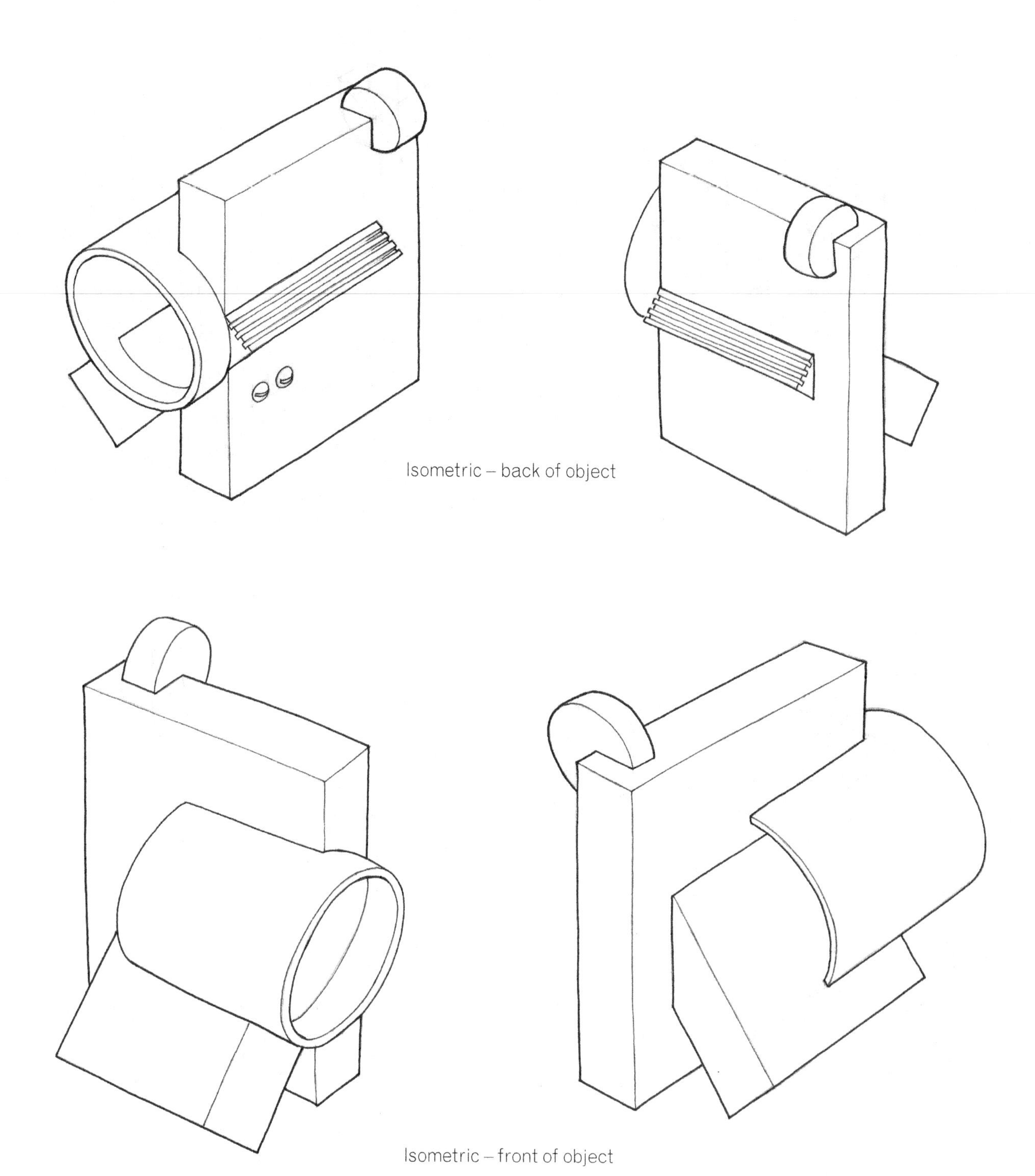

The photographs below show the actual object that has been sketched.

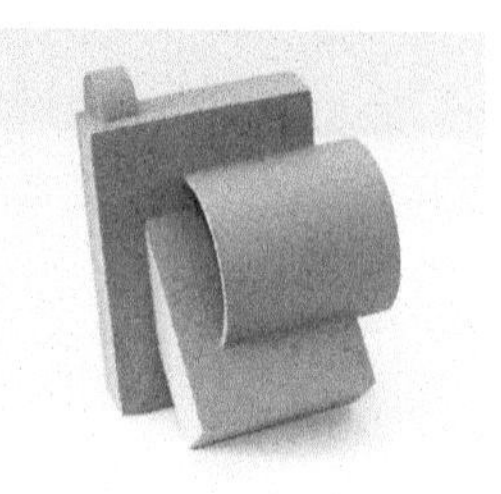
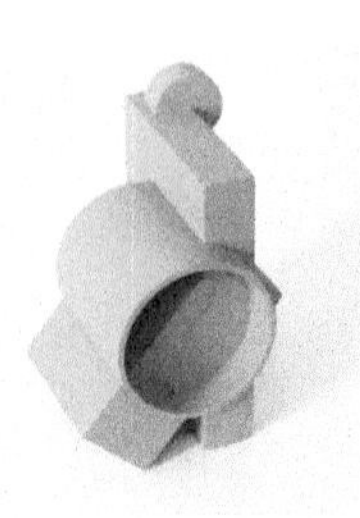
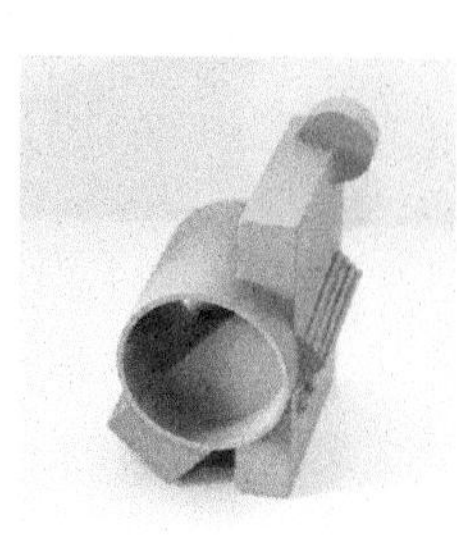
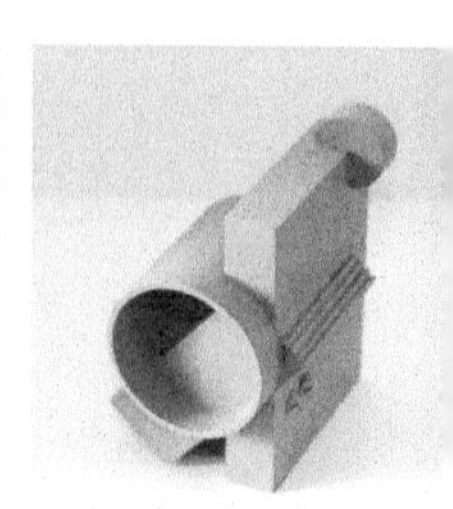

ISBN: 9780170233279

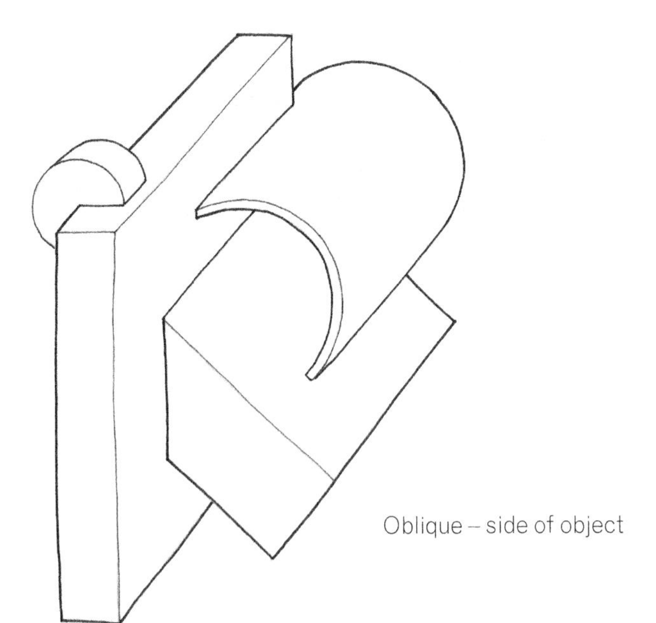

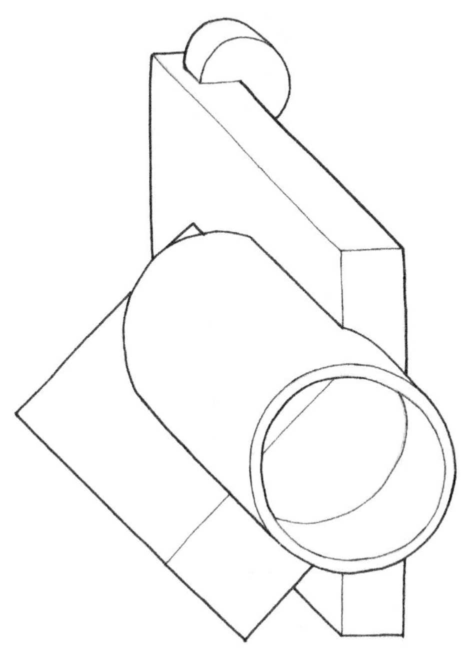

Oblique – side of object

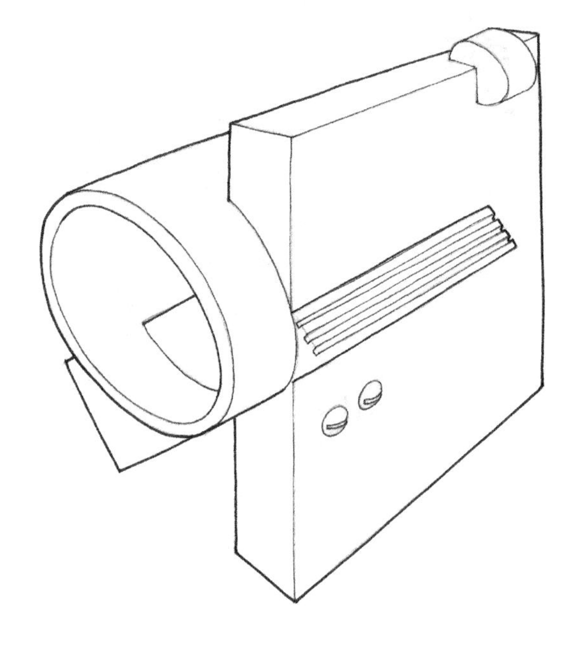

Perspective – looking down on the object from the back – below the horizon line

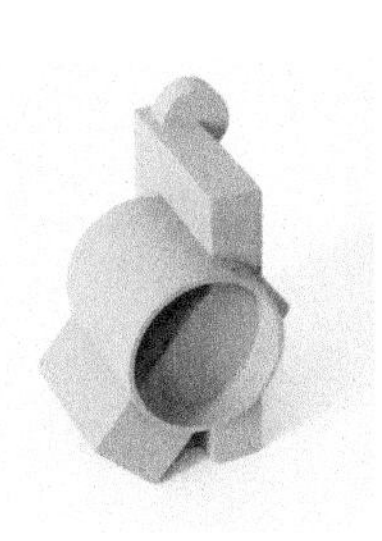

ISBN: 9780170233279

Perspective – looking down on the object from the front – below the horizon line

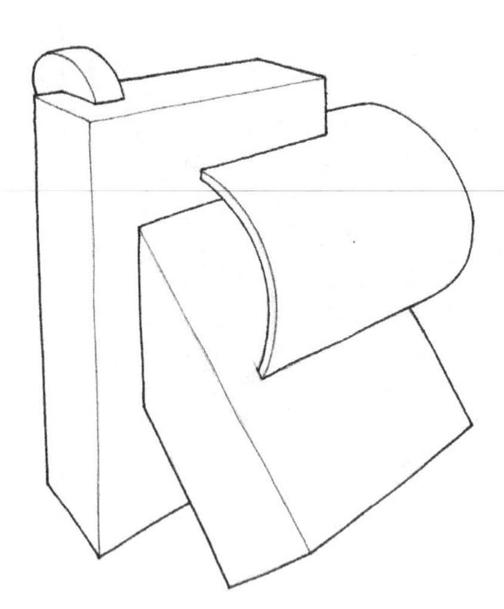

Perspective – looking up on the object – above the horizon line

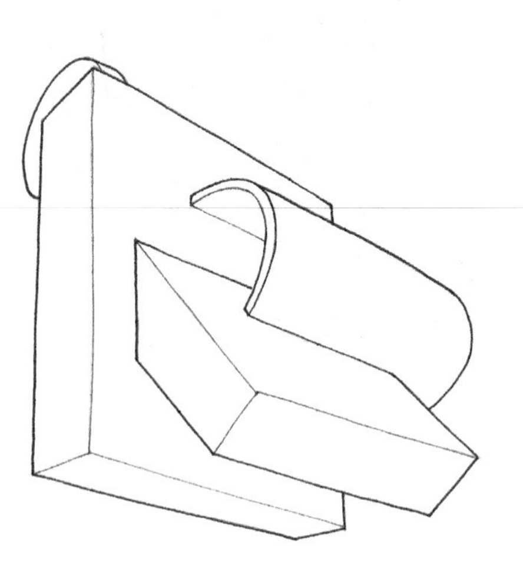

Perspective – looking up on the object – above the horizon line

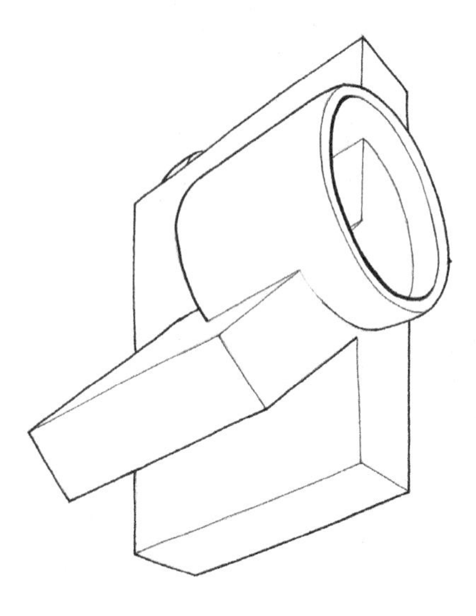

Perspective – object sketched above and below the horizon line

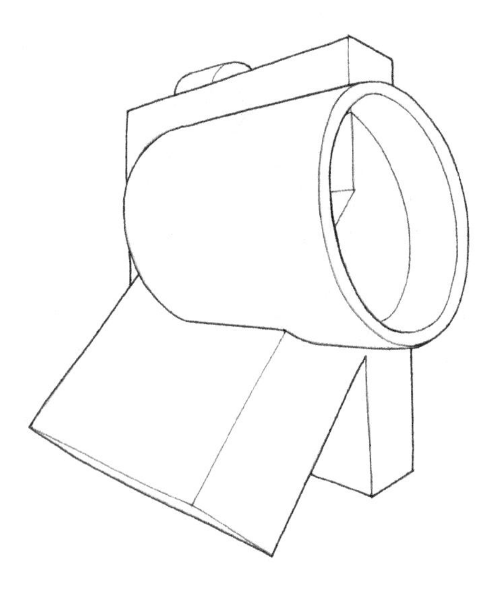

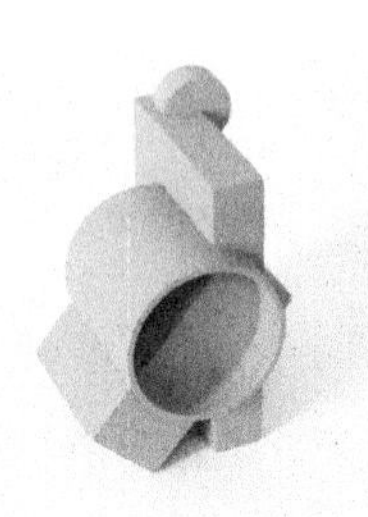

ISBN: 9780170233279

Christian Burgos — St Peter's College

The following work produced by Christian Burgos demonstrates detailed 2D and 3D freehand sketching. He has turned the seat of the vehicle in different directions while exploring the shape. Christian has presented a 2D sectional view and a 3D exploded sketch to demonstrate how the components fit together. He has also shown the relationship of the human body to the object.

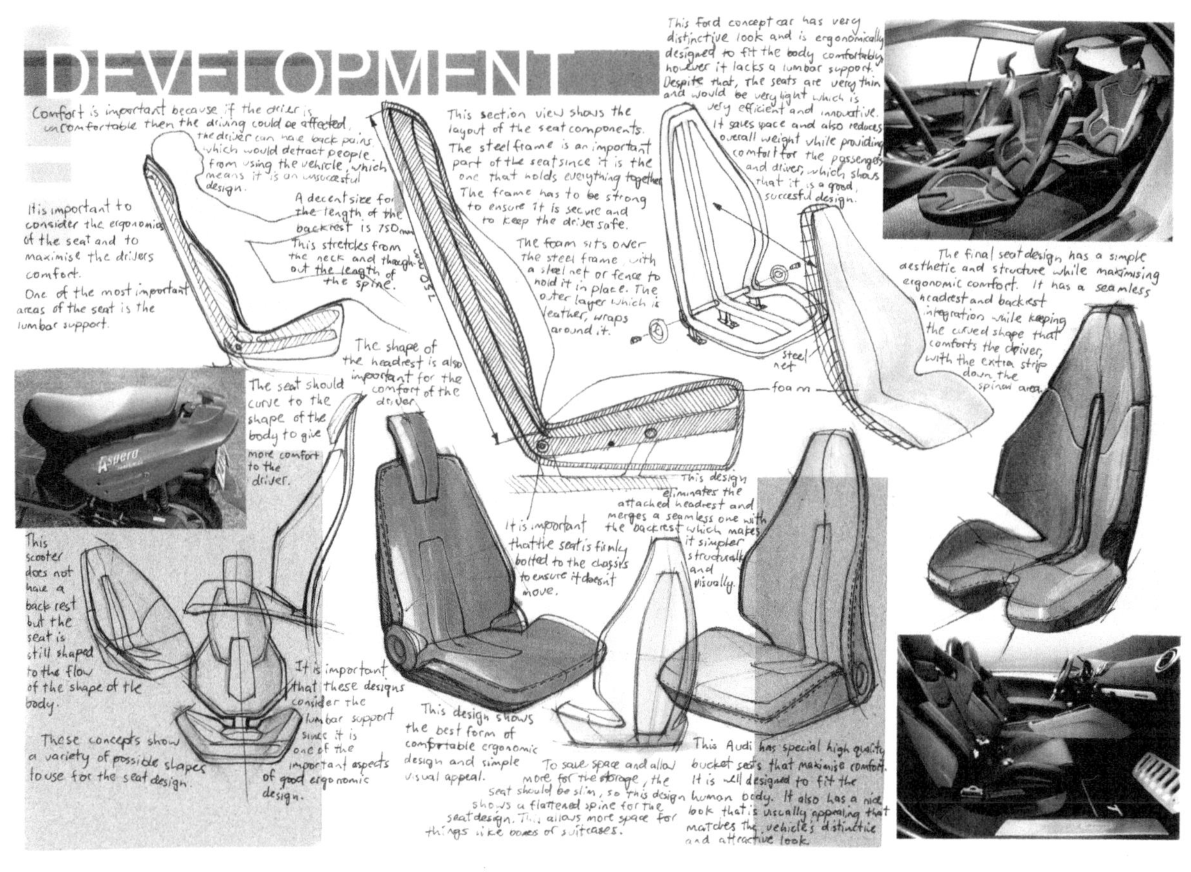

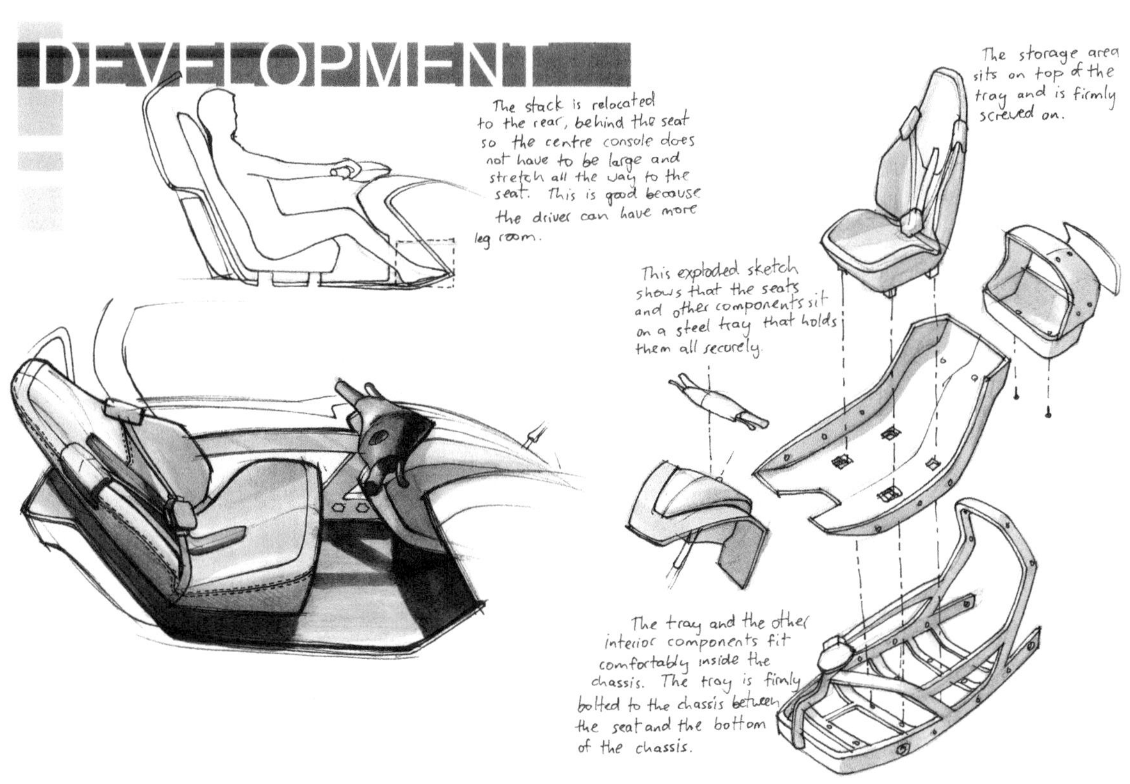

ISBN: 9780170233279

3 rendering techniques

Rendering techniques are used to communicate the form of design ideas. The form is the definition of an object to show the shape, material, texture and finish. Design ideas are your responses to a design brief.

Rendering involves indicating the tonal qualities produced by an identified light source and its three dimensional effects on the object's shape and surface qualities.

To effectively communicate the form of your design ideas you will need to skillfully apply rendering techniques, enhancing the realistic representation of the design qualities.

There are six basic points to remember when rendering. These are: the light source, tonal change, highlighted edges, a background outline, shadow and material value. These points need to be applied to all rendering regardless of the type of media used. Rendering is used throughout the development of ideas and for final instrumental drawings.

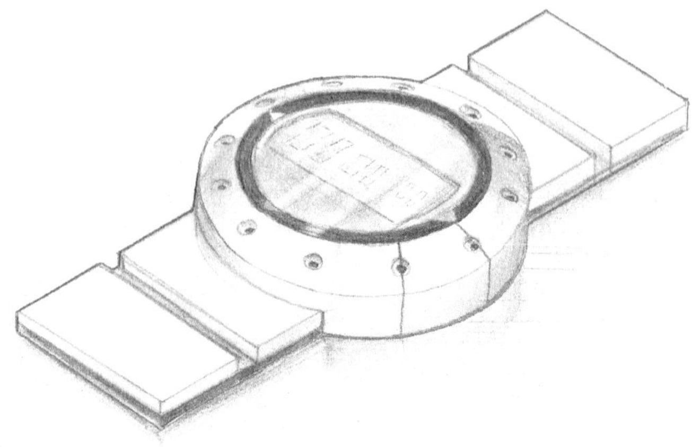

Christian Burgos • Instrumental Rendered Pictorials

ISBN: 9780170233279

The following examples explain the techniques required to apply different graphic media.

The first thing to consider when rendering is where the light source is coming from. The shading of the dark and light areas on the object gives the appearance of form which is essential when explaining a design idea.

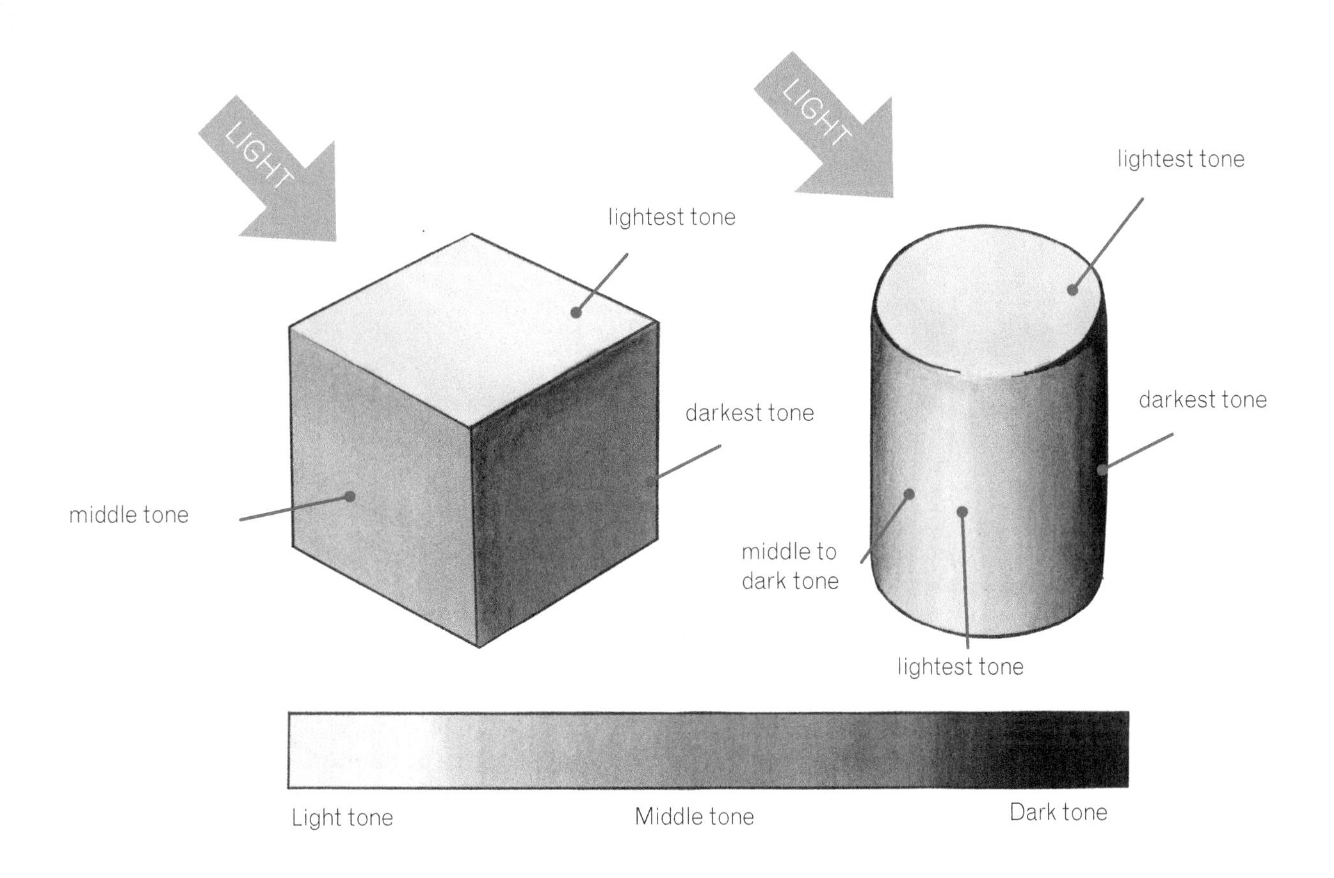

Once you have learnt how to render basic geometric shapes you will then be able to apply these techniques to more complex objects.

Pencil rendering

Rendering a cube

Step 1 • Shade the surfaces with the different tones. The top should be the lightest tone. The side surfaces are the middle and darkest tones depending on where the light source is. So that the application of the pencil is consistent apply the pencil methodically. Move the pencil in one direction across the whole surface, then work across the area in the opposite direction. If you move the pencil in many different directions randomly the surface will look inconsistent and you will be able to see the direction of the pencil. To avoid smudging put a scrap piece of paper under your palm.

ISBN: 9780170233279

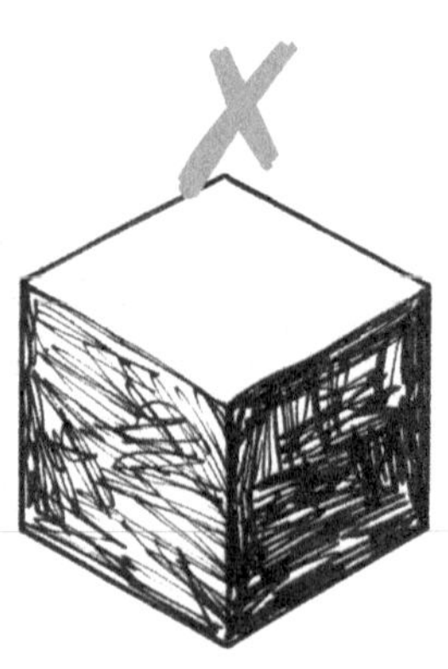

Incorrect use of pencil

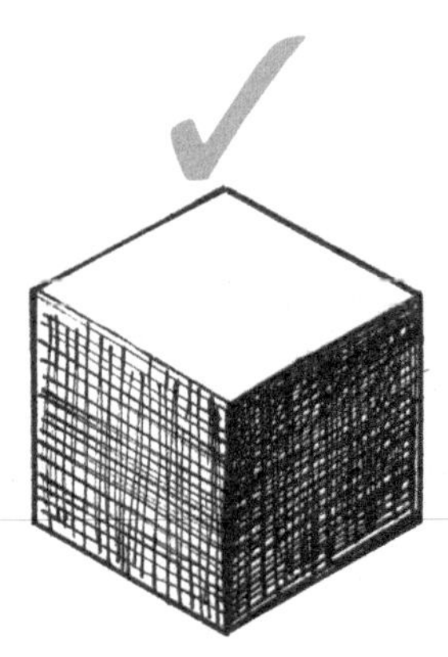

Correct use of pencil

Step 2 • Draw a dark line around the background of the object. DO NOT darken the front edges.

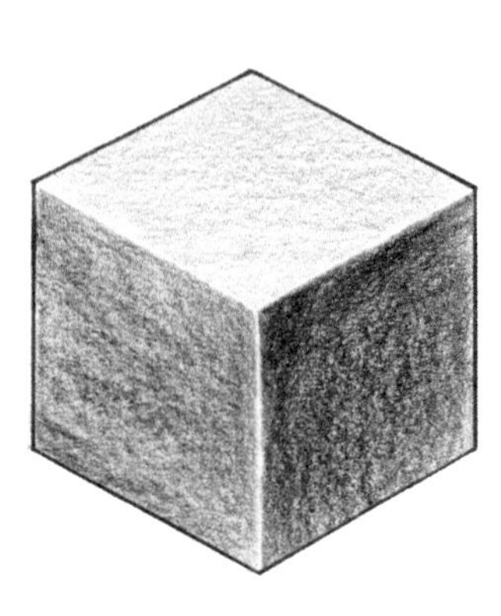

Step 3 • Render a shadow to make the object look more realistic. The shadow should be lightly shaded and no outline of the shadow should be clear.

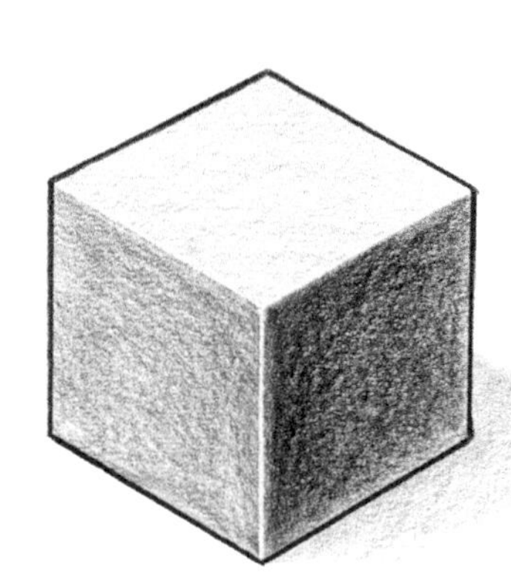

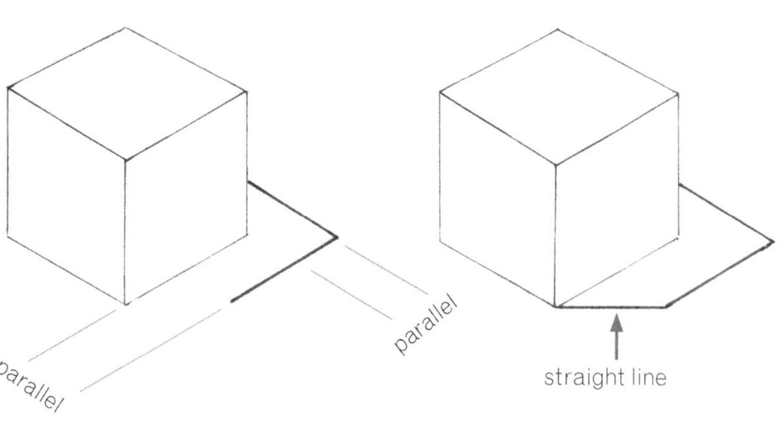

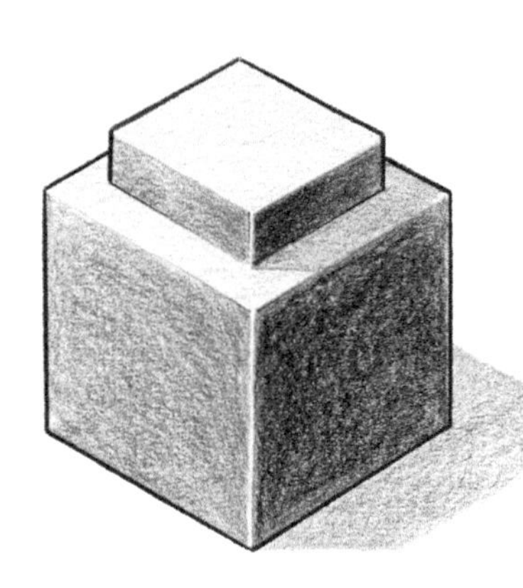

The same method applies when rendering a more complex shape. A shadow should also be reflected onto the object if required.

ISBN: 9780170233279

Rendering a cylinder

Step 1 • To show that the cylinder is a curved surface the rendering should graduate from dark to light from each outside edge. The highlighted space does not have to be in the middle, it can be off-centre according to the light source. To gradually change the gradient successfully work the pencil straight up and down along the cylinder. Do not try to follow the shape of the cylinder when rendering.

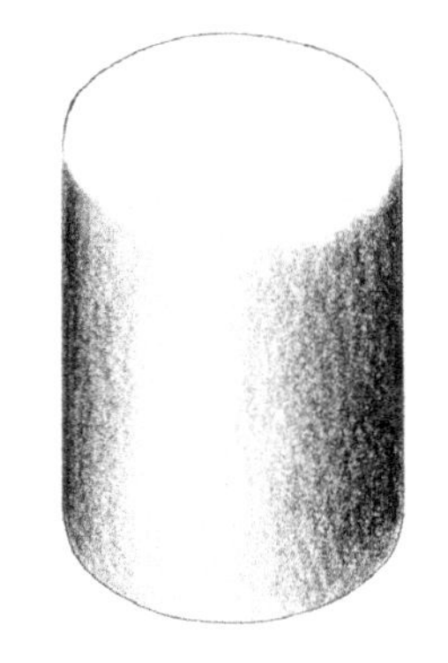

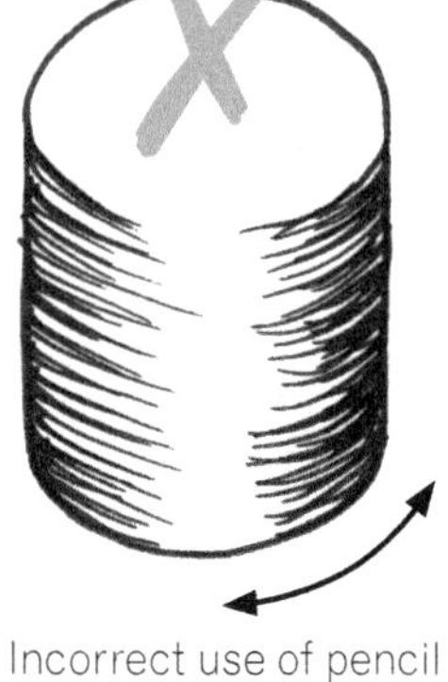

Incorrect use of pencil

Step 2 • Apply the shading to the top surface. Work from the front edge to the back. Render the top as a flat surface. Next draw the background outline. Blend this line into the front edge. There should not be a full line running around the front edge.

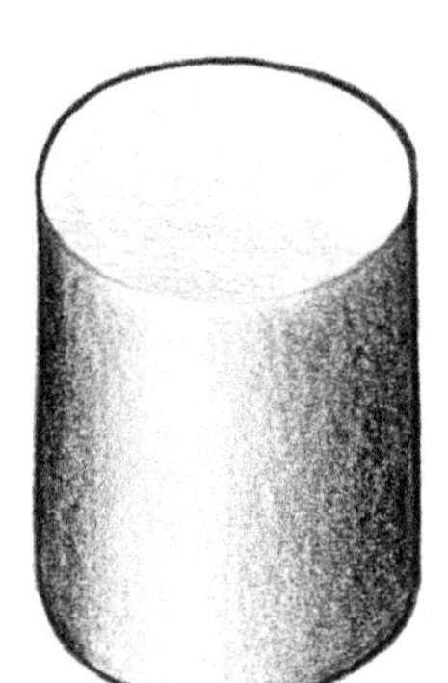

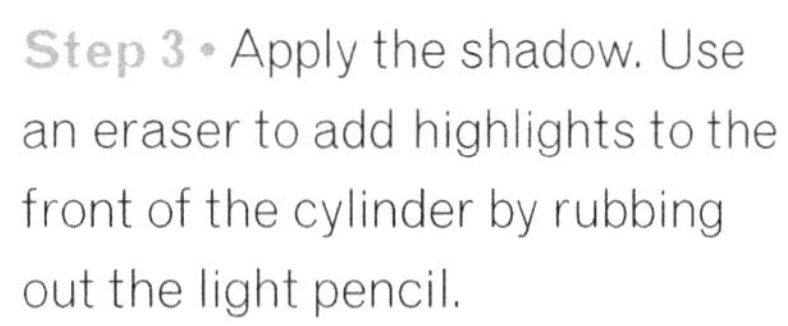

Step 3 • Apply the shadow. Use an eraser to add highlights to the front of the cylinder by rubbing out the light pencil.

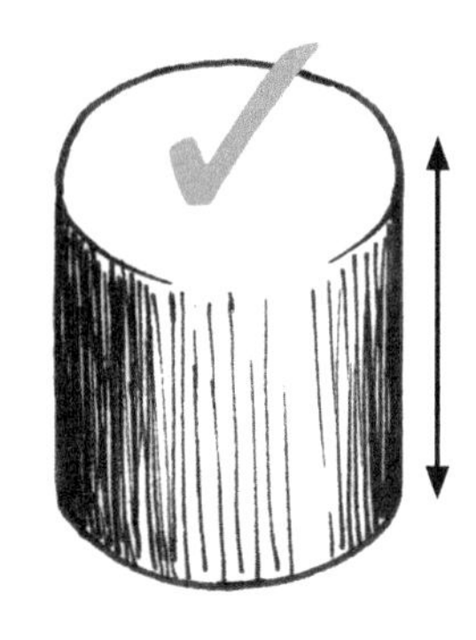

Correct use of pencil

Rendering curved edges

This example demonstrates how to present curved edges on an object. When rendering, as with a cylinder, you need to blend one side into the other.

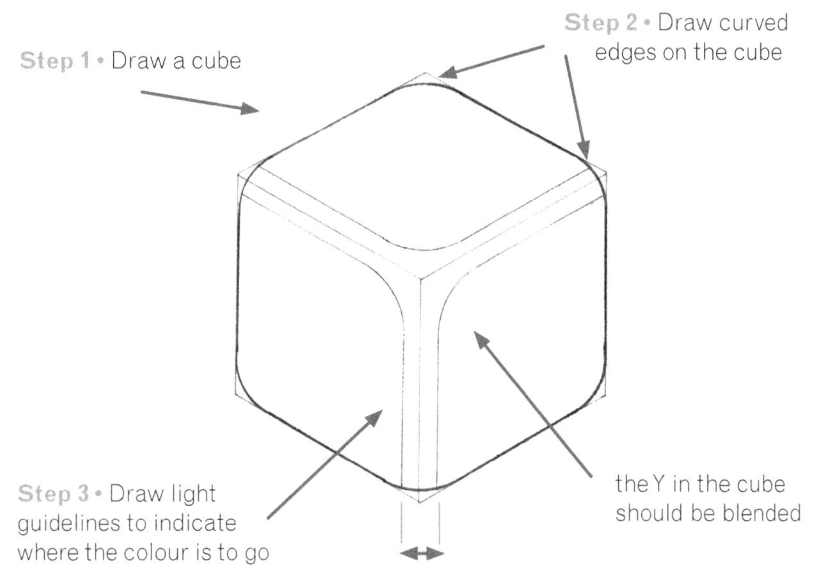

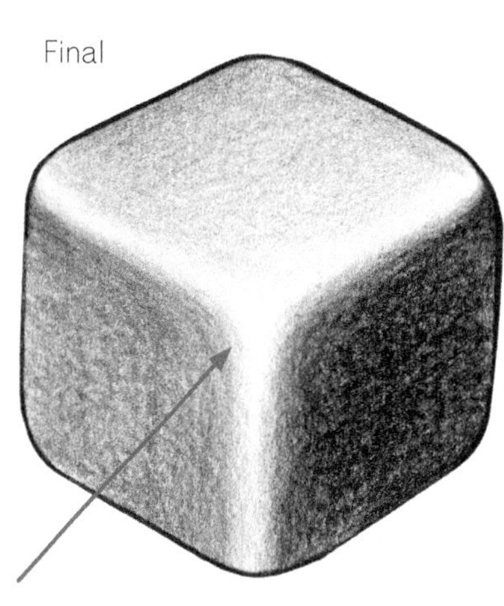

Graphics media may include pastels, airbrush, colour pencils, collage, marker pens, paint, gouache, card and digital media. The following examples present the rendering of basic solids using chalk pastel, colour paper, colour pencil and colour marker. They are all rendered using the same step-by-step process but with different media. This shows the use of a broad range of media options to present design ideas.

Chalk pastel

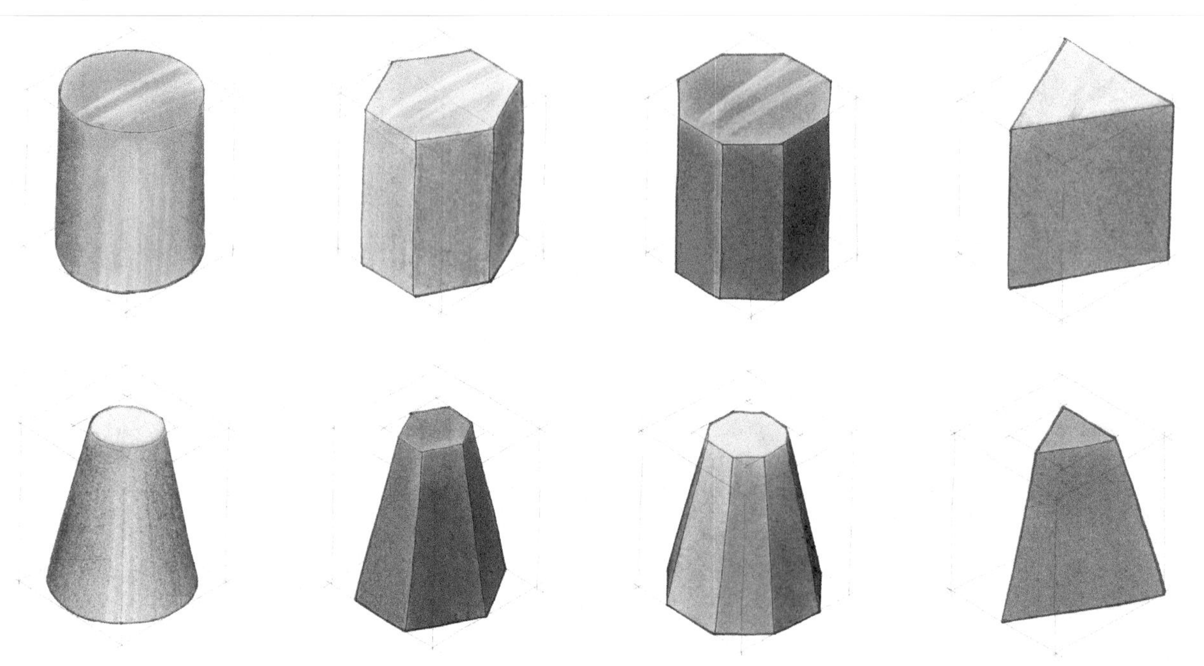

Colour paper

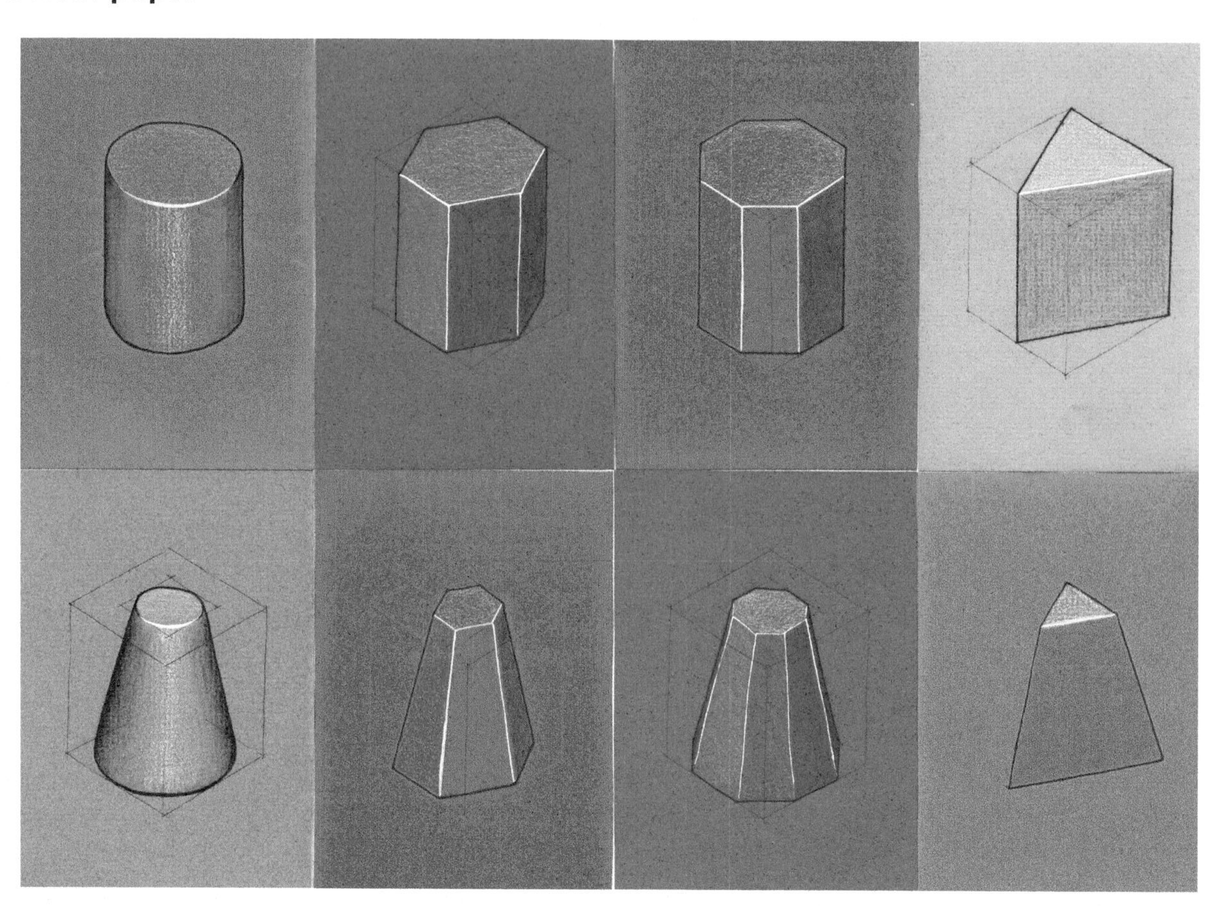

ISBN: 9780170233279

Colour pencil

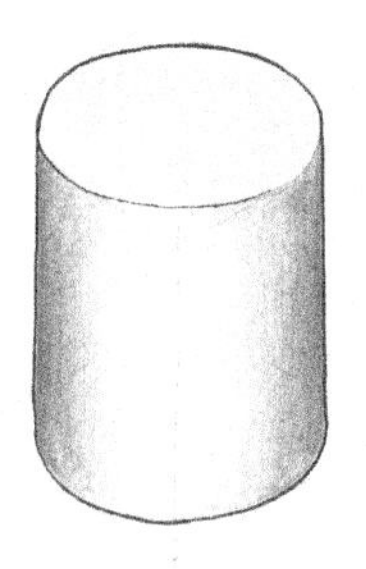
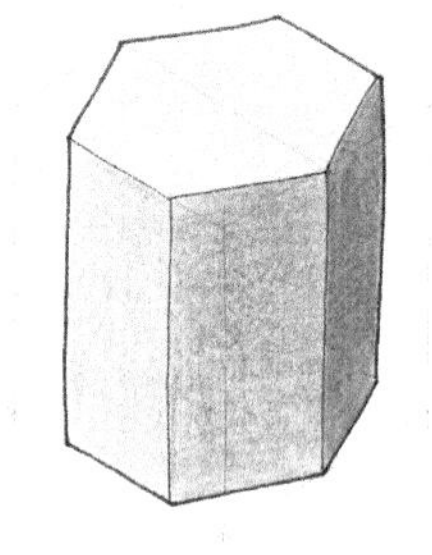
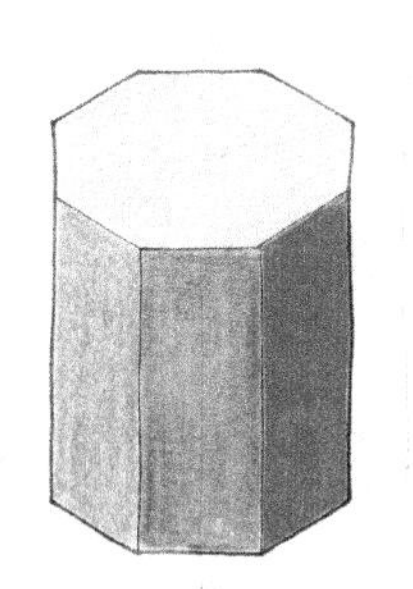
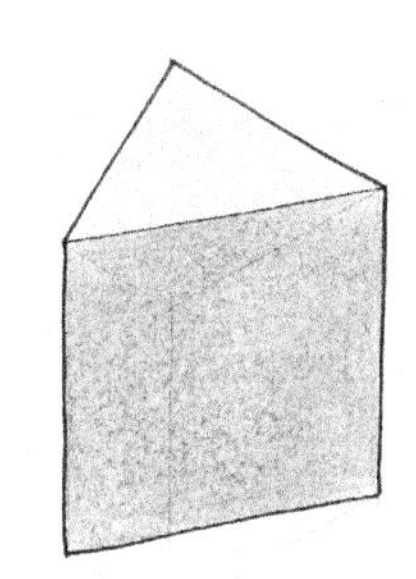

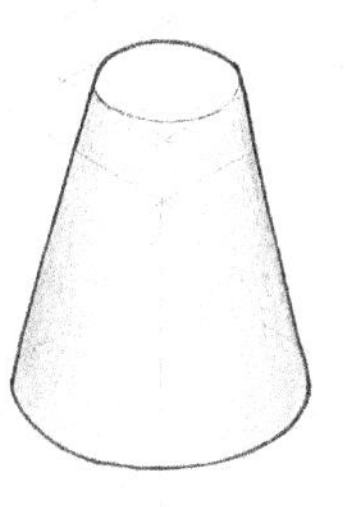
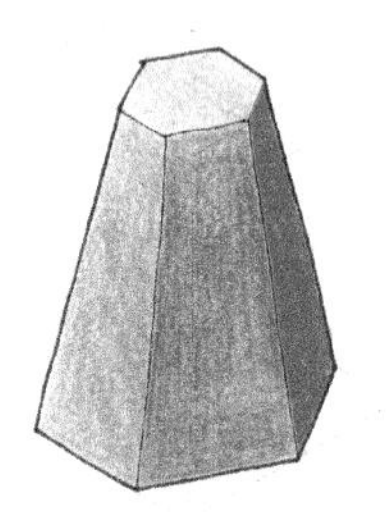
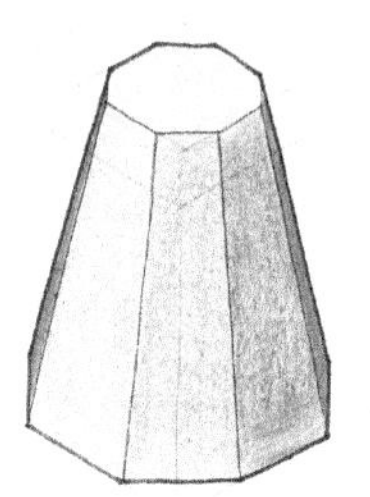
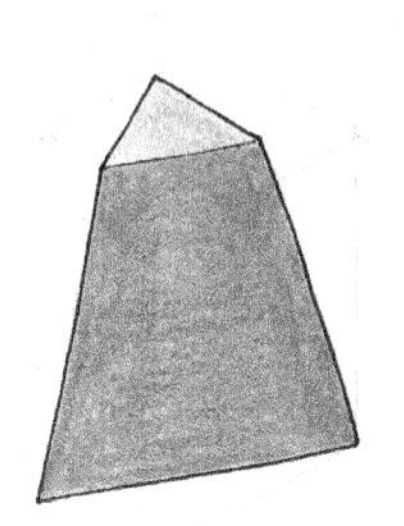

Colour marker

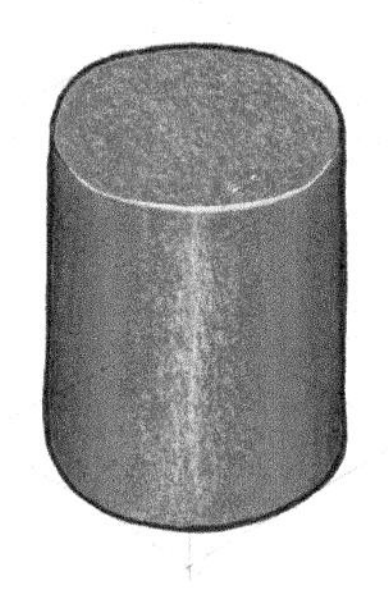
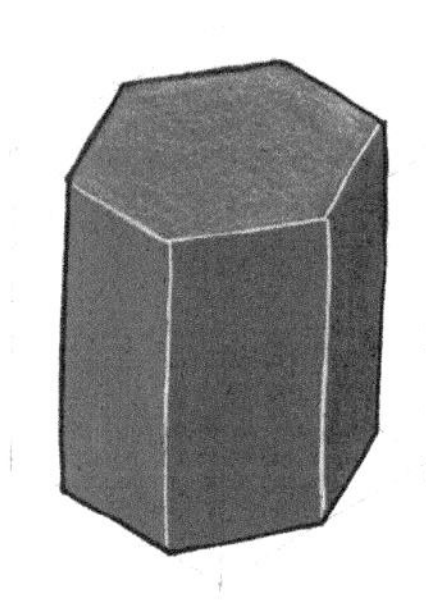
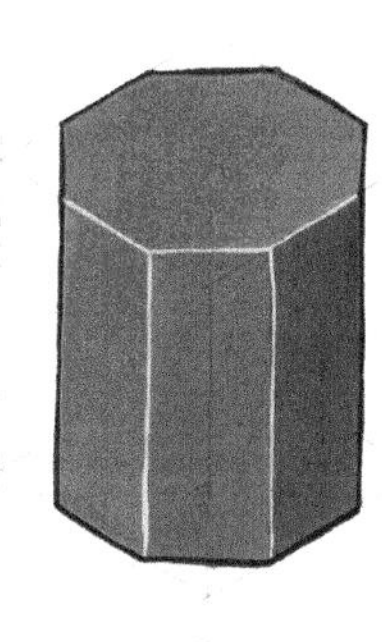
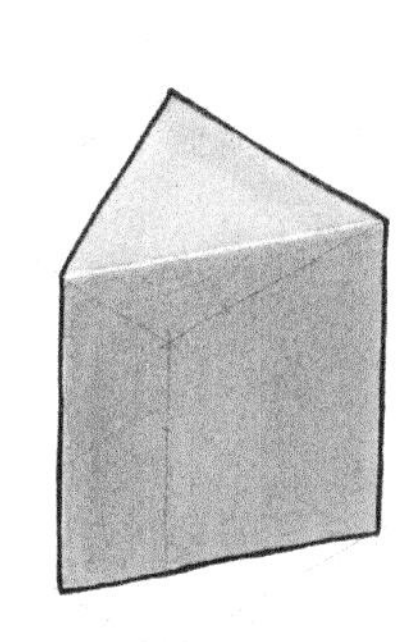

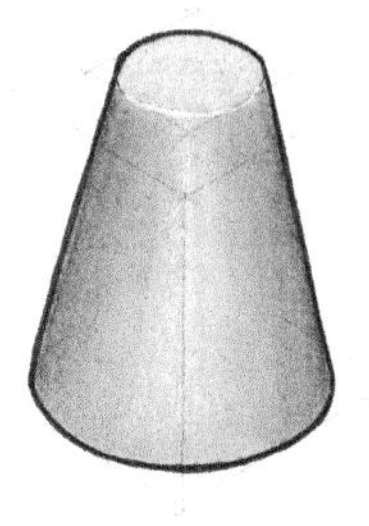
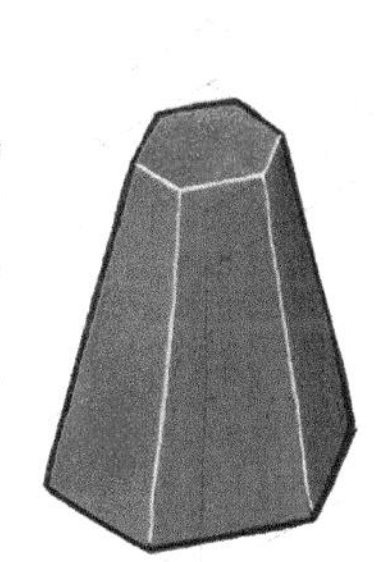

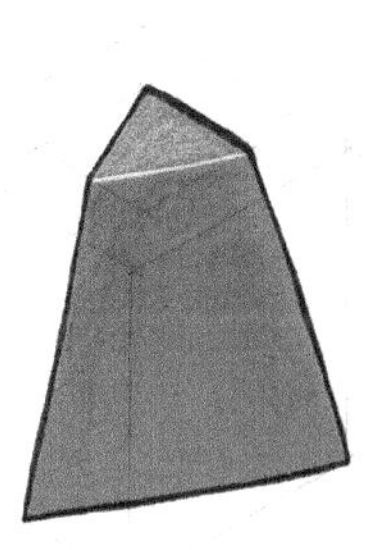

Geometric shapes

The same rules apply when rendering ALL geometric shapes. All you need to do is remember the six basic steps:

1. The direction of light on the object.
2. The tonal change which defines the shape, form and structure.
3. Highlighted edges and in some cases surfaces.
4. A background outline.
5. The shadow so that the object looks grounded.
6. The material value and texture of the object.

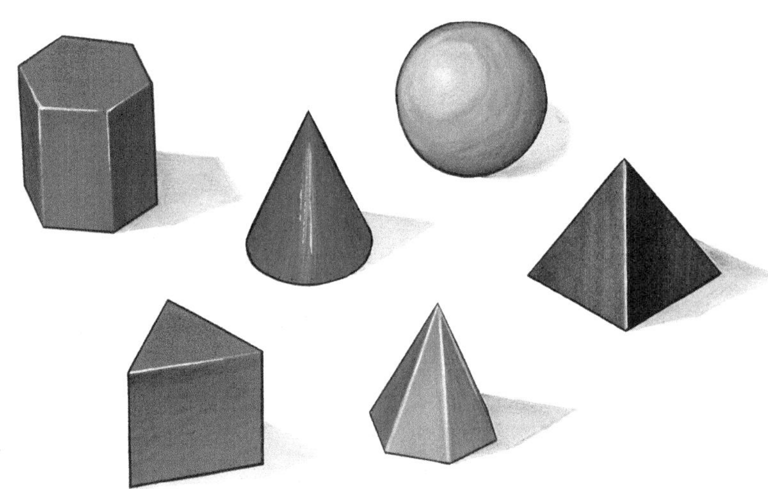

ISBN: 9780170233279

Sam Matijevich — St Peter's College

The following work produced by Sam Matijevich demonstrates rendering for a shelter design. He has used markers and pencil to present the effect of material value and texture for concrete, timber and plastic.

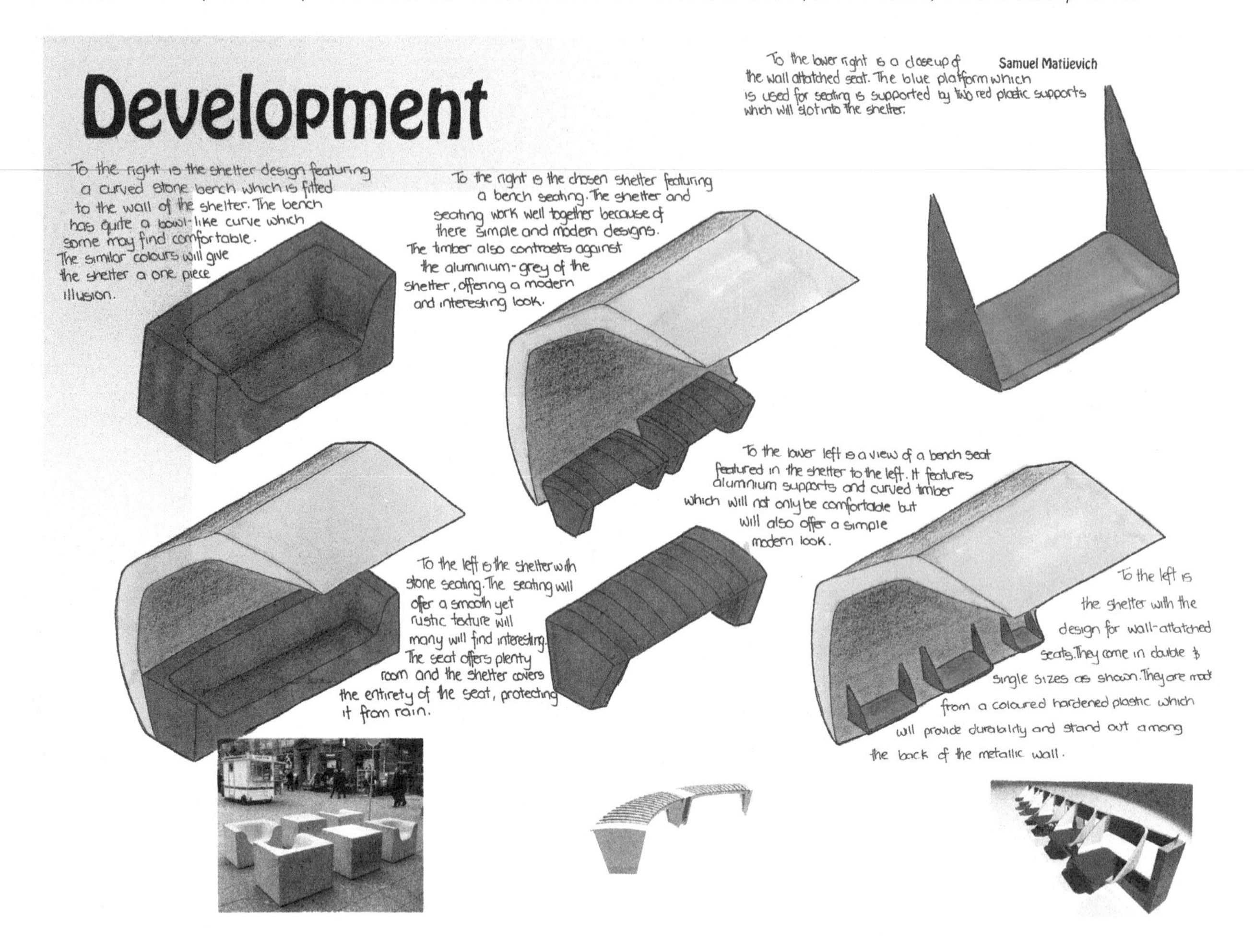

Marker rendering

Rendering a cube

Step 1 • Apply a base colour of marker to the object. Use the marker quickly and as evenly as possible to avoid streak marks.

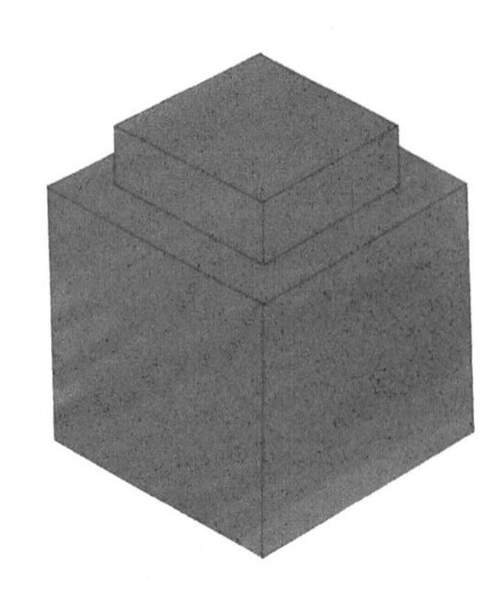

ISBN: 9780170233279

Step 2 • Apply grey marker to the darkest side of the cube. Again this is just to add a base shade. Be aware that the marker will bleed so don't colour right to the edges. Apply as quickly as possible so you get as few stroke marks as possible. Don't worry if some of the edges do bleed a little or that colours overlap as they will be tidied up when you apply the white and black pencil. Test out the colour marker and the greys you are going to use before working on your final pictorial.

Step 3 • Draw highlights on the front edges with a white pencil. Draw a dark outline around the back edge of the object with a black pencil. Do not darken any of the lines before this stage. Add a little white pencil to the top surface, starting from the back corner. Add a little black pencil to the two darker surfaces, starting from the main corner working to the opposite corner. You only need to add a little pencil to make it look effective. DO NOT over do it.

Step 4 • Add a shadow using grey marker and black pencil.

Rendering a cylinder

Step 1 • Apply a base of coloured marker over the area of the cylinder. Once the marker has dried add grey marker to the edges and carefully blend into the highlighted area. The marker will not be as easy to apply as the rendering of the cylinder with the pencil.

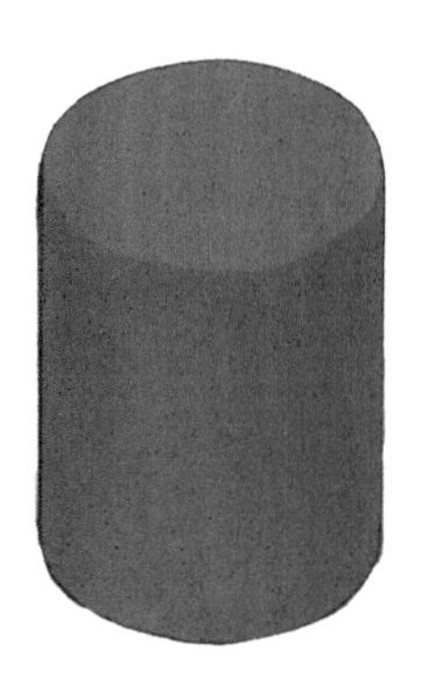

Step 2 • Blend black pencil from the edges to the highlighted area. Apply white pencil to the highlighted area. The curved front edge should be highlighted. The back edge should have an outline.

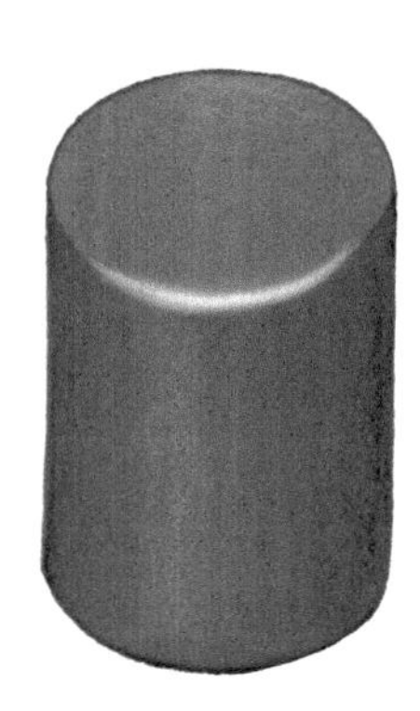

ISBN: 9780170233279

Step 3 • Apply Twink to the highlighted edge. Add the shadow with grey marker and a black pencil.

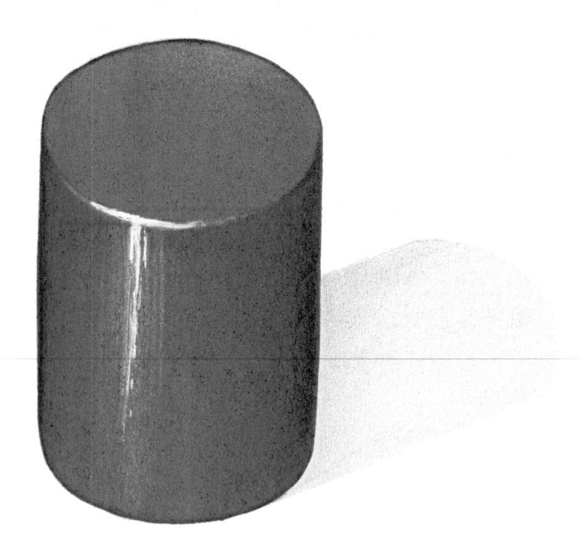

Colour paper rendering

Colour paper can be used very effectively to present your design ideas. Colour paper rendering could be used to present a final rendered instrumental pictorial or for rendering drawings at the development stage of the design process.

To render on coloured card you apply markers and pencil using the same method as in colour marker rendering. The colour paper is used as a base instead of the colour marker. Using paper that has a texture will give an effective finish. Colour paper rendering is a quick and effective way to present your ideas. It is ideal for the development of design ideas when you want to be able to put ideas down quickly. Colour paper can also be used for collage throughout your design work.

ISBN: 9780170233279

Thanya Chansouk — St Peter's College

This work produced by Thanya Chansouk demonstrates rendering for a shelter design. He has used colour paper as the base colour and added the tonal change and highlights with marker, pencils and black pen.

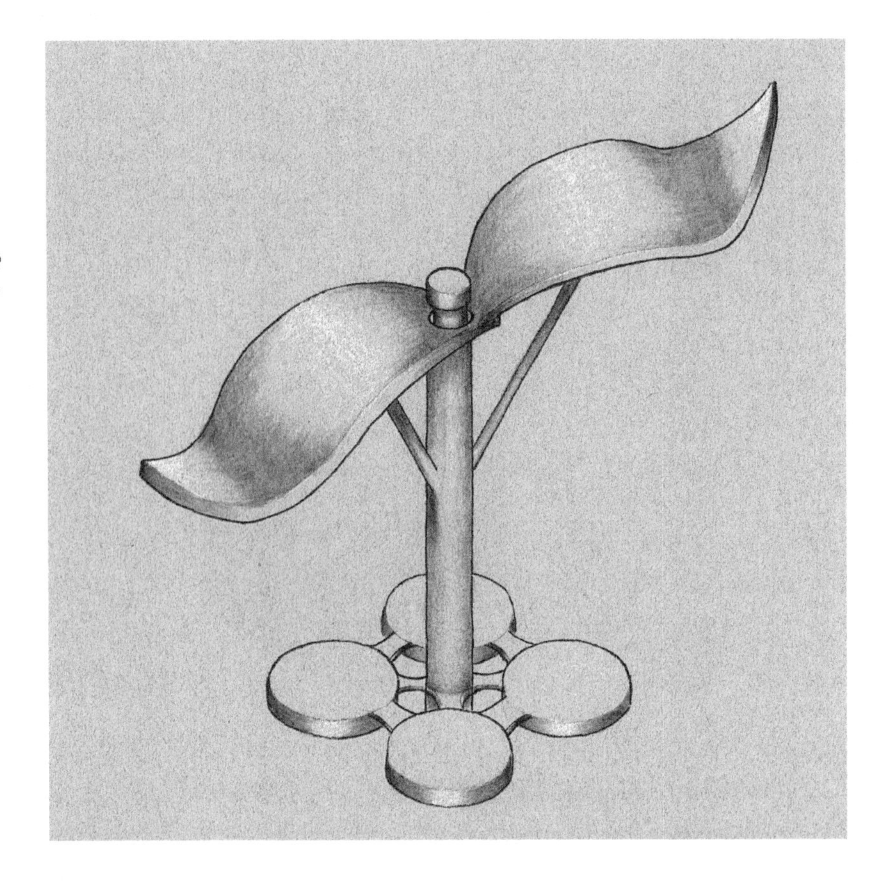

The following work produced by Thanya demonstrates rendering for a shelter design. He has used markers and pencil to present the effect of material value and texture for timber, plastic, glass and metal.

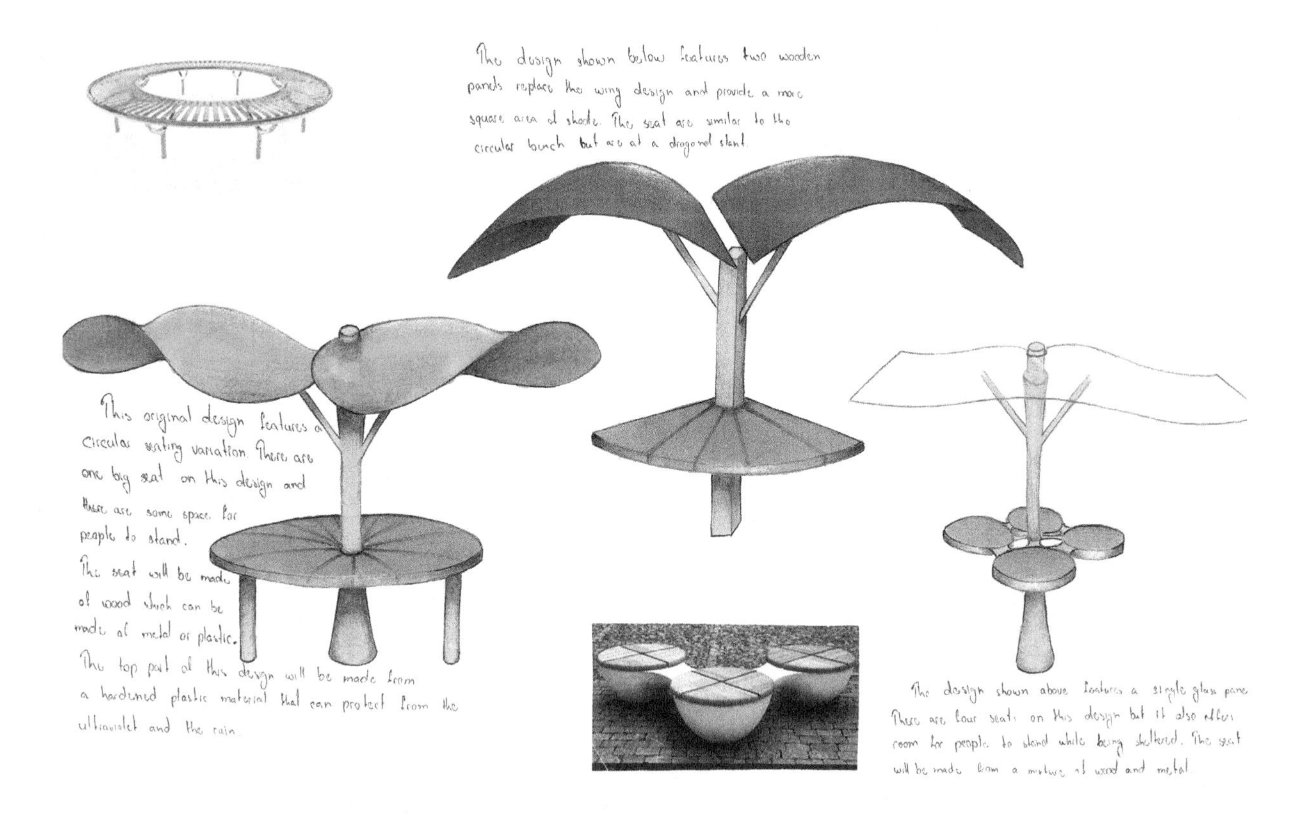

ISBN: 9780170233279

To present your renderings as realistically as possible and your ideas in the best way, you need to add material value. This will also add texture to your rendering and it allows you to explain your product in more detail. The following explains techniques to represent plastic, metal, glass, concrete and timber.

Plastic (Gloss)

Render the top surface with chalk pastel. Chalk pastel is used as it is easy to rub out lines for reflection on the top surface. Render with the markers as shown in previous examples e.g. highlighted edges, dark outlines, Twink the highlighted edges.

Plastic (Matt)

Matt plastic is rendered using the same process as gloss plastic but without any of the highlights. A white pencil should still be used on the edge but give the edge a softer finish.

Techniques using chalk pastel

Scrape the chalk pastel stick with a craft knife. This forms a fine dust which can then be used for rendering.

Use the cotton wool to spread the chalk pastel onto the drawing.

Rub away the unwanted edges of the chalk pastel. Continue on the other surfaces using marker.

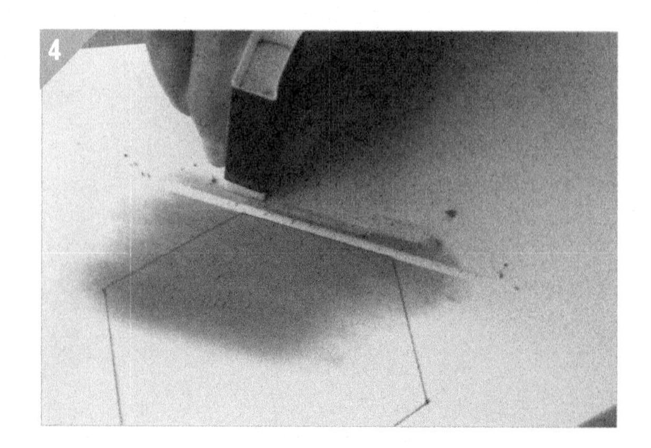

ISBN: 9780170233279

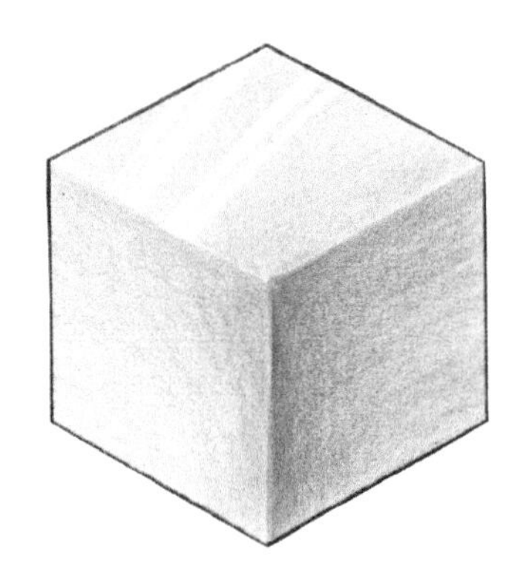

Chrome

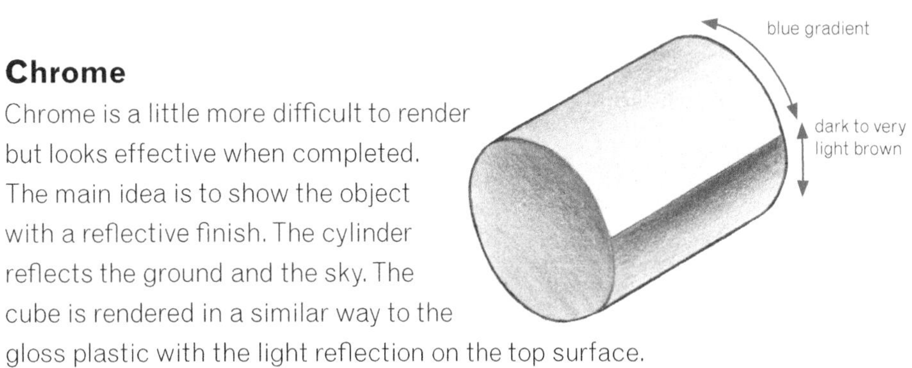

Chrome is a little more difficult to render but looks effective when completed. The main idea is to show the object with a reflective finish. The cylinder reflects the ground and the sky. The cube is rendered in a similar way to the gloss plastic with the light reflection on the top surface.

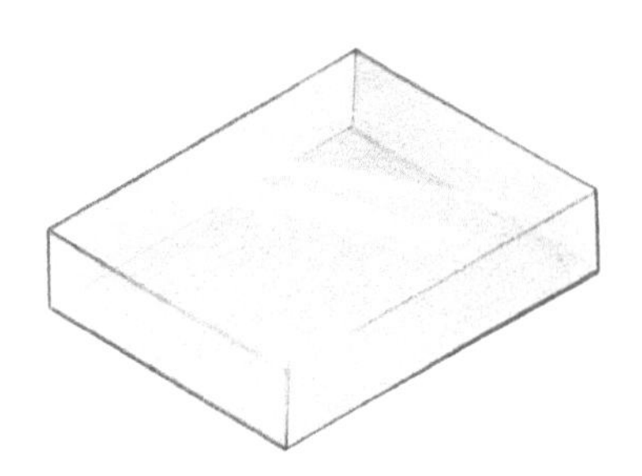

Glass

Apply chalk pastel, light blue or green pencil or marker lightly to the area. Render each surface (including the surfaces in the background) separately. Draw the back outline in blue. Rub out lines on the top surface to give the indication of reflection.

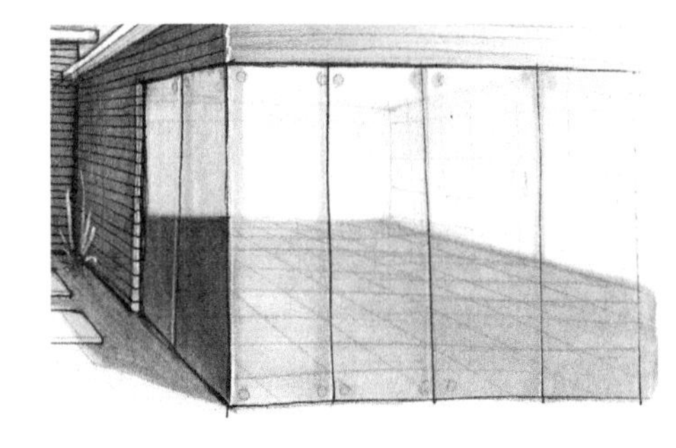

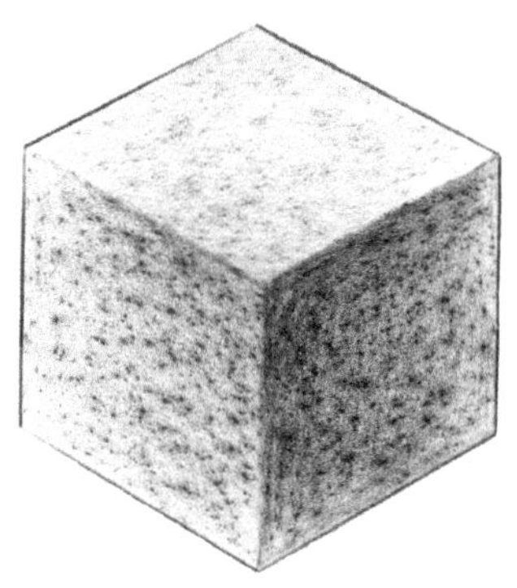

Concrete

Use a grey and black pencil to represent the concrete. Before rendering the object put a rough piece of sandpaper under the image and then follow the main rules of rendering. The texture of the sandpaper will come through and give the appearance of the stones in the concrete.

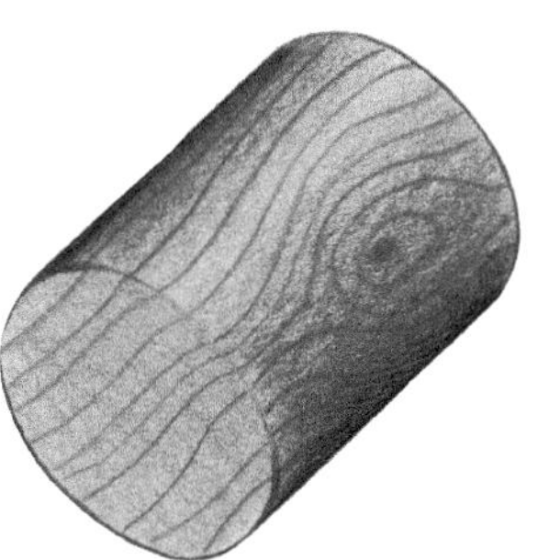

Timber

Apply a base colour to the area that is to represent timber. Draw the grain of the wood onto the area with a dark brown pencil. Render with a dark and light pencil to get the tone.

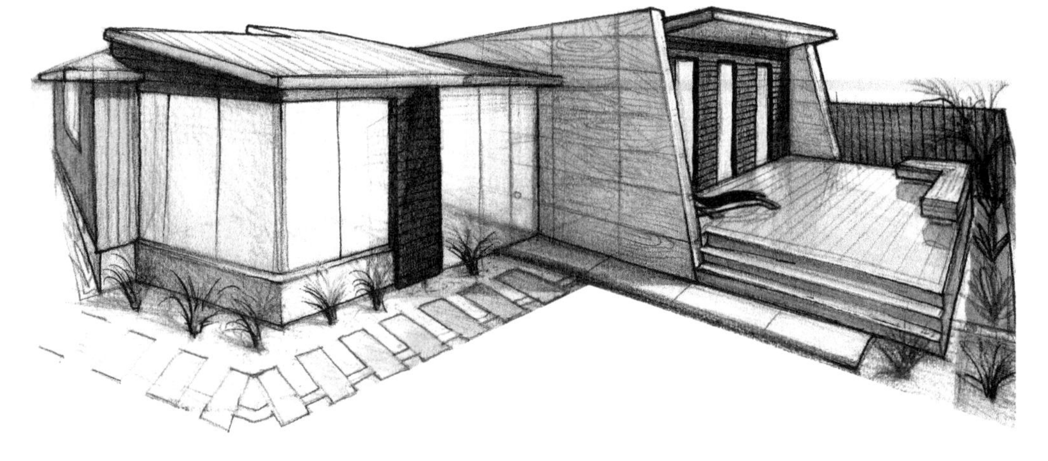

ISBN: 9780170233279

Liam Matheson — St Peter's College

The following work produced by Liam Matheson demonstrates rendering for a shelter design. He has used markers and pencil to present the effect of material value and texture for timber, plastic and glass.

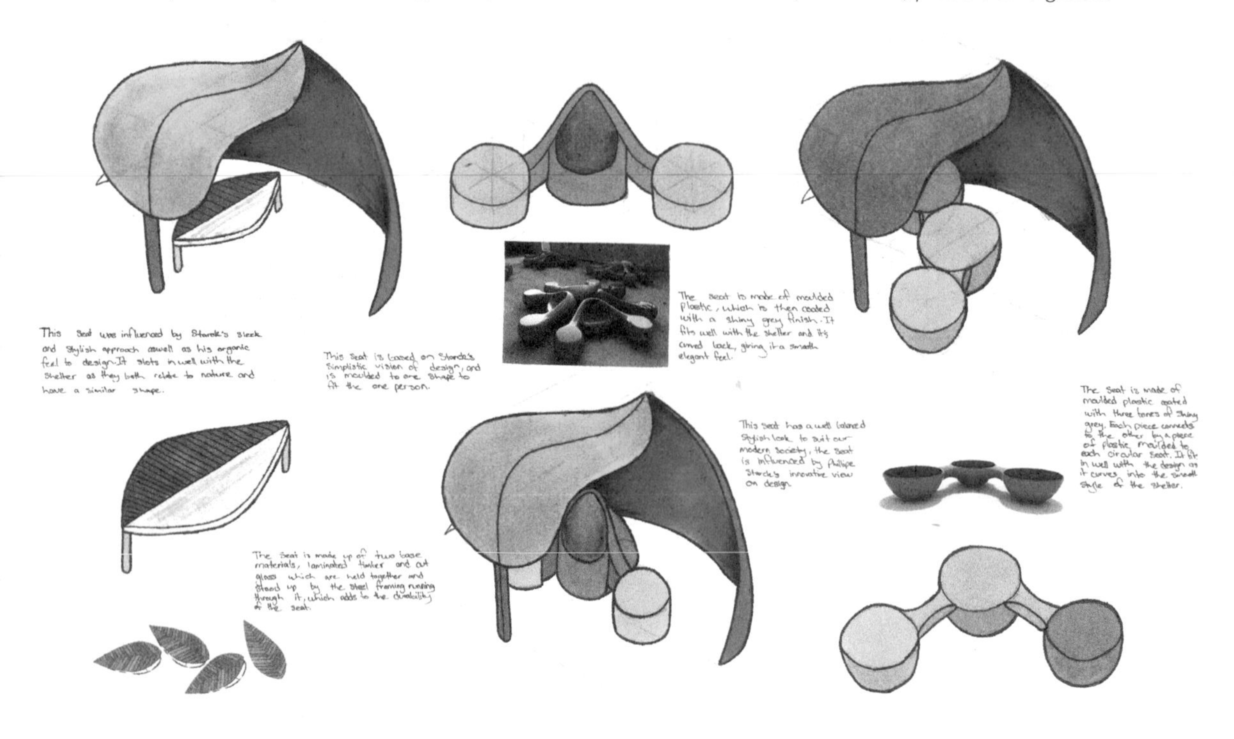

Christian Burgos — St Peter's College

The following work produced by Christian Burgos demonstrates a high level of understanding about where to add tonal change and highlights. At times he does not render the entire object for effect but visually communicates clearly the material value of the product.

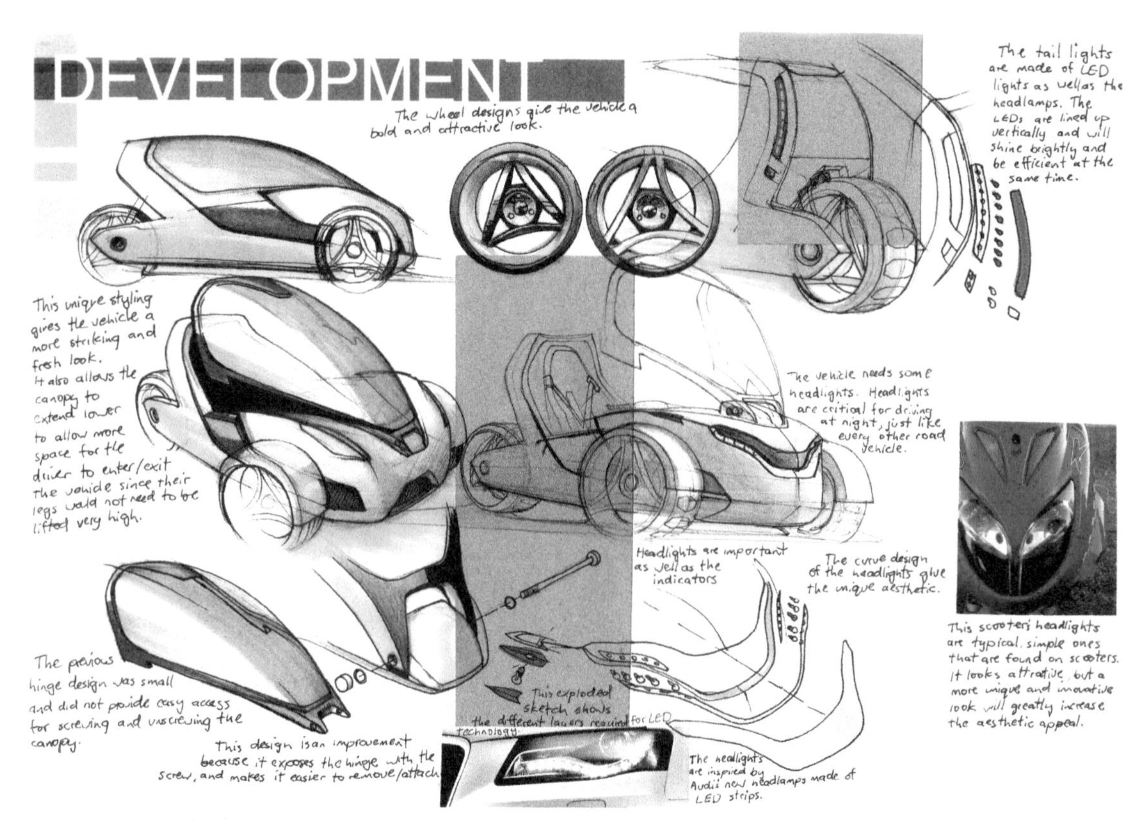

ISBN: 9780170233279

The following demonstrates two ways to construct a shadow.

A indicates a 1 point light source. The three steps to construct this type of shadow are shown below.

B indicates a cast shadow using parallel lines.

A Step 1 • **1 point light source**

A Step 2

A Step 3

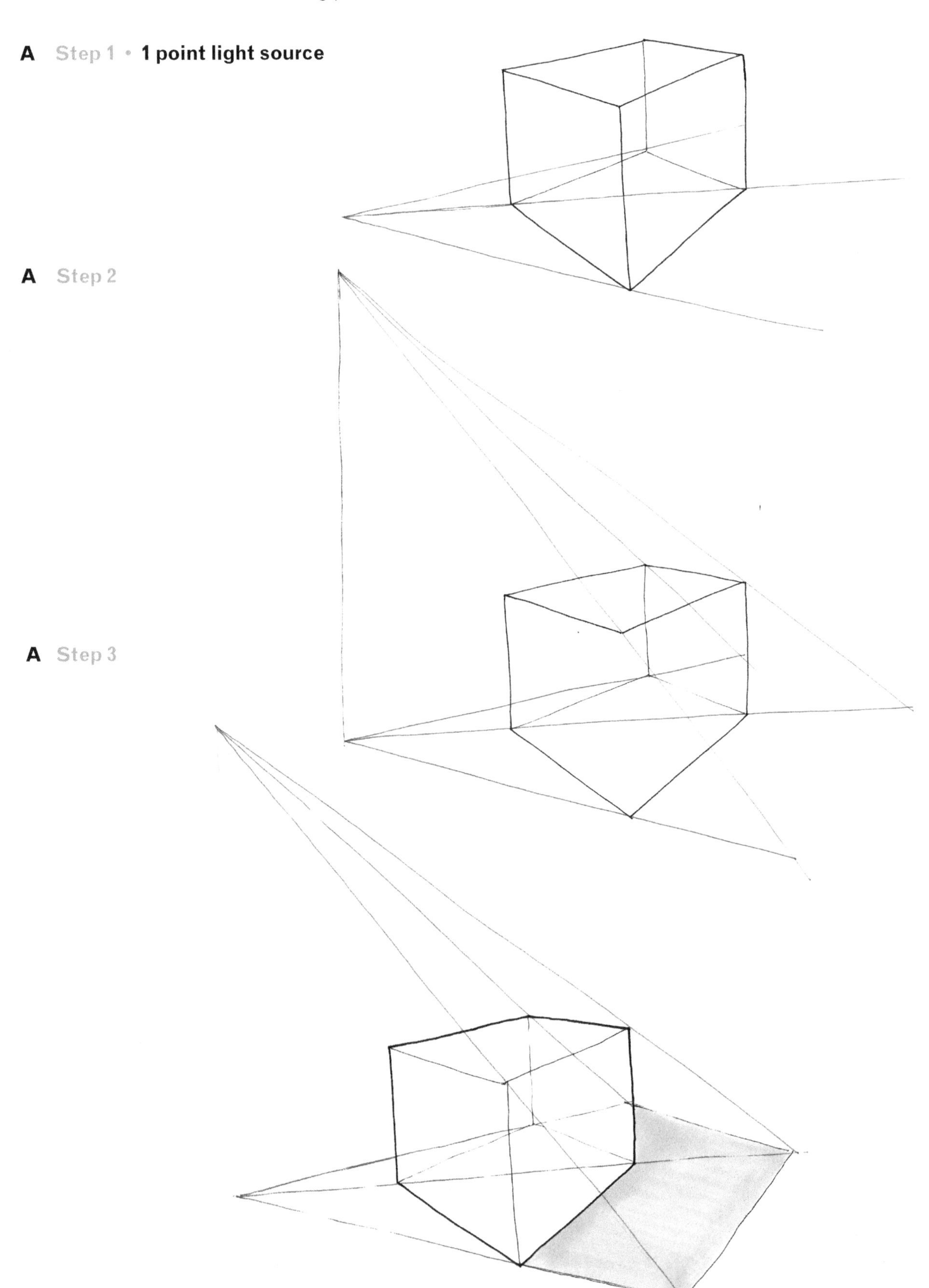

ISBN: 9780170233279

B Cast shadow

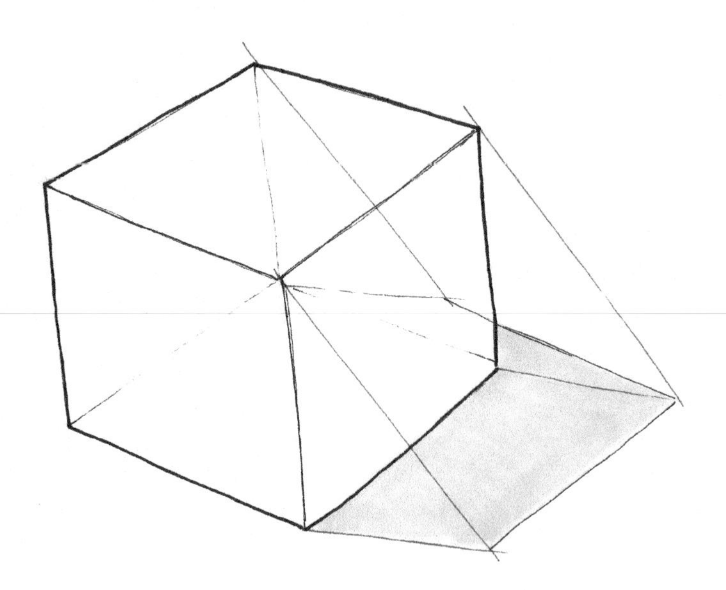

Block model rendering

The rendering of the block model demonstrates the application of shadows.

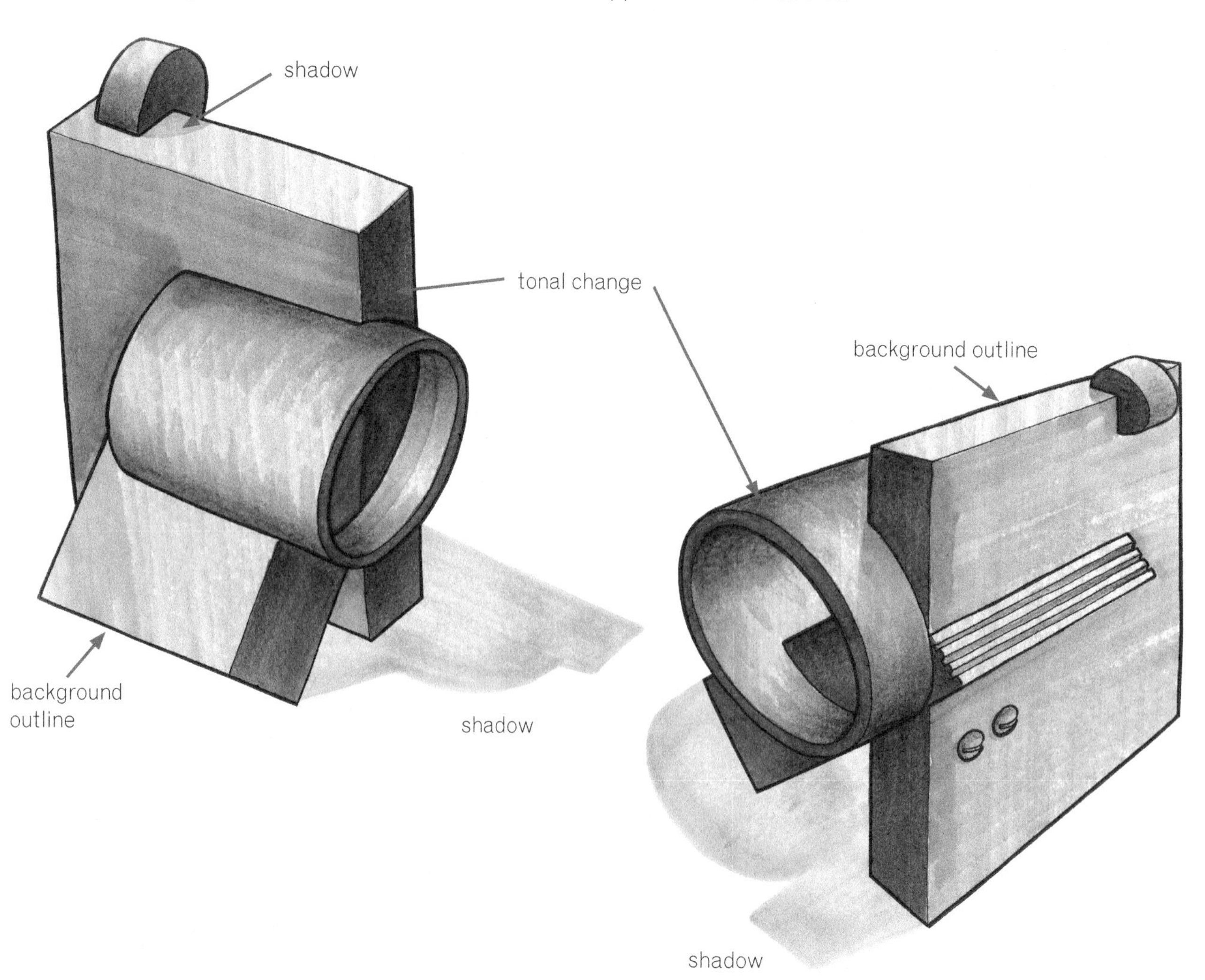

ISBN: 9780170233279

Thick-thin line

The thick-thin line technique is used to define an object. The outside lines of an object are to be defined with a thicker line and the inside or front lines of an object are left with the thin lines.

These examples of the barbecue, clasp, watch and torch show how the products stand out a lot more clearly with the use of the thick-thin line technique.

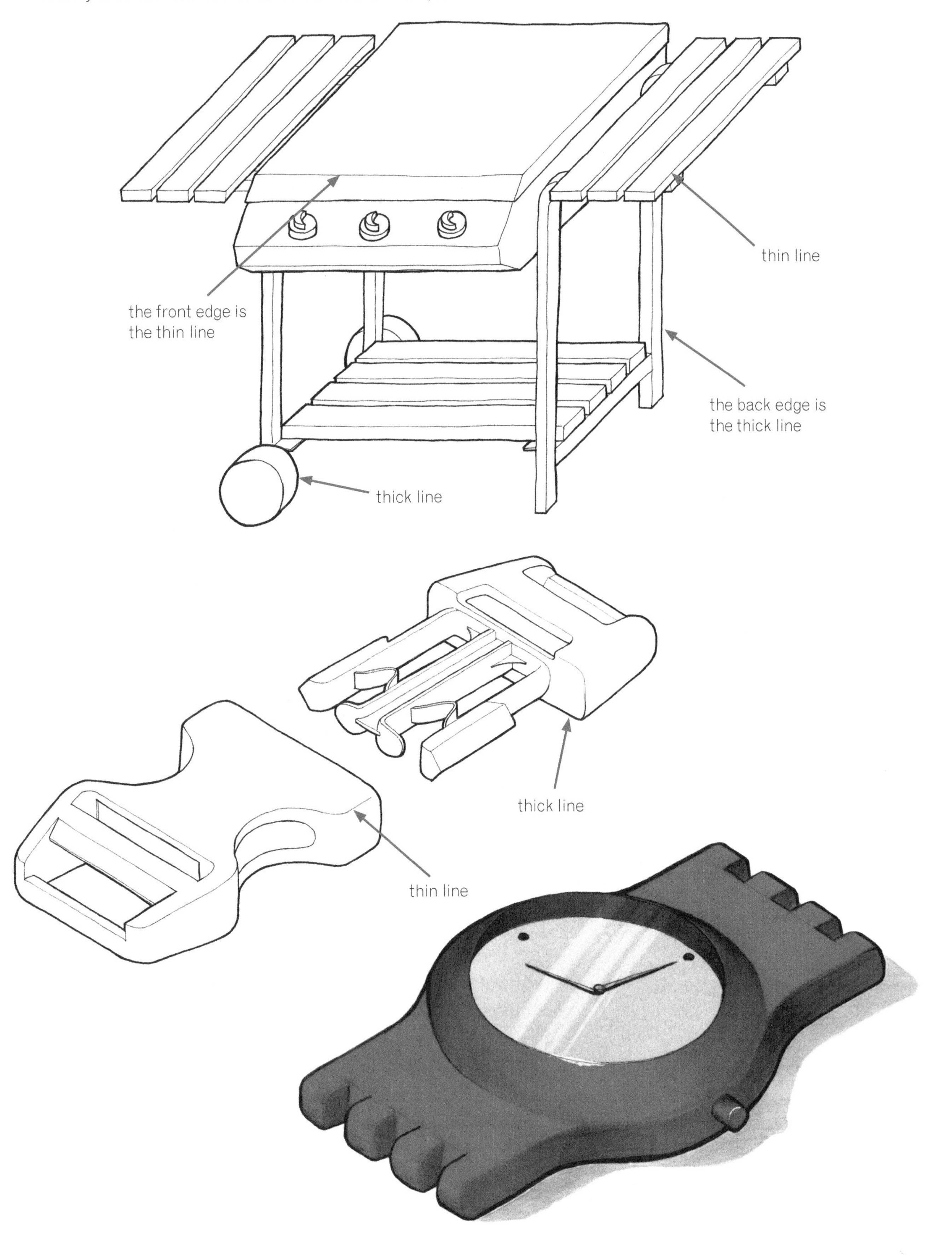

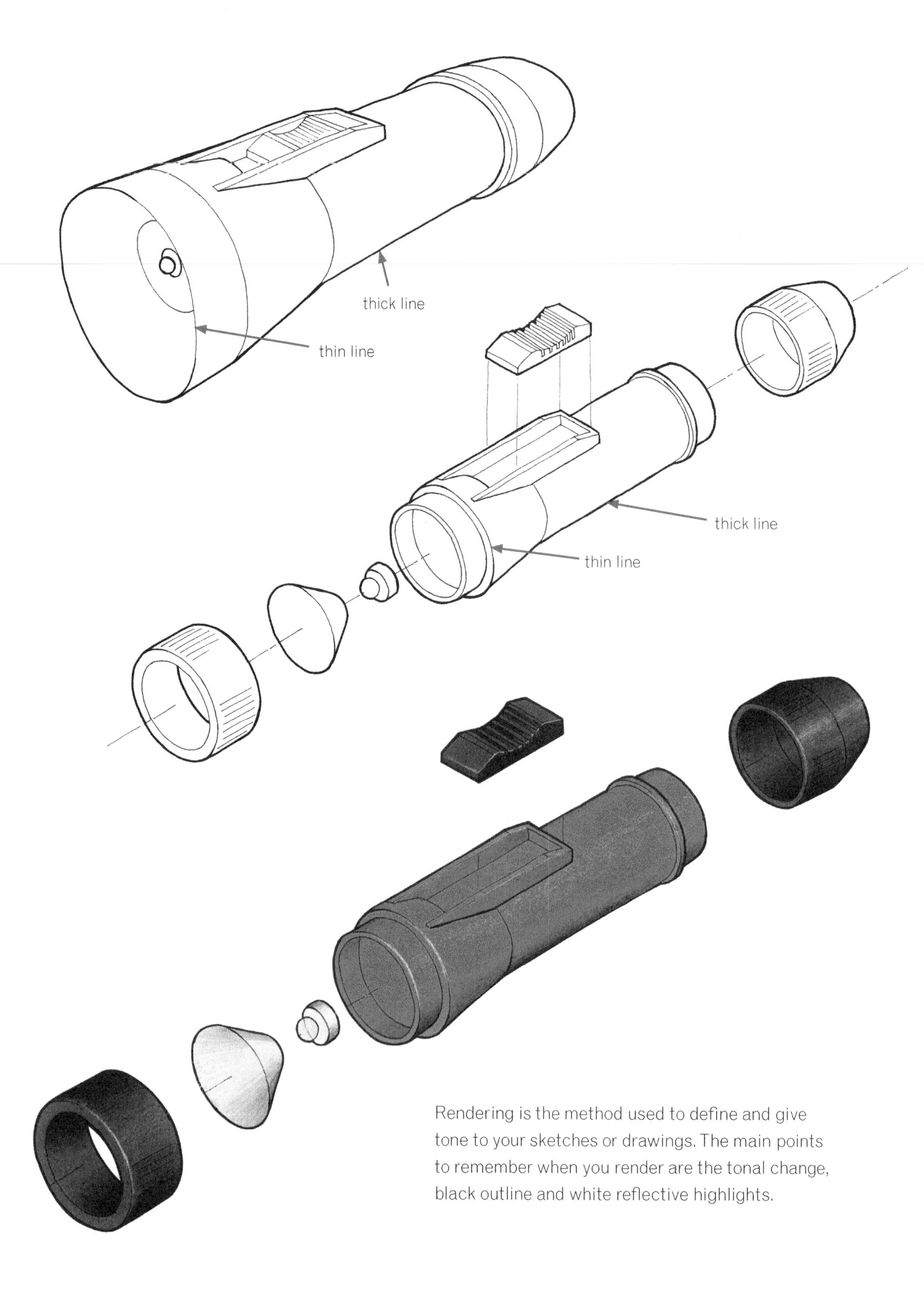

Rendering is the method used to define and give tone to your sketches or drawings. The main points to remember when you render are the tonal change, black outline and white reflective highlights.

ISBN: 9780170233279

Torch rendering

The rendering has been completed using a grey and orange marker, a range of colour pencils and fine tip black pen.

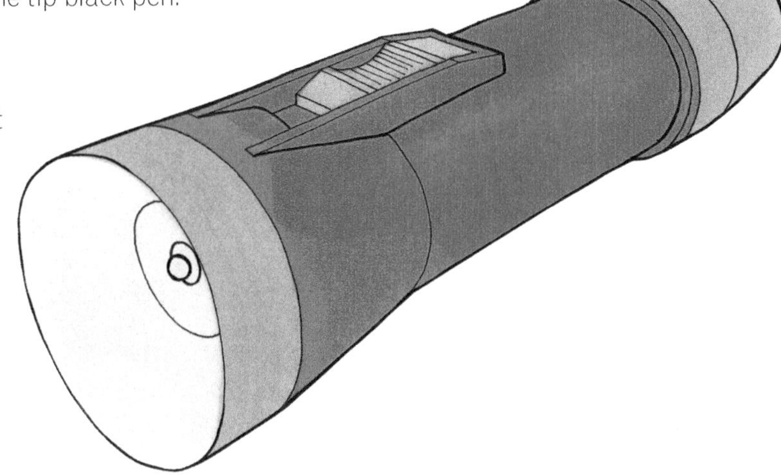

Step 1 • The base colour is applied first with a grey and orange marker.

Step 2 • Using the same markers, tonal change is added in the darker areas on top of the first application once it has dried. Markers can be used in this way and do show up darker on a second or third application. The marker can be blended to a certain extent.

Step 3 • Then, more definite tonal change is added using black, white, orange and red pencil. Hint: start with the white pencil and do the white highlights; then, you can work out where to add the dark outlines and tonal changes.

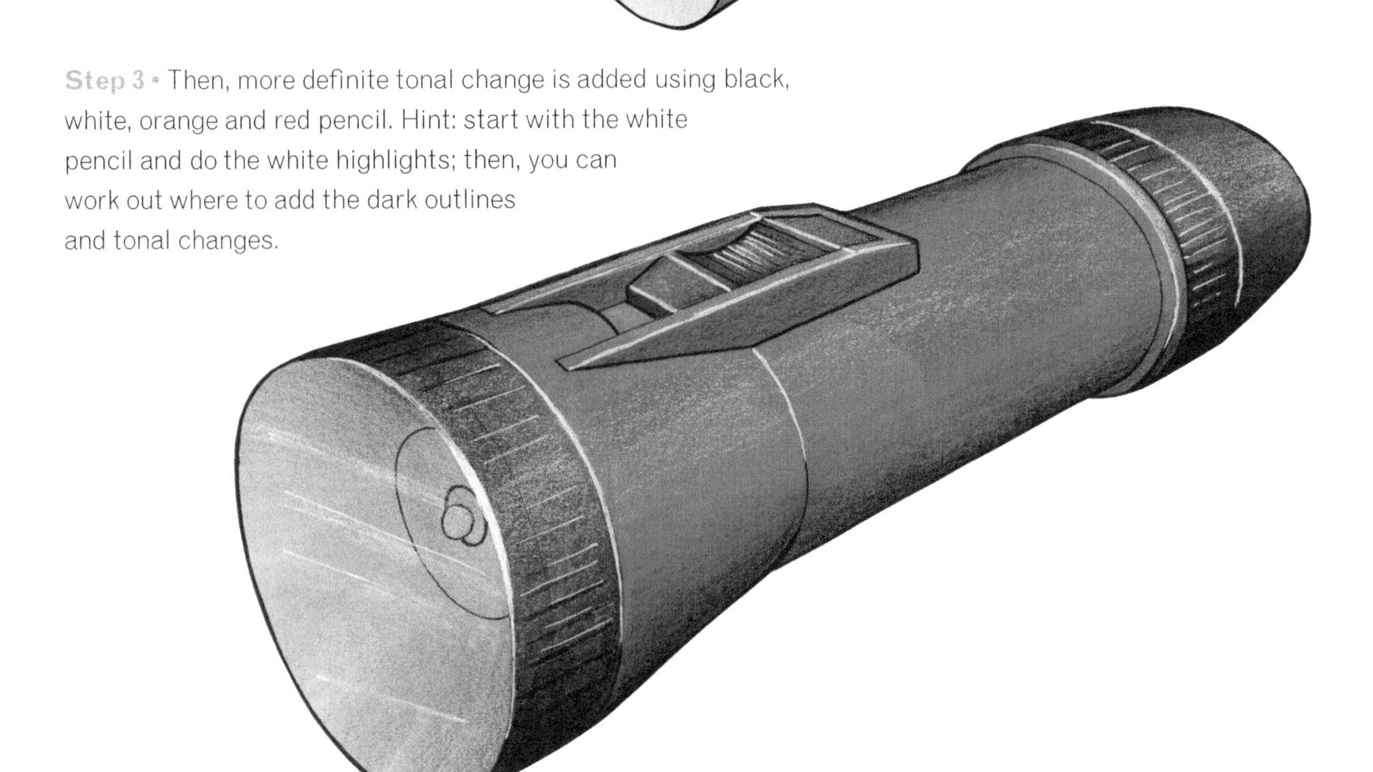

ISBN: 9780170233279

Clip rendering

The rendering has been completed using blue marker, black and white pencil and fine tip black pen.

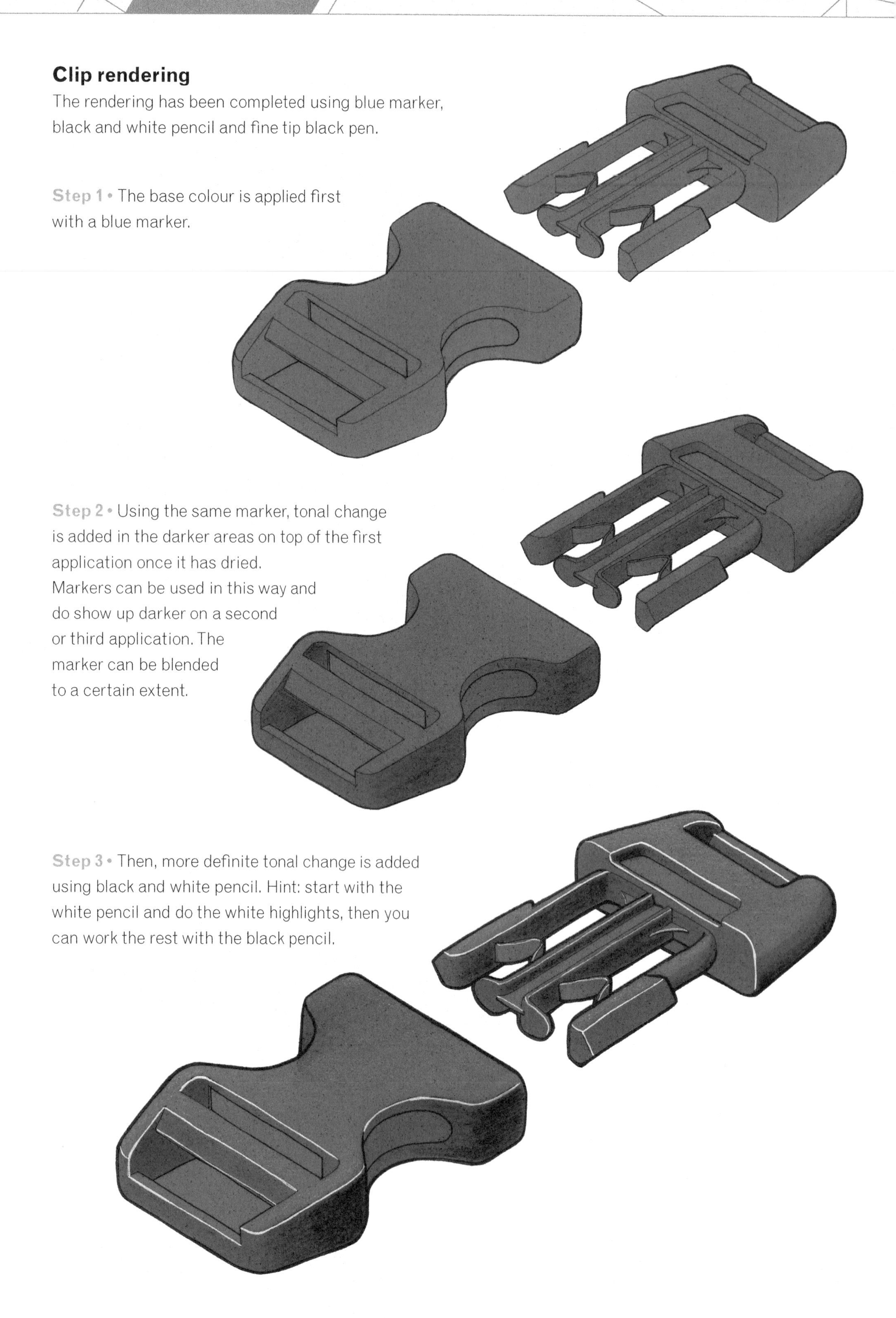

Step 1 • The base colour is applied first with a blue marker.

Step 2 • Using the same marker, tonal change is added in the darker areas on top of the first application once it has dried. Markers can be used in this way and do show up darker on a second or third application. The marker can be blended to a certain extent.

Step 3 • Then, more definite tonal change is added using black and white pencil. Hint: start with the white pencil and do the white highlights, then you can work the rest with the black pencil.

ISBN: 9780170233279

Rendering products and spatial design

Bike rendering

The rendering of the bike demonstrates the application of tonal change, a background outline and shadows. The texture and material of the seat and frame have been visually communicated with the use of colour marker and pencil. The handle bars, seat and wheels are plastic. The tubular frame is metal.

Furniture rendering

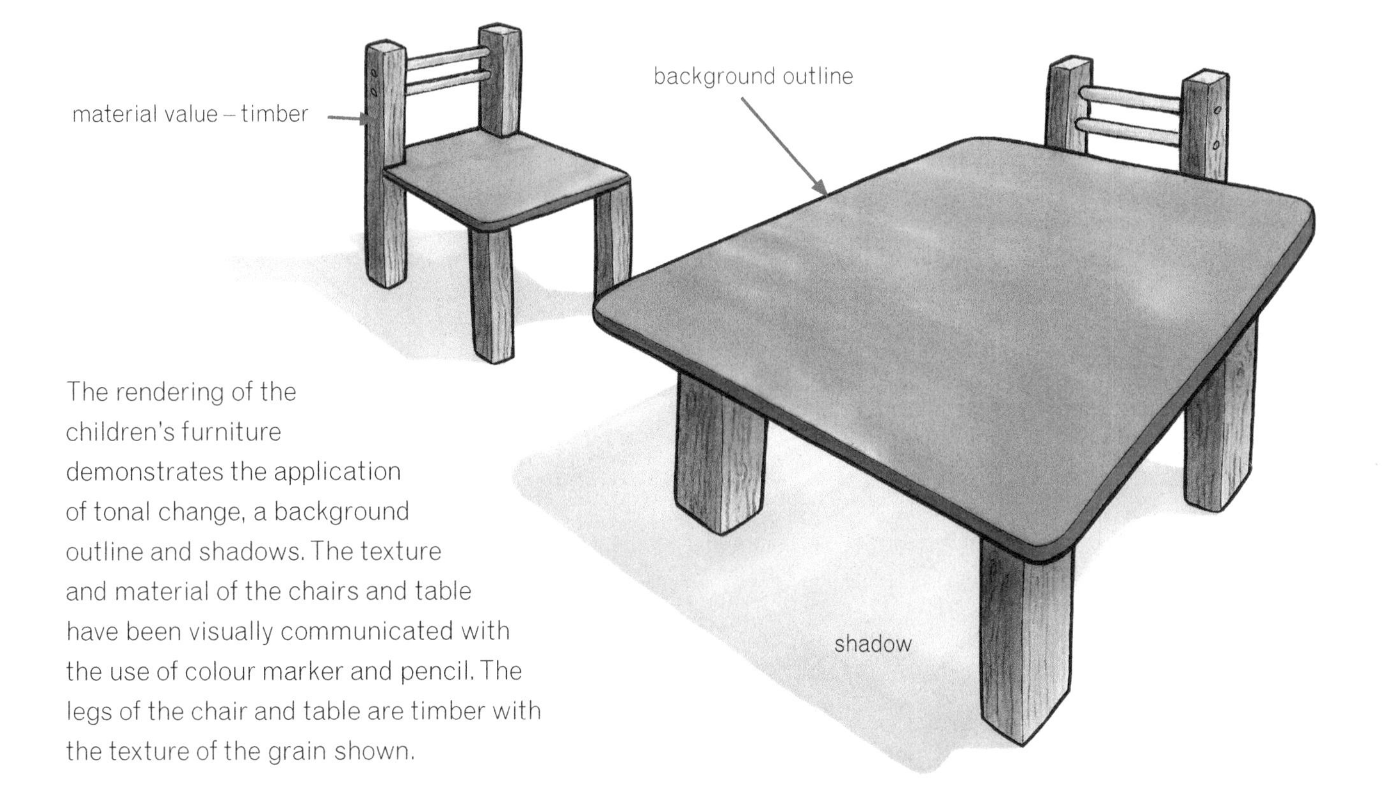

The rendering of the children's furniture demonstrates the application of tonal change, a background outline and shadows. The texture and material of the chairs and table have been visually communicated with the use of colour marker and pencil. The legs of the chair and table are timber with the texture of the grain shown.

ISBN: 9780170233279

Lamp rendering

The colour paper rendering of the lamp demonstrates the application of tonal change, a background outline and shadows. The texture and material have been visually communicated with the use of colour marker and pencil. The highlights and reflections indicate a glossy metal.

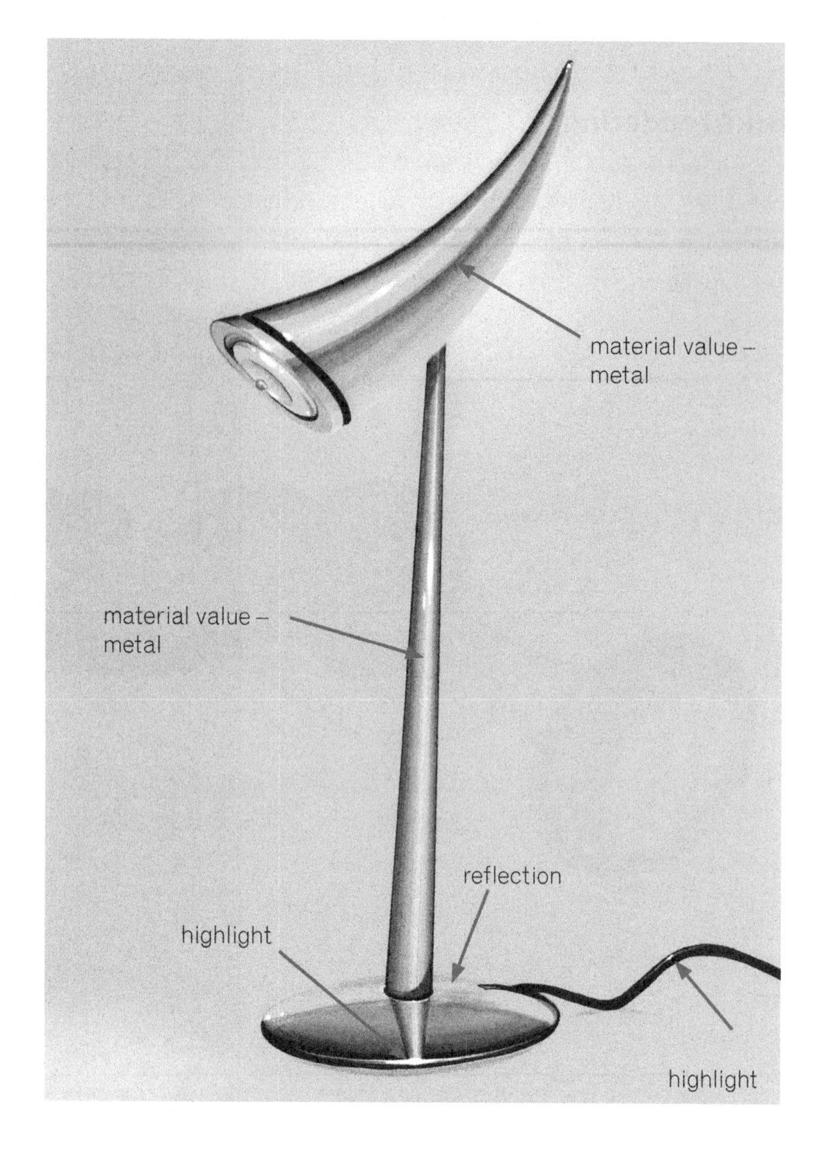

Container house rendering

The rendering of this container house demonstrates the application of tonal change, a background outline and shadows. The texture and materials have been visually communicated with the use of colour marker and pencil. The texture of the exterior of the containers has been shown. The use of highlights with a pastel green marker indicates reflections of the window and glass railing. The decking and stairs is timber with the texture of the grain shown.

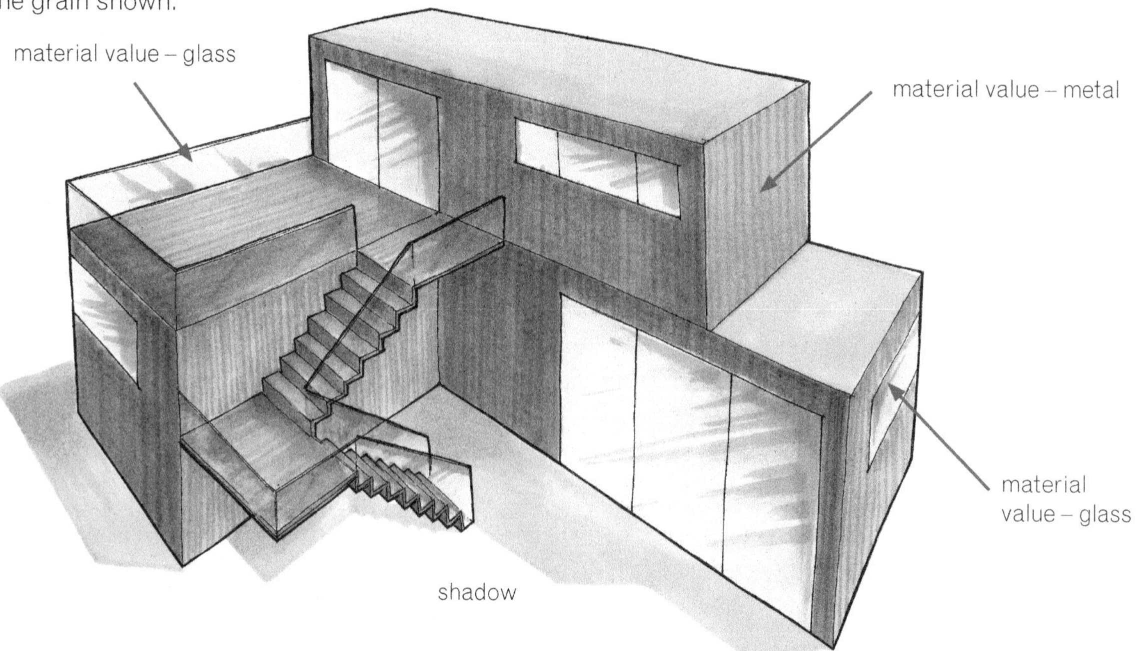

ISBN: 9780170233279

Christian Burgos — St Peter's College

The following work produced by Christian Burgos demonstrates a high level of understanding about where to add tonal change, highlights and shadows. The shadows give a realistic feel to the interior and exterior and add another dimension to the rendering.

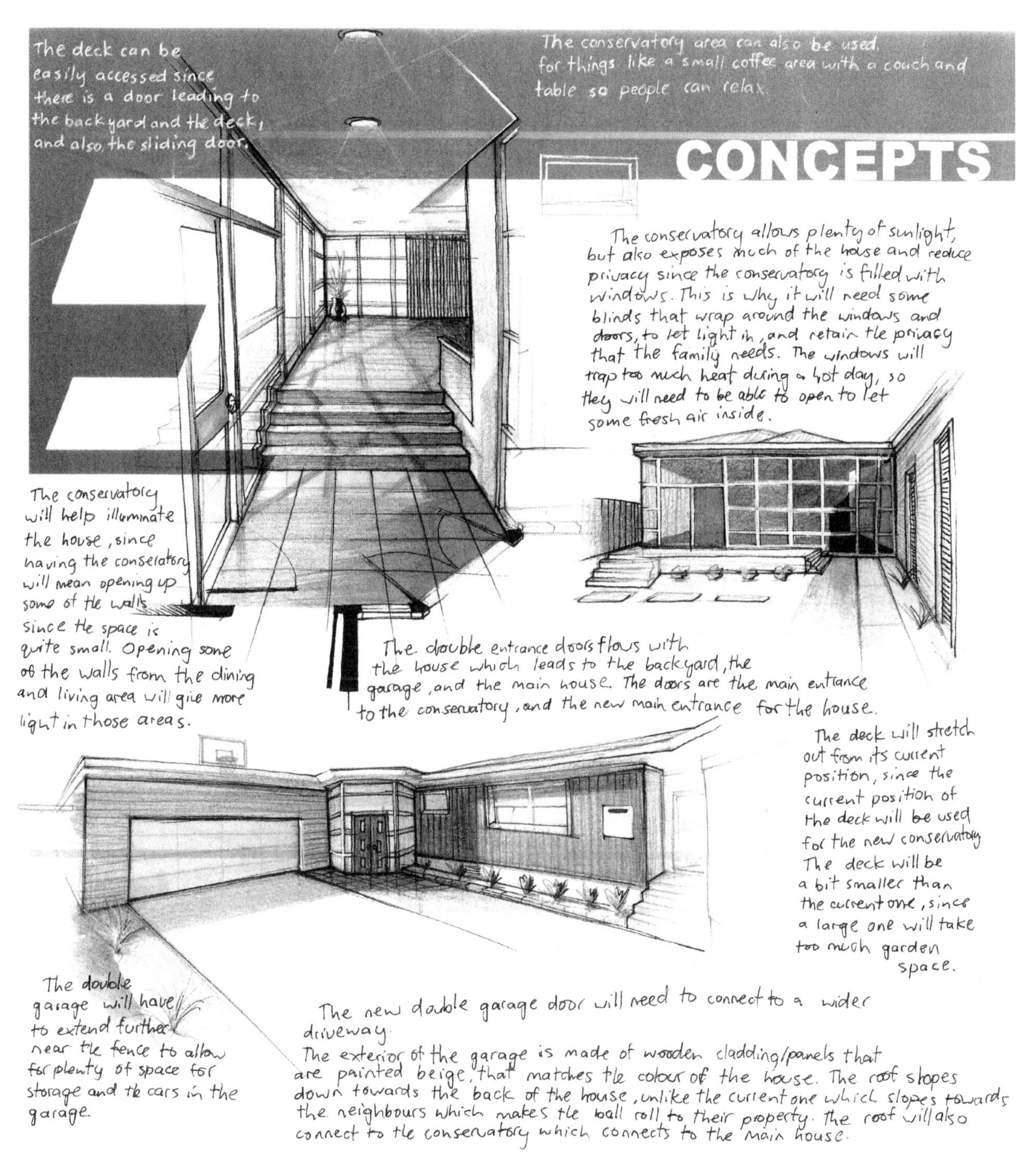

Christian Burgos

Christian Burgos — St Peter's College

The following work produced by Christian Burgos demonstrates a high level of understanding about where to add tonal change and highlights. He has used markers and pencil to present the effect of material value and texture for timber, cladding, windows, tiles and decking.

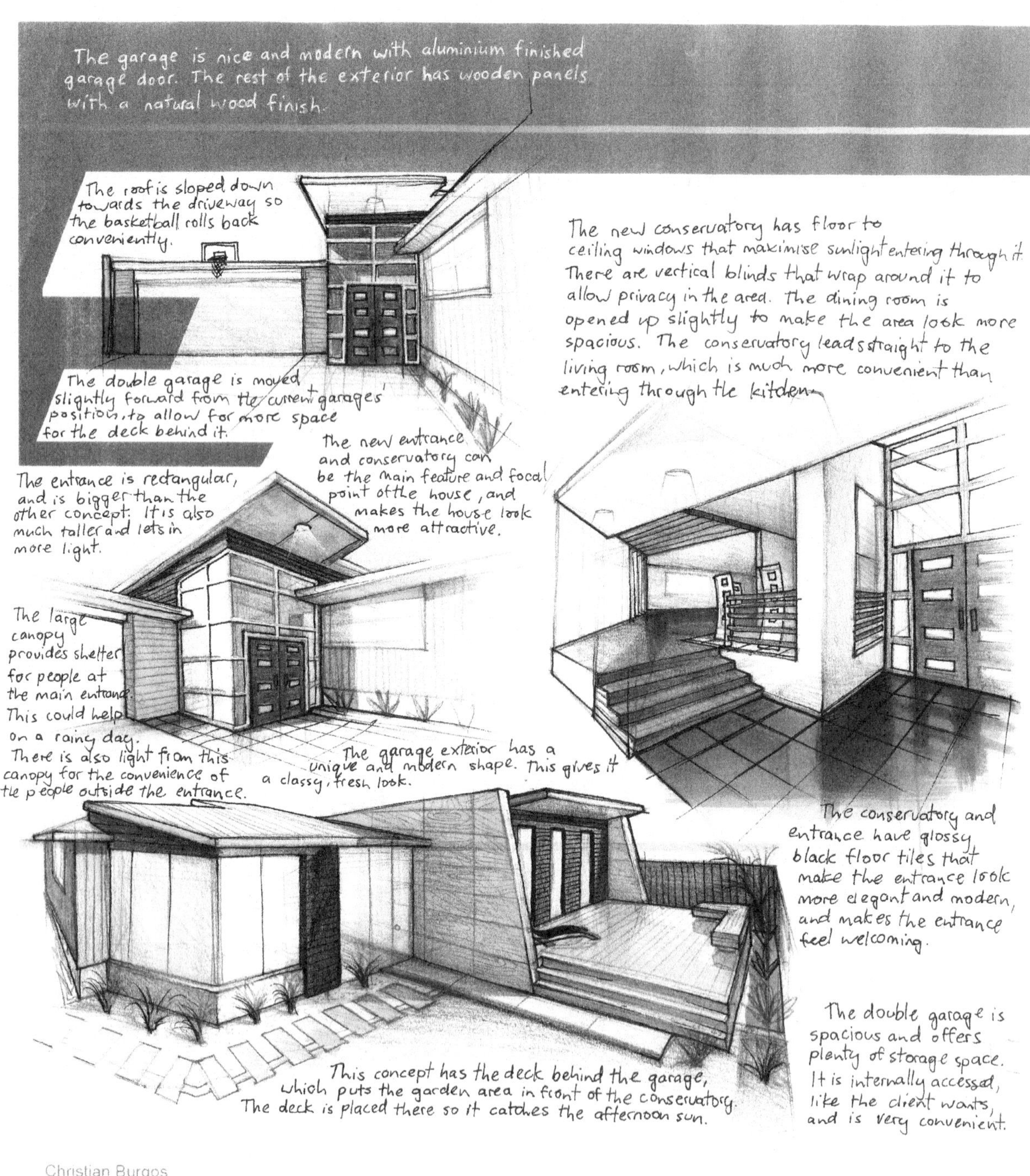

Christian Burgos

ISBN: 9780170233279

A range of media

The following pages of development present design ideas using a range of different media.

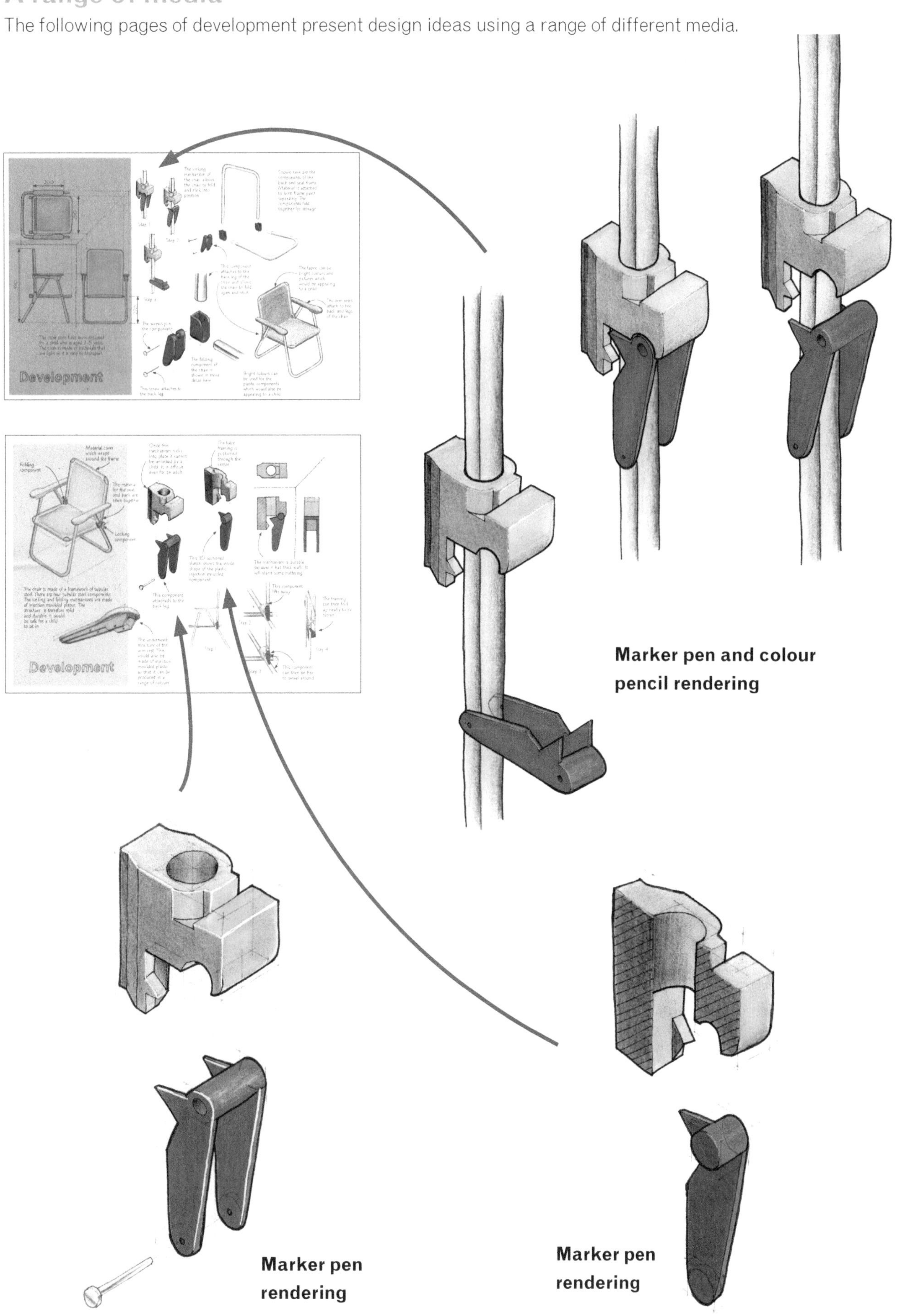

ISBN: 9780170233279

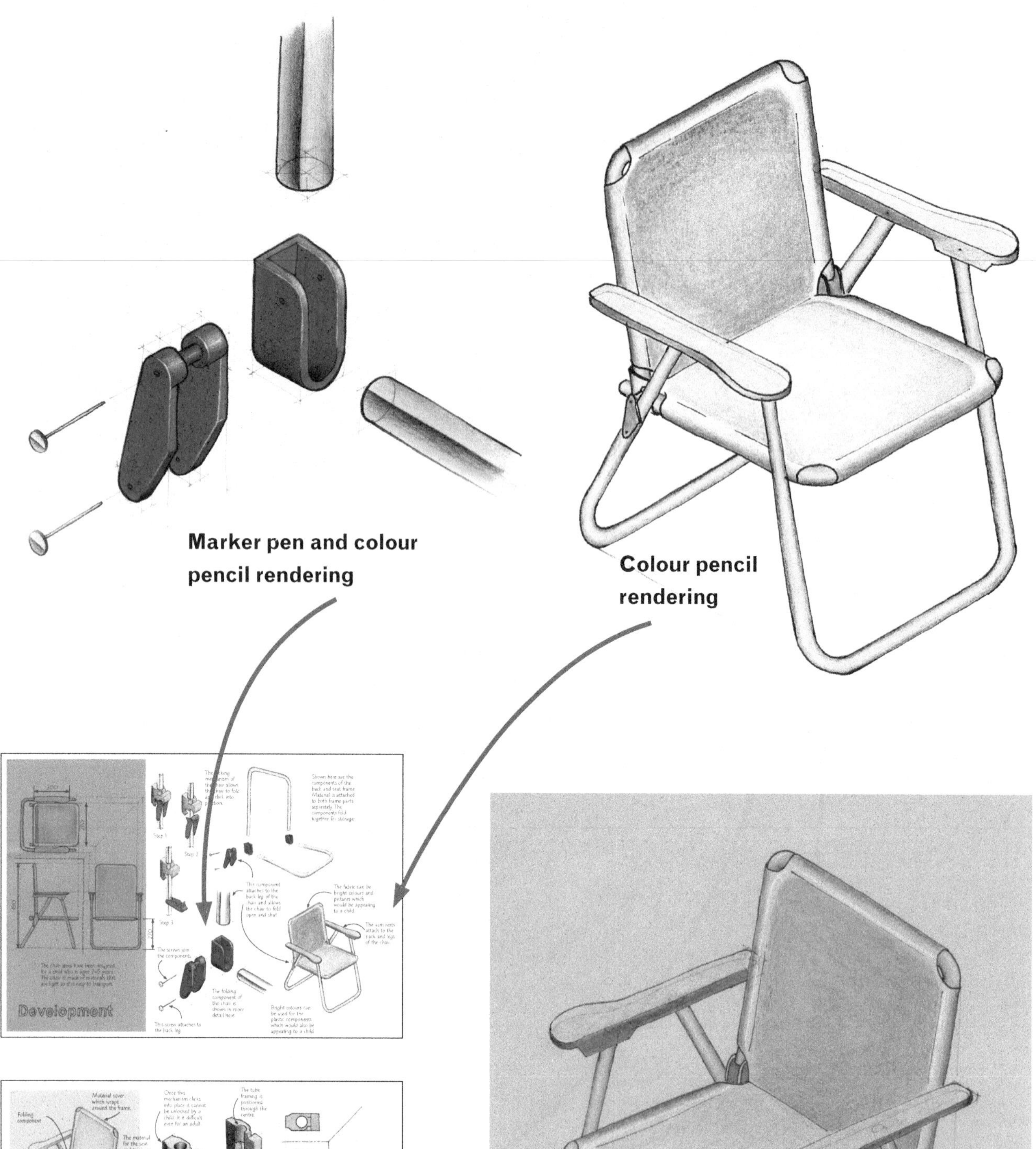

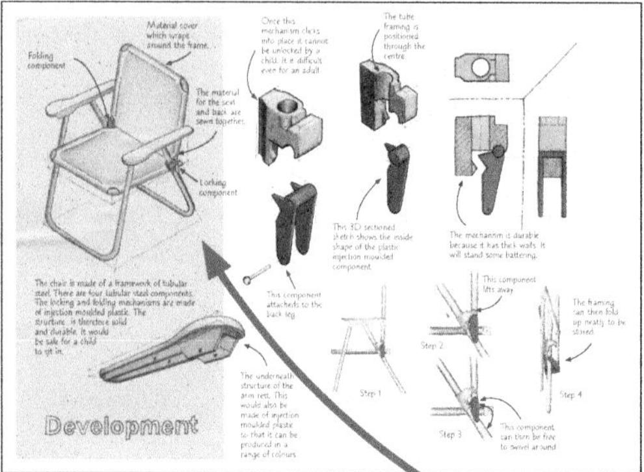

Colour paper rendering

ISBN: 9780170233279

To undertake the development of design ideas through graphics practice you will need to explore, refine, make decisions and visually communicate ideas in response to a brief. You will need to explore design ideas by considering possible alternatives. To refine design ideas you will need to consider the details. The consideration of aesthetics and function in response to a brief involves making design judgements that consider qualities of design ideas. You will need to visually communicate the details of the design idea.

Graphics practice involves using a visual literacy such as sketching, drawing, model-making or digital media, through the development of a design idea. Visual communication techniques are explained in all other chapters – Idea Generation, Freehand Sketching, Rendering Techniques, 2D and 3D Instrumental Drawing and Promotion of Design Work. These tools are used for aiding design thinking and the visual communication of the features of design ideas.

Design judgements, which are decisions made, are supported through research. Supporting research can be evident in a visual, oral or written form. Examples of research used to generate ideas has been demonstrated in Chapter 1. Within this chapter the integration of research at the development stage has been demonstrated, as well as initial research.

Spatial design

Spatial design is the design of exterior and interior spaces, which could be architectural, interior design or landscape architecture.

Design approaches that can be used for spatial design are tools used for the development of ideas such as bubble diagrams, symbolism and paper architecture. Technical knowledge that can be applied to spatial design is building materials and details, processes, sustainability and environmental considerations such as climate, space and light. Visual communication techniques relevant to spatial design are architectural sketching, drawing and rendering, and photo manipulation.

ISBN: 9780170233279

Product design

Product design is the design of objects and artifacts which could be related to fashion, packaging, media, consumer and engineered products.

Design approaches that can be used for product design are tools used for the development of ideas such as research, ergonomics and anthropometrics, mockups and models.

Technical knowledge that can be applied to product design is materials, joining, fitting, assembly, fasteners, finish, sustainability and environmental considerations.

Visual communication techniques relevant to product design are product design sketching, drawing and rendering, prototypes, models and animation.

Research

Existing products

Initial research may be undertaken to identify the design features of existing products in terms of the aesthetic and functional factors related to appearance and use. You will be able to source images from the internet, magazines, books or from products or spatial design around you. Having an actual product or spatial design that you can interact with will give a better understanding of the good and bad points or the aesthetic and functional features. You could then photograph or sketch the existing product or spatial design and reflect on these images.

Robert Gorrie — St Peter's College

The following work produced by Robert Gorrie and Samuel Matijevich demonstrates the use of the internet to gather images of existing shelter designs. Robert and Samuel have then analysed these products with reference to the aesthetic and functional features of each of the designs.

Robert Gorrie

Existing Product Research - Shelter Design

This is a very radical take on a bus stop. It not only provides a shelter for you while waiting for buses but is in fact made of bus parts. This bus stop is found in Athens, Georgia, United States and was created with a design that resembles a school bus. Every part of the shelter is made from either lamp posts or old school buses that used to stop there. Seeing as it is made from bus parts, it has a yellow metal roof and frame. The windows from the bus act as windows for the shelter allowing for a view to be seen outside. The seats are of average size and so a wide variety of people could be accommodated by this shelter although there are only three passenger seats and so the function is reasonably minimal.

Sjolander da Cruz designed the Wolves shelter to be used by children/teens to hang out with friends. There are two parts that make up this shelter, the ground and the roof. The ground was custom made by a group of people using bricks and cement. Located at a skate park in Wolverhampton, the Wolves shelter features undulating wave-like steel cantilevers over a brick terraced seating plinth, giving it a modern but at the same time natural design. There is plenty of space for seating which are in different sizes for different body types. The steel roof stretches over every part of the shelter and so it is protected from the elements. In conclusion this shelter has a terrific fitness for purpose.

This design is not only a bus stop, but is also a soccer goal. It is located in Sao Paulo, Brazil where soccer is talked about everywhere. This modern design is made from metal that acts as the frame of the goal. This is followed up with the netting to finish of the aesthetics of this shelter, both of which are white. Along with the stunning appeal of the shelter, there is a fair amount of seating on offer. Up to four people can occupy this bus stop at once, as well as having some poles for others to lean on. Though there may be a good amount of seating the practicality is lacking. There is only a base to sit on leaving the occupant having to sit up straight or against the back wall.

Designers Barry Sheehan and Gregor Timlin have come together to create this strange but unique shelter. This cart is not only used for carrying goods but is also to be used as a shelter when flipped upside down. These designers created this piece to help awareness towards the issue of urban homelessness. It does provide some shelter but is limited to one person per cart. It is not adjustable and depending on how tall you are, you will have to bend over while sitting in the cart due to a lack of height. Despite this, when upside down and fully extended, the cart can be used to lie down and even sleep in. It has a yellow coat and all of the aspects of a regular cart (wheels, push bar etc). Due to all of this, the carts have a reasonable fitness for purpose.

ISBN: 9780170233279

Sam Matijevich — St Peter's College

Protection Against The Elements

Samuel Matijevich

Blue Spots Of Santa Monica

The Blue Spots are a competition winning design for the 'Big Blue Bus' company operating in Santa Monica.
The LOHA was in charge of making these shelters with the idea of breaking apart the typical bus shelter and turning it into several components. the design was made to provide more flexibility, shading and visibility for the Santa Monica environment. The shelter was designed to use alternate energy and long lasting products so the shelter will have extended life spans compared to ordinary bus stops. an example of this is the LED illumination. this is easy to maintain and efficient. Besides the functionality of the Blue spots the are also quite bright and vibrant. they stand out among other buildings making them easily recognisable in there blue colouring. They have also been describe to be 'street art' in the fact that they a variation of shapes and shades. With this in mind various people have shared their opinion of the shelter many commenting on that it does not seem as though it would offer much protection against elements such as rain. in a country such as New Zealand this may not be as practical. Overall this is a great design for a country in which it is often sunny. This shelter is average in terms of being comfortable to the occupants. it offers little seating and benches which seem to be made of a hardened plastic. probably not meant for long durations of sitting. although the shelter offers plenty of shading for those people who prefer to stand

The Drop Spot

The image in the top right is The Drop Spot designed by Jonas Elslander and Jeroen Robberechts. this shelter was designed to be an inflatable shelter that offers an instant and cost-effective alternative to bus-stops and rain shelters. The Drop Spot is a collapsible bus-stop and is inexpensive to produce and easy to install which provides instant shelter for travellers in any situation. The shelter is coloured bright yellow which allows it to stand out amongst crowds. yellow is also thought to be the common colour of buses making it obvious and easy to recognise. The majority of the shelter is made out plastic which helps make the design cheap and cost effective. plastic is also an excellent material to be used for protection against the rain. while it is not as strong as other materials plastic does not absorb or soak up water and instead water just slides off. Another material which is also used is the small portion of metal. the metal is used as supports and can easily withstand small amounts of damage and lots of rain. metal is a good choice of material for supports as it is sturdy and long lasting. The fact that the shelter has a simple design makes it stylish and modern among the flashy stores and streets of cities. As for seating. this shelter looks as though it would be rather uncomfortable. I do not think this shelter was made to provide people with good seating because it is more a emergency shelter so seating is the last of the designers worries. it is only to provide shelter. The choice of colour and design makes The Drop Spot stand out among the crowd and the choice of materials easy to use capability makes the shelter functional.

Raytree Bus Shelter.

This shelter is designed with particular reference to trees. in terms of its shape. lines and the general look of the supports. This shelter was made to cope with the weather conditions of Singapore particularly during the summer months. The roofing is made out of a heat absorbing plastic which suits the harsh sun in Singapore. so it general provides excellent shelter against the sun. the shelter also provides protection against the rain. it achieves this in the way roofing of the shelter is shaped. The roofing has a very wavy shape is tilted on an angle so as the rain can flow to the back of the shelter once it hits the roof. if the occupants a sitting or standing in the shelters protective zone they should not get wet. The shelter. as mentioned before. is designed to resemble trees. you can see this in the way the supports are shaped. with branch-like ends coming out of them and at the top the support comes of onto another long branch-like support and the roofing begins to flow. making it seem like a top layer of leaves.
Aside from the shape and inspiration of the shelter. the use of the colours silver and blue work quite well together in this instance making the shelter quite attractive to look at. The shelter offers plenty seating and quite enough standing room. the seats a slightly curved in their appearance and have a very straight at the back. this means people will have to sit with their backs upright with not much room for slouching. so the back support of these seats is quite good. This shelter is quite good looking with its curves and shapes and is quite functional in its purpose.

Eco Friendly Beach Tent

The bright green image on the bottom right is the solar powered beach tent. This beach tent was designed by Emma Harris and was designed for the British coast where the sun is lacking and wind is strong. The tent has been designed to be able to withstand the brutal winds by making the design aerodynamic which allows the wind to flow over the tent towards the back where there is a wind turbine. this helps provide power within the tent and protect it from the brutal sea side winds. Continuing with the eco-friendly design aspects the tent is made from a photovoltaic fabric and helps to harness solar energy which is fed into a storage generator inside the tent. this provides the tent with a clean and cost-free way of providing power while outdoors. This power can make this tent the social arena for you and your friends because this energy concept is revolutionary for outdoor tents and will be the talk of all camping or outdoor events. Besides the tents functionality this beach tent is also quite stylish. with its smooth curves and semi-dome shape this tent can make and impact on any scene. the tent is really only curved with minimal straight lines which give the tent its own unique look. the tent also stands out with the choice of the bright green colour. while this can also help with the tent being in the forest. this tent is used primarily at the beach so it will stand out and be noticeable. The functionality and the general aesthetics of the tent are new to the tent shelter world and are surely the start of a revolution.

Zack Kite — St Peter's College

The following work produced by Zack Kite demonstrates sketching of an existing torch as research. Instead of finding a range of different products to analyse he has focused on the components of the existing torch. From this Zack has gained a sound understanding of the parts and how they fit together, the materials, aesthetic and functional characteristics. This not only is useful as a starting point when designing a product but also improves sketching skills by doing these observational sketches.

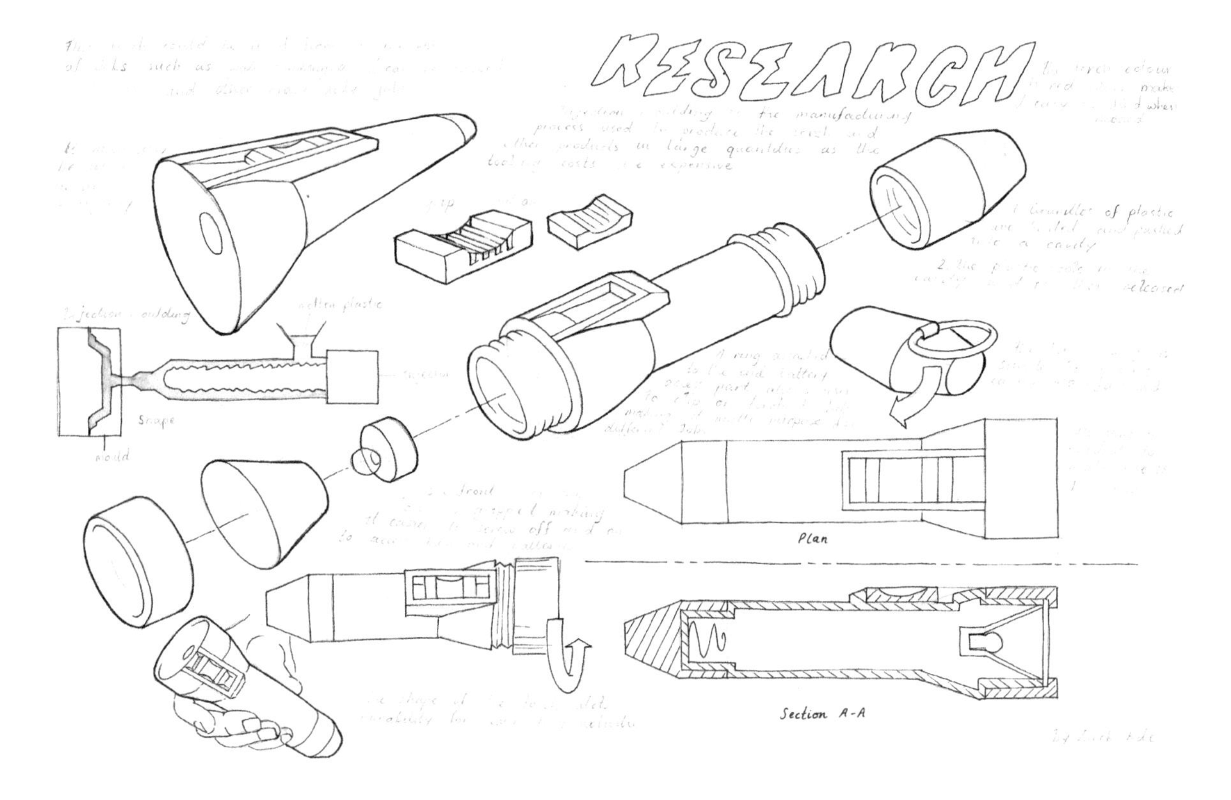

ISBN: 9780170233279

Christian Burgos — St Peter's College

The following work produced by Christian Burgos demonstrates the use of studying existing products that relate to his spatial design brief while also outlining characteristics of the site for the architectural design.

RESEARCH

Existing products

The clients want a new conservatory area on their current deck. The conservatory area will become their new entrance to the house, which is much more handy than their current entrance - the kitchen. The conservatory area will be the main foyer of the house which connects the integrated garage to the house, which gives it internal access, which is what the client wants. The conservatory area will take up the current deck space, so there will have to be a new deck in front of the current one. The conservatory area will wrap around the dining room, linking the living room to the garage, to the new main entrance. The conservatory area will have to be ful of glass to provide plenty of sunlight in the house. The house is actually elevated about half a meter above ground, so the conservatory area will have to be elevated also, and have steps that go down to the garage and main entrance. The new conservatory area will be quite large, and can have some seating.

The conservatory area will have steel frames, and will stretch around from the living room to the garage. It will be full of glass which retains the current amount of sunlight that enters the living room, and can increase the amount on the dining room if the dining room wall between the conservatory area is cut away. This can help open up the house a bit, which saves from moving the dining and living room around.

The entrance/conservatory will need to have a small canopy outside, like most modern houses have, much like the one in the picture above.

Also, the conservatory will have some blinds across it for privacy. The conservatory will cause the house to be more exposed, so blinds will help keep the privacy that the house needs. The blinds could either be vertical or horizontal, or those plain sheet ones, depending on the clients preferences. A good colour would be white or beige, which will make the place look modern.

This folding side up garage door is good since it does not end up scratching the cars and does not bump into people and things. It is safe and also reliable, which is something that is useful and something that the client wants for the new garage.

A good option for the material of the garage door is steel, as it is practical, and very common and reliable. A nice colour that would suite the garage exterior is grey, white, silver or green.

This garage door is quite stylish and distinct. This would look good on the new garage, if the client wants the garage to look modern and sophisticated. Although, this option is visually appealing, it might be too expensive.

These storage cabinets are really good for storing many things. The new garage will need plenty of these in the garage so they can store some of the things in their garage right now. They have many things in the garage that need storage right now, so they might also need some sort of mini storage room. These cabinets also keep the things safe. The garage gets quite moist sometimes, and this is a good way to keep their belongings safe from damage.

Christian Burgos

SITE PLAN

The site plan shows the current position and size of the house and garage. There is plenty of space around the garage for expansion, especially the side next to the fence wihch is currently used as sort of a shed to store some things like firewood. There is also plenty of space behind the garage, which is currently just a large grass area.

The current garage is quite small, and does not provide enough spaec for the clients storage. The new garage should be much larger than the current one since it will need to fit the two cars and plenty of space for their storage. There is enough space beside the garage to expand it slightly further to make it wider and make better use of the space.

The current garage has a single garage door which only makes room for one car in the garage at a time. It is also made from a cheap corrogated steel material that becomes rusty over time which can look unpleasant, and can be quite dangerous. The garage floods when there is a storm, so the floor will need to be heightened slightly to prevent the water from sliding down the inside of the garage and damaging the things that are stored inside.

The client wants a new conservatory area. This is quite clever since it is a good way to connect the house to the garage and can also form a new entrance, which saves the client from rearranging the layout of the interior such as shifting the living room to the kitchen etc. The conservatory area is a nice simple solution to connect the garage to the house and create a new, convenient

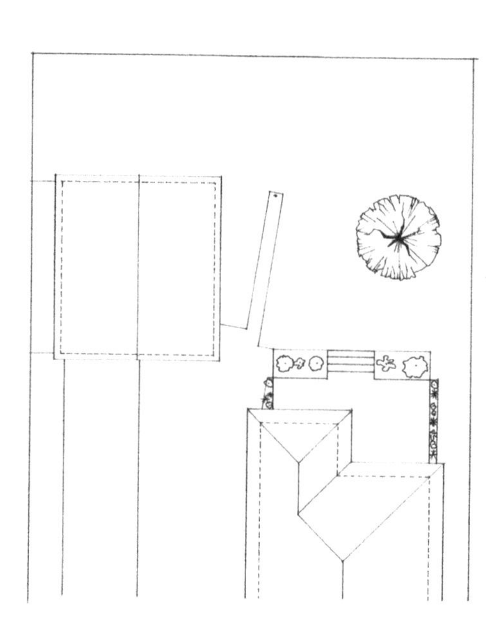

The new conservatory will need to have a ramp or some steps that lead up to the house, since the house floor is raised. It will need to lead back down to the ground level so it can provide acess to the garage and the outside. It will also need to connect to the new deck, which will have to expand into the garden since the current deck is will be used for the conservatory.

The current garage interior has no gib walls which mean that the structure of the houes is exposed. The new garage will have to be properly gibbed up so it looks neat and presentable, and can provide for insulation.

The roof of the garage will need to slope down on one side, somewhere towards the property since the basketball often rolls to the neighbours property. When the ball goes on the roof, it can roll back somewhere on the garden, and is easy to reach.

Christian Burgos

ISBN: 9780170233279

Sheldon Carr — Onslow College

The following work produced by Sheldon Carr demonstrates the use of photographs instead of drawings to analyse the site for his architectural design.

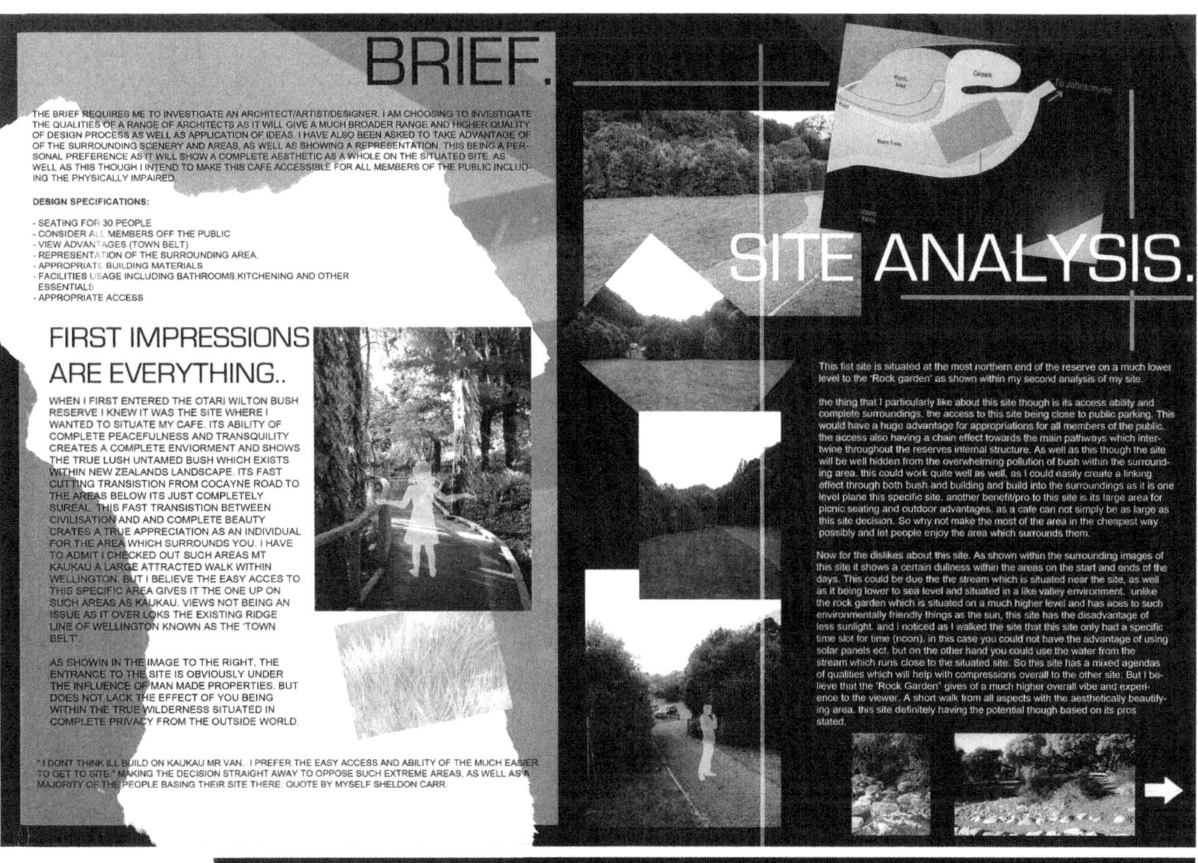

Christian Burgos — St Peter's College

The following work produced by Christian Burgos demonstrates the use of the internet to gather a range of existing products to analyse the shape, style and function of a range of vehicles.

RESEARCH engines

ELECTRIC
Electric engines are clean, as it does not directly produce emissions, however since it requires batteries, it needs to be charged, and the batteries would be have harmful acids that are bad for the environment.
The electric motor can also be charged at home, but then since the electricity comes from large power stations, it will still cause pollution. If the electric engine was to be connected to solar panels, but they may not be large enough to power the electric engine. The best way t0o ues an electric motor is to plug it to a solar panel that replaces the battery.

BIOFUEL
Biofuel is the use of waste products or vegetables like corn. It is good since it uses natural fuel, however it still produces carbon dioxide which is bad for the environment.

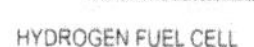

HYDROGEN FUEL CELL
The Hydrogen fuel cell technology is growing, and many automotive and moped manufacturers starting to use it as an alternative to petrol, and even electric engines. Hydrogen fuel cells convert hydrogen into electricity. By having hydrogen fueling stations, which are currently underway around the world,hydrogen fuel cell vehicles can recharge at those stations. The fueling stations are powered by solarpanels which collects energy from the sun and converts the water into hydrogen.Hydrogen fuel cells are a great way to power a vehicle with zero emissions, which is the most important part of this technology as it is environmentally friendly. They are very clean and are easy to dispose of, and are very light which means that it requires less power since the overall weight is quite light.
However, the disadvantages of hydrogen fuel cells are, that the hydrogen molecules are very sensitive, which means that the slightest puncture on the hydrogen tank would release the gasinstantly, and also, most hydrogen fuel cells at the moment are not even strong enough to last as long and go as fast as a petrol engine, but since this technology is for a moped, it is not much of a problem sincethe moped will be light and compact which means that it requires much less power than a car.

PETROL
Petrol engines are very fast and lasts longer than hydrogen fuel cells and electric ones, but are very bad for the environment since it produces a lot of emissions and requires a battery which is also harmful for the environment. A majority of the vehicles in the world are petrol because it is the most practical. Petrol engines are easy to use but are bad for the environment. There are already enough polluting vehicles in this world, and creating a new petrol vehicle is going to make it worse.
The world is running out of oil, and it is predicted that it will run out in a few years if we keep using it to run our cars. Sooner or later manufacturers will eventually be forced to use other means to power vehicles, like electricity, solar power, or even hydrogen fuel cells, which is one of the best options right now.

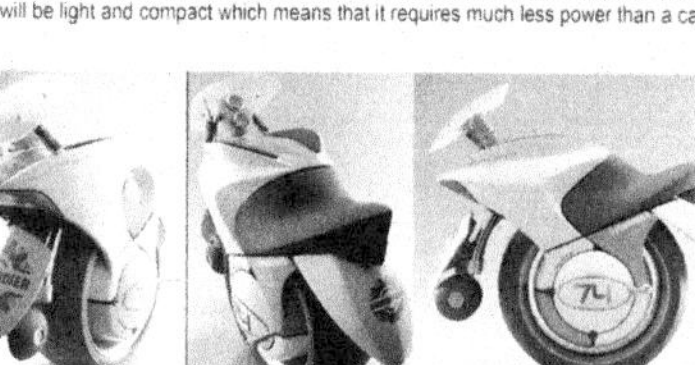

Also, fuel cell life is about 5-10 years, but is very expensive to replace. A battery takes 3 years before it needs to be replaced.

RESEARCH shape and form

The typical shape of a motor scooter is quite common, and in no very unique. However, the detailing such as the lights and the wheels, and some of the aerodynamic lines are still attractive and modern which makes this classic design aethetically apealing.

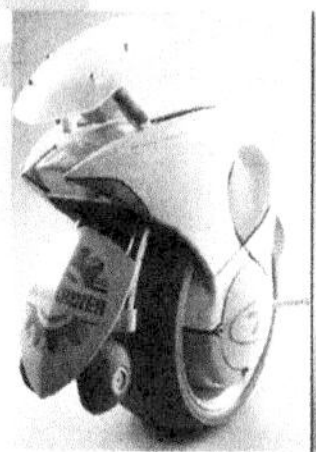

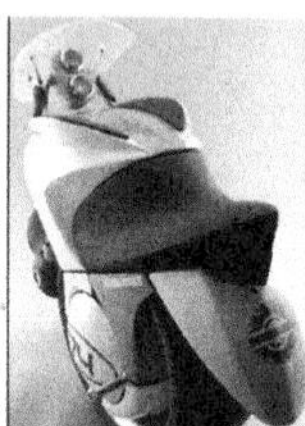

This is the Embrio gyro-scope fuel cell concept.
The shape is very unique and uses features from motorbikes like the seat and the controls. It has a unicycle look to it since it balanceson one main wheel. The extra wheels are small and act like 'landing gears' when it travels less than 20 km/h. It uses smooth flowing lines thatcreate an aerodynamic look and has colours like shades of grey and white with some blue accents which makes it look modern and appealing.The diagonal line that runs above the wheel stretches to the back and the shape of the upper body give the look of an animal, which gives the moped a unique characteristic. It has a leaping forward kind of a look which resembles a motorbike.

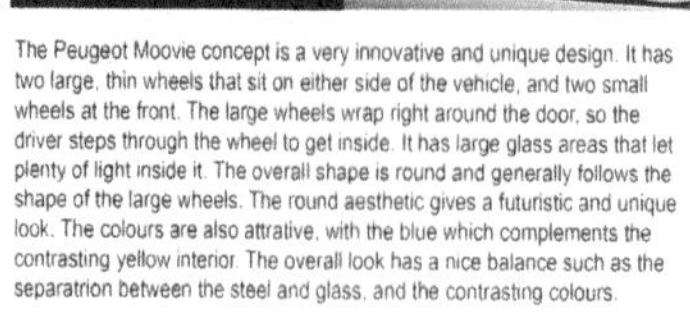

The Peugeot Moovie concept is a very innovative and unique design. It has two large, thin wheels that sit on either side of the vehicle, and two small wheels at the front. The large wheels wrap right around the door, so the driver steps through the wheel to get inside. It has large glass areas that let plenty of light inside it. The overall shape is round and generally follows the shape of the large wheels. The round aesthetic gives a futuristic and unique look. The colours are also attrative, with the blue which complements the contrasting yellow interior. The overall look has a nice balance such as the separatrion between the steel and glass, and the contrasting colours.

The Peugeot HYmotion3 has typical scooter proportions but has the unique roof that wraps around the driver. It has three wheels that make it look quite different from a normal scooter. It has many designs similar to peugeot cars like the head-lights and the overall body lines. The headlights make it look like a car, and also the sporty looking wheels, make it look attractive. The all-glass roof gives it a simplistic look and at the same time gives a comfortable amount of light on the driver.

This Can-Am Spyder is unique as it has 3 wheels with a quad bike aesthetic. It has flowing lines that rise towards the back which gives it a forward looking stance and also shows speed. The vehicle is avaiable in different colours as a two tone with black. It gives a unique look that makes it look sporty and fast. The shape of the vehicle is unique and appealing, while also being safe, as it has two wheels at the front and the overall width is quite wide which reduces the risk of falling over. It uses characteristics of a sports car, such as the wheels.

This volkswagen concept has a very round appearance which makes it look compact and distinct. It uses some typical car like designs with shoulder lines that run above the wheels, and the shape of the bonnet and the cabin. The rear drops down like a sports coupe which also give a look of a car.

The k-007 Scooter has very muscular motorbike proportions. Its wide rugged looking wheels make it look fast, alog with the long body which is very low to the ground. It is available in differ-ent colours that are bright and make the bike look attractive. It has 4 wheels altogether, but are paired in doubles on the front and the back, which gives a look of two very wide tires, and makes it look unique.

ISBN: 9780170233279

Research

Influence of a designer, design movement or era

As identified in Chapter 1, research about a designer, design movement or era can be undertaken as a way to generate ideas. The ideas, shape, style, materials of the designer, era or movement can then used to influence your own design ideas.

Sarah Stevenson — Onslow College

The following work produced by Sarah Stevenson demonstrates the study of Santiago Calatrava, the era, his style, his ideas, aesthetic and structural characteristics of his design. The pages shown are only a sample of her work. Sarah has taken images of his structures to inform her design ideas for a chair to produce a range of effective intial ideas.

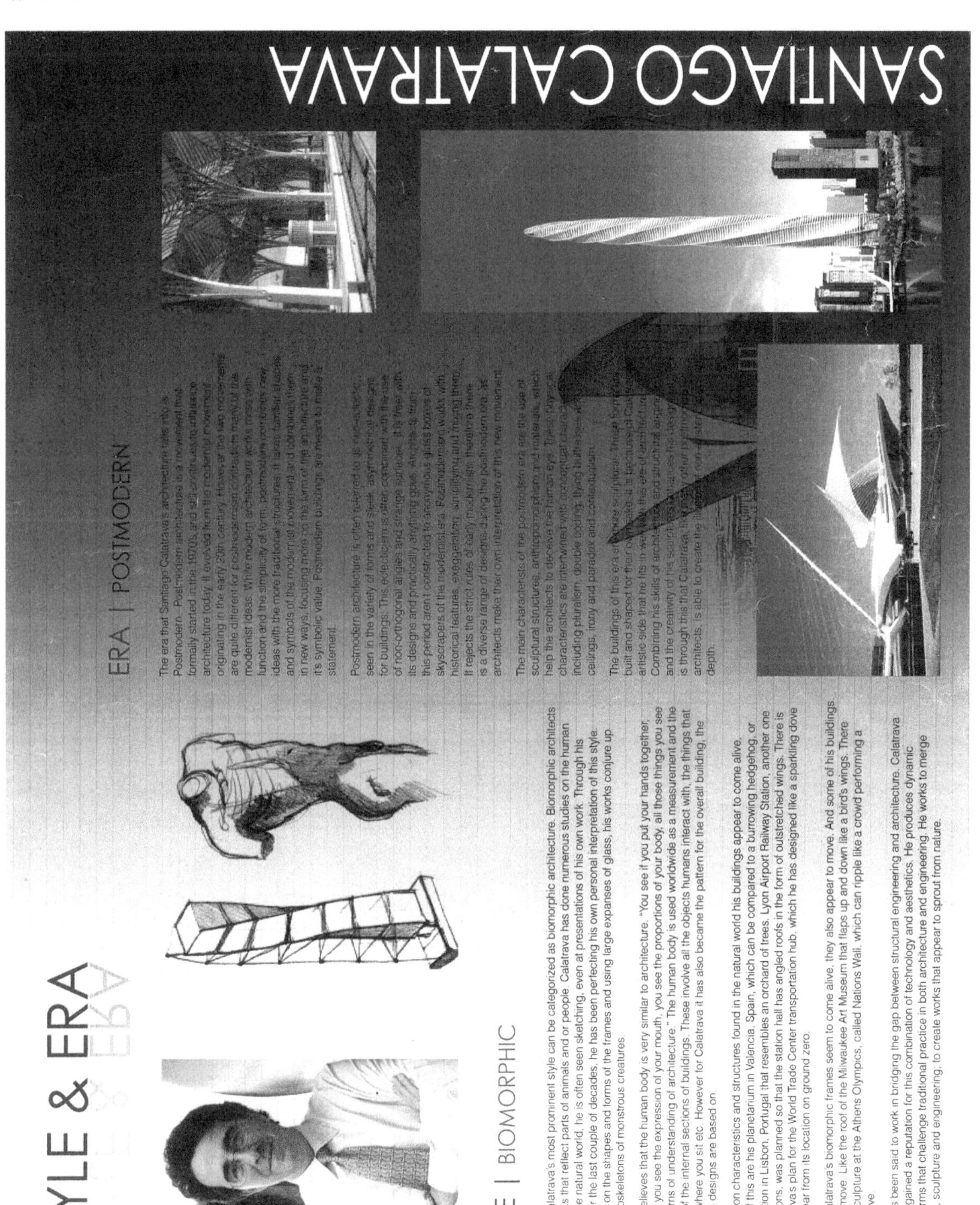

ISBN: 9780170233279

CALATRAVA AND HIS VISION

PHILOSPHY

Santiago Calatrava describes himself as an "architect and engineer by trade, sculptor and painter by vocation" and this is reflected in his work philosophy. He claims that the [illegible] forms into one. Architecture is art in the way that it affects the viewer. It, like art, has a symbolic value and a certain meaning. Whether it is seen as beautiful or ugly it was [illegible] that by getting closer to the frontier between architecture and sculpture we are able to understand architecture as an art. He relates it to reading. He argues that nobody reads a book just because of [illegible] the book, they read it because of the symbols and the meanings behind it.

His philosophy is that architecture is art. Calatrava uses the creativity of his paintings and sculptures, incorporating them into his architecture to create unique designs. [illegible] worldwide. They reinforce his main ideal for his architecture and the metaphor of his work. Calatrava states that his principle is to make works of pure engineering that are inspired by [illegible] innovative buildings he wants to give people a break by experiencing a different environment, even just for a couple of minutes. It creates a link that [illegible] bring people together.

Geometry is vital in the field of architecture, according to Calatrava. The shapes that are used in the architecture and how they interact with each other can have a fundamental [illegible] with or in contrast to their environment. The environment plays an important roll in Calatrava's work philosophy. He is known to use the motto 'nature is both mother and teacher' [illegible]

"If you look to our environment and you see the importance that architecture has in our artificial environment you immediately can understand that the control of it and the quality of it is fundamental [illegible] and the impact or the part that the architects have in this built environment is enormous. So this makes maybe open the mind of the people how important it is the quality in this particular part [illegible]

Calatrava's theory is that you can increase the quality of life in certain areas by creating a focus. In an interview a few years back he talked about the intervention in the Expo in Lisbon. When [illegible] buildings were built he said it almost looked like a desert. However after six or seven years it had transformed and become one of the most attractive living areas of Lisbon. [illegible] more identity. "They can re-enter themselves, they can say I live close to this building. It also produces a facility, if this is a transportation facility, if it is a bridge or a station or whatever, [illegible] everyday lives because they are living close to a facility."

"The most touching thing that anyone can say to me is that I have done something beautiful for the community." Even though he has designed many extravagant works he still believes that even [illegible] natural place, a winery in a delicate setting, can also move your sensibility and show you how important it is that architecture does not become a predator on the landscape, and rather [illegible] environment. Even in the most modest circumstances, there is the possibility for emotion and poetry."

HISTORY

Santiago Calatrava was born on the 28th of July 1951 in Valencia, Spain. He grew up in a family involved in agricultural exports, the principal industry for his coastal town. His father loved art and would take Santiago to experience Spain's greatest Museum, the Prado in Madrid. This played a vital roll in sparking Santiago's interests in the arts.

Calatrava aspired to be an artist and at the age of 17 he planned to move to the French capital and immerse himself in the prosperous art community of Paris of the 1960s. However this never occurred, due to the student protests of the late 60s that resulted in the cancelling of classes. Instead he chose to study at the Escuela Técnica Superior de Arquitectura (Technical University of Architecture) as an undergraduate. After he graduated in 1975 he moved to Zurich, Switzerland and enrolled at the Swiss Federal Institute of Technology (ETH). During the next four years Calatrava was able to gain a PhD in civil engineering and technical science and in 1981 he completed his doctoral thesis on the "Foldability of Space Frames".

It was in Zurich that Calatrava met and married his wife Robertina, a law student at the time. They settled in Zurich, which naturally became the logical location for his first office.

At first Calatrava struggled as a new architect and he spent the next few years competing in various competitions in an attempt to get his name out. He finally got his ticket into the limelight in 1983 when his design for the Stadelhofen Railway Station in Zurich was chosen. This opened doorways for him into the world of professional architecture.

L'Hemisfèric

THE CITY

The L'Hemisferic is one of the buildings that make up the City of Arts and Science, designed by Santiago Calatrava and built on the dry riverbed of the River Turia in Valencia, Spain. It took fours years to design and was under construction during 1996-1998. The City of Arts and Science incorporates an Opera house, a Science museum, an Auditorium, a walkway and garden, an open-air oceanographic park, an IMAX Cinema and Planetarium. The L'Hemisferic, containing the cinema and planetarium, was the first of these elements of the 'City' to open to the public.

STRUCTURE

Calatrava's L'Hemisferic eye can also blink. A steel and glass shutter is operated by hydraulic lifts, moves the eye's 'lid' giving the impression of a blink. The caged lid that encloses the center of Calatrava's 'eye' can be raised above the water as well as fold.

The structure of the eye is reasonibly strong. Calatrava has used croncrete walls smaking the building more stable. Aluminium beams were a

THE EYE

The L'Hemisferic resembles a gigantic eyeball floating in the water. Calatrava has used this water surrounding the City of Arts and Science to enhance the building's impact. On a still day the mini lake acts as a giant mirror reflecting the building, revealing a perfectly symmetrical eye. In this design Calatrava has used the movement of the sweeping curves (seen in the upper lid and roof of the eye) and the 'pure ' shape of circles (pupil) to draw the viewers attention to the center. By sculpturing it so the ends peak off, it gives the building a sense of depth. By simplifying the shape of a human eye and sticking to realistic proportions Calatrava is able to give a life-like quality to the eye.

The shapes used in the eye harmonize together and in conjunction with the white and grey of the building puts off a calm and relaxing vibe. Calatrava really uses bright and contrasting colours to catch attention and this design is no exception. The blinding white doesn't distract from the natural beauty of the eye. It actually enhances and modernizes the buildings. The light tint of grey for the roof complements the changing colours of the water and helps break up the sections of white.

The 'pupil' of Calatrava's L'Hemisferic is the IMAX Cinema, its perfectly spherical exterior is a mosaic of fragmented tiles. This attention to detail makes sure that not just those who are seeing the building from a distance are impacted by the design. Calatrava makes the pupil a focal point of the L'Hemisferic using the creation of positive and negative space through the light that is reflected off the billions of tiles. The pupil glows due to these titles creating a stunning effect on Valencia's night horizon.

The socket and the lid of the eye are made out of concrete and contain elongated aluminum awnings of differing lengths. Calatrava uses the repetition and rhythm of these white beams to portray a cage like structure entrapping the pupil of the eye. His use of lines is also shown in the corrugated panels that construct the roof of the L'Hemisferic.

SANTIAGO CALATRAVA

SANTIAGO CALATRAVA

ISBN: 9780170233279

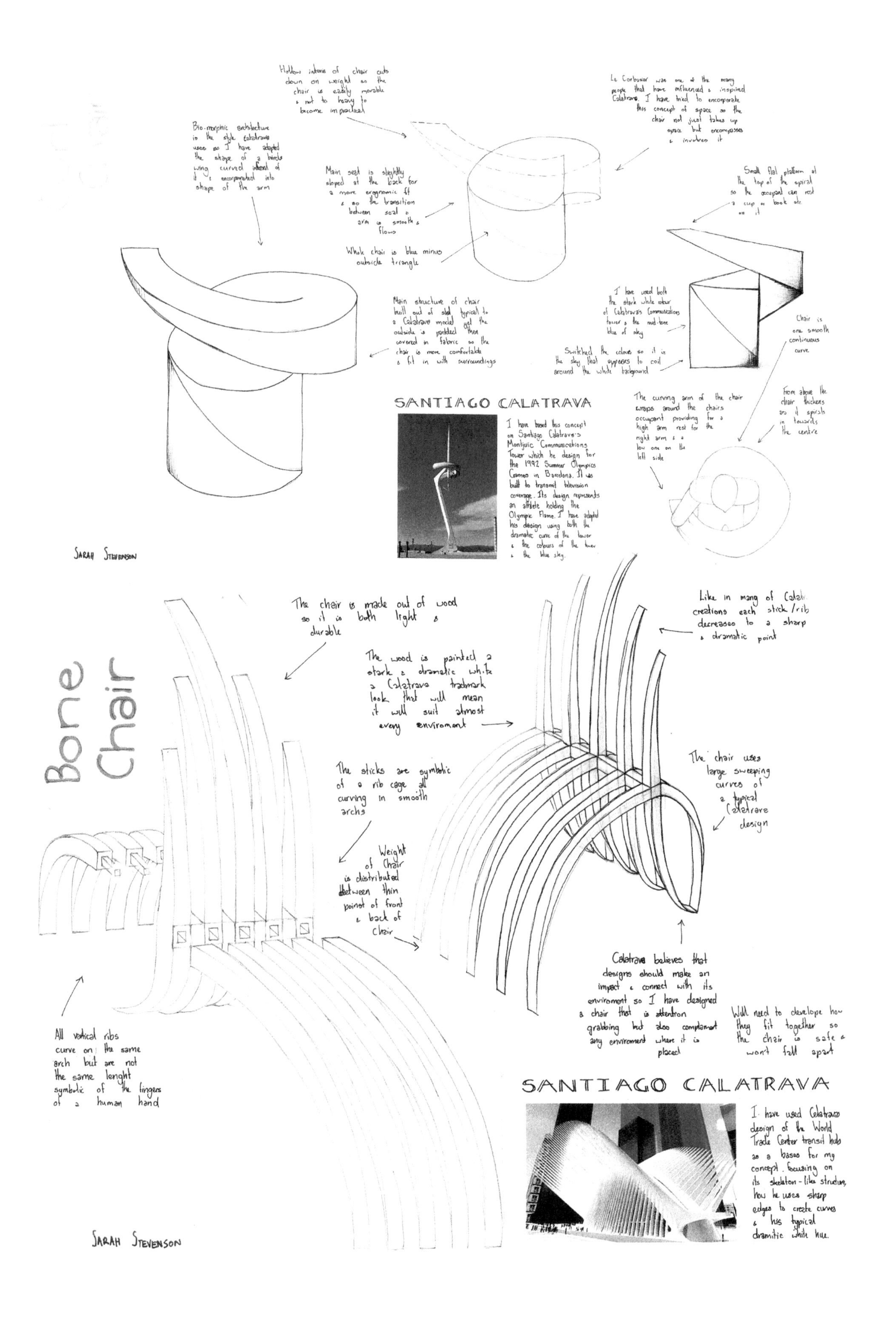

ISBN: 9780170233279

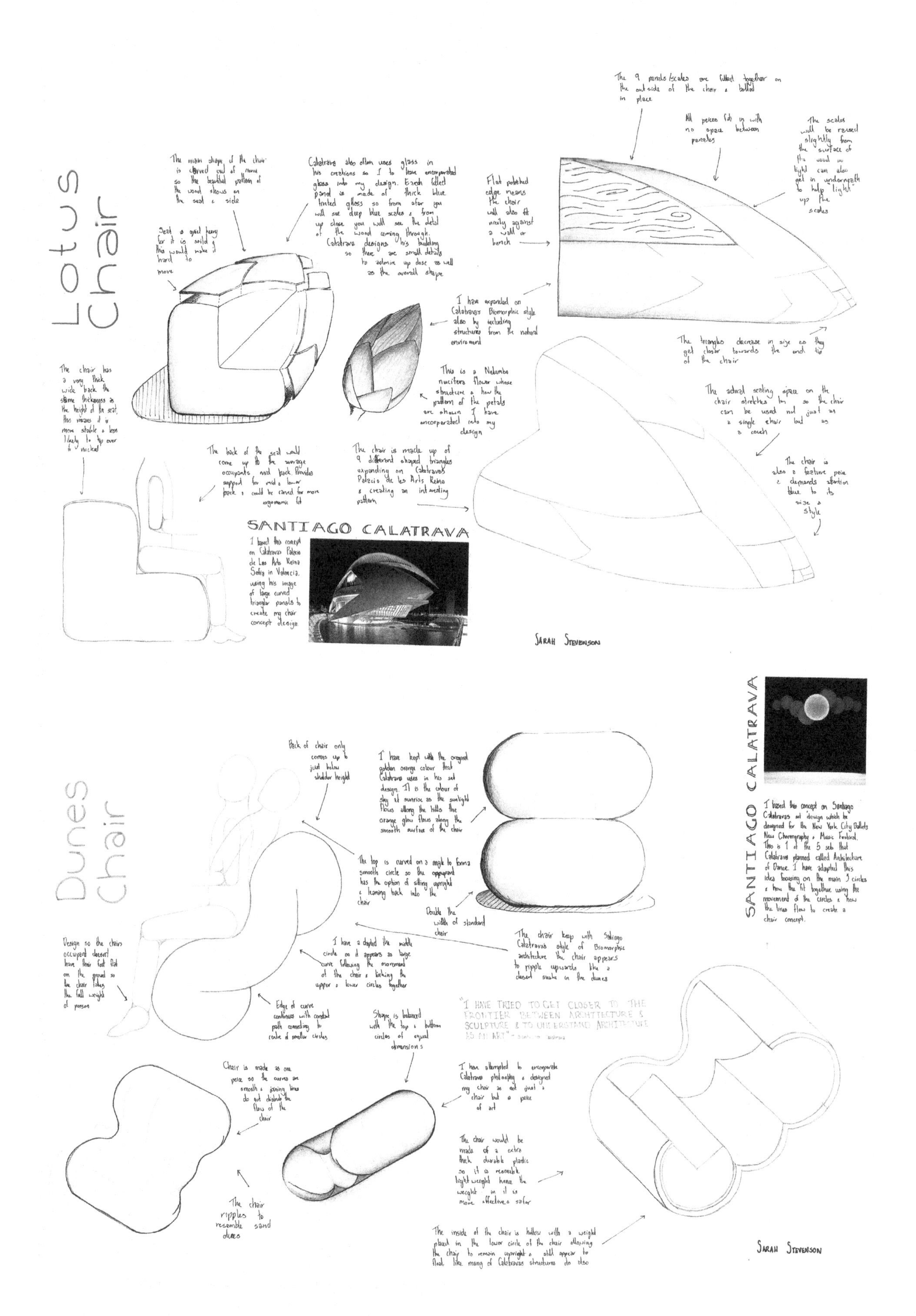

ISBN: 9780170233279

Initial sketches / ideas

Generate a range of initial ideas using sketches and/or mockups. Initial sketches are your first ideas that could be inspired by research of a designer, design movement or era, existing products or spatial design, the context or environment you are designing for and your own ideas. This is a starting point for your design development.

Yiqiu Hong — Epsom Girls Grammar School

The following work produced by Yiqiu Hong demonstrates the use of a bird form to influence her ideas for a product. Her initial ideas are unrestricted, playing around with different forms.

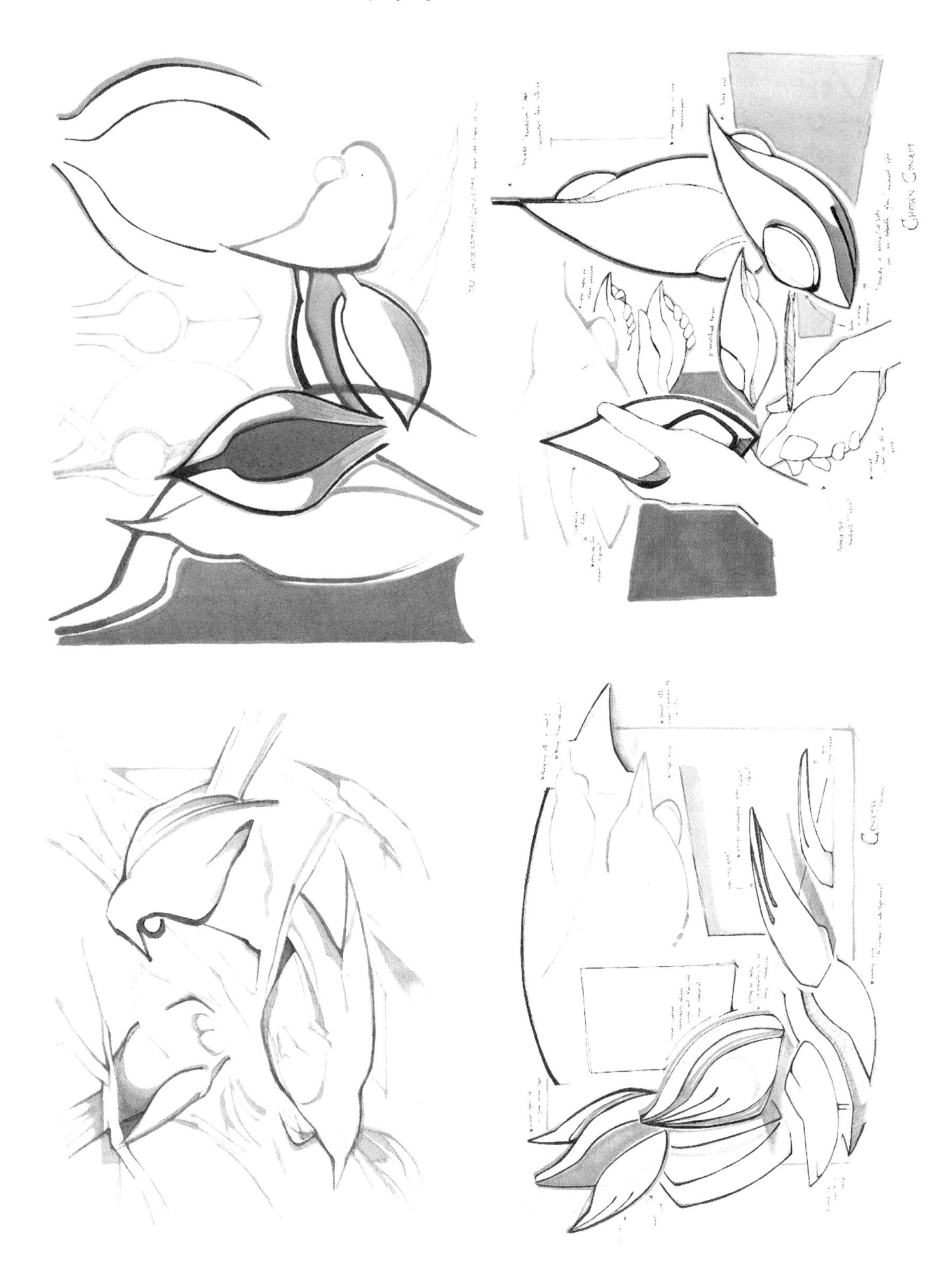

ISBN: 9780170233279

Bronwyn Kan — Epsom Girls Grammar School

The following work produced by Bronwyn Kan demonstrates the use of origami as inspiration for lighting design. She visually plays around with the idea of folds, concertina of paper and repetition.

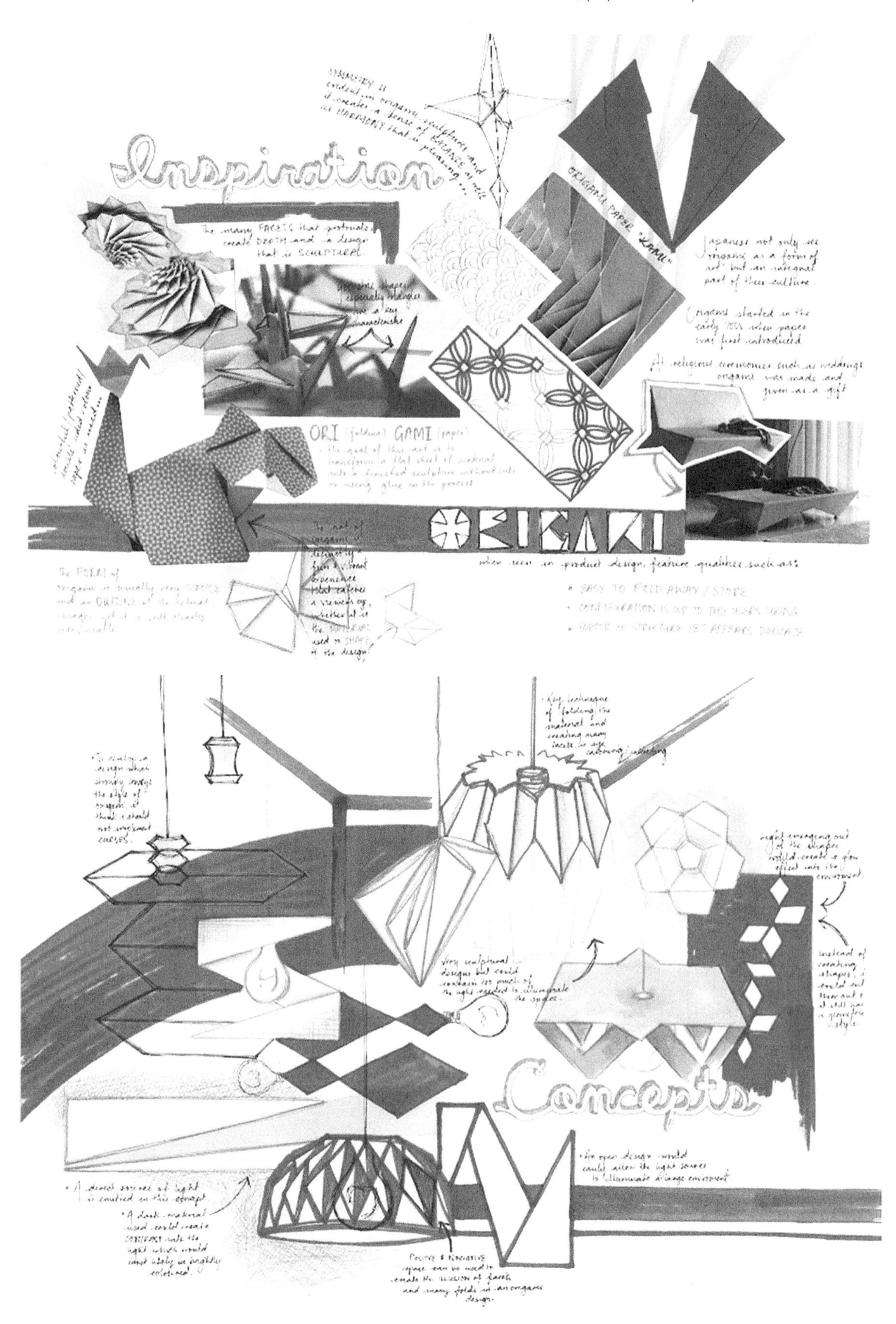

ISBN: 9780170233279

Using forms from the era Art Deco as an influence, initial sketches have been produced for lighting ideas.

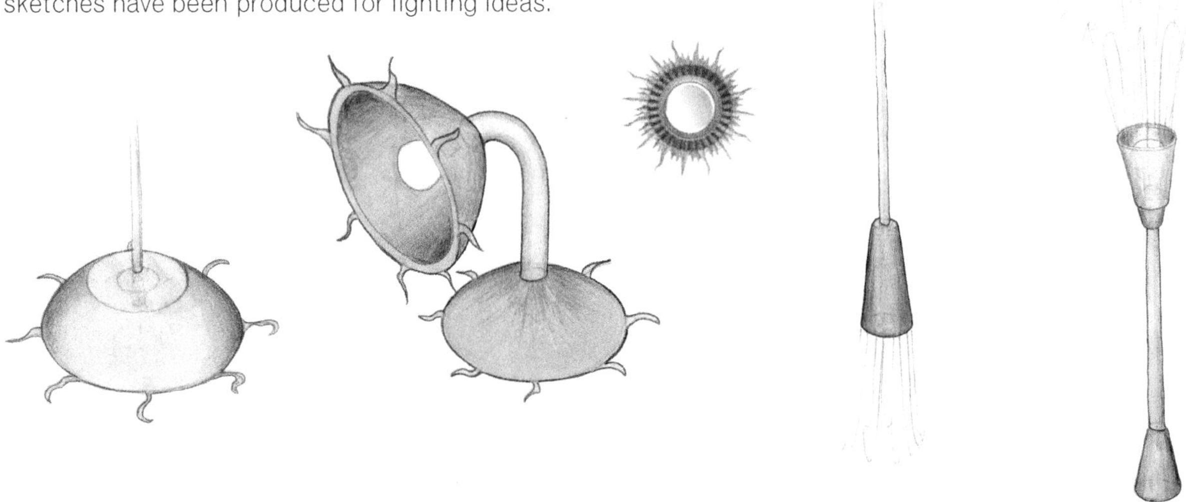

Julie Lam — Epsom Girls Grammar School

The following work produced by Julie Lam demonstrates the use of a model to inform design ideas. The model is a starting point for the exploration of many different ideas.

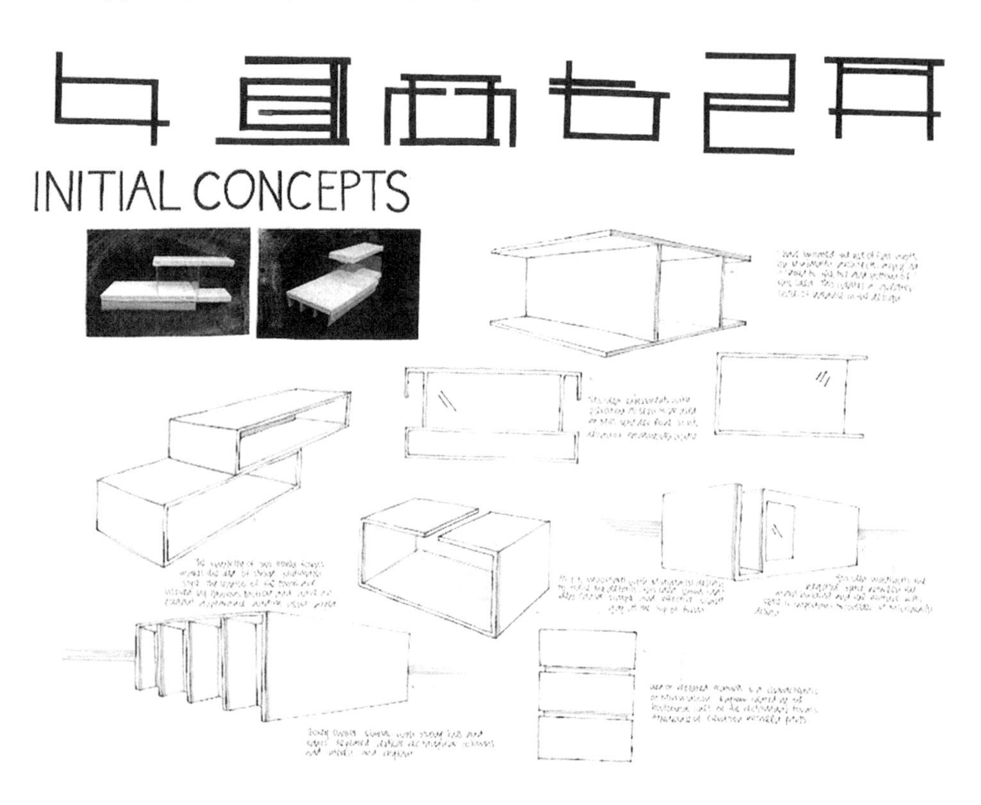

Further development

Further develop the best idea or ideas. You may choose one idea or combinations of several ideas. Make sure you explain your choices.

The following are different examples of exploring, reviewing and refining design ideas using visual communication techniques that explain design thinking. The following techniques and approaches explore alternatives so that ideas are well-considered.

Exploring design ideas – Consideration of aesthetics

Exploring the shape and form of your ideas. Exploration could be in the form of sketching, rendering, models or mockups.

ISBN: 9780170233279

Julie Lam — Epsom Girls Grammar School

Further exploration of initial ideas demonstrated on the previous page.

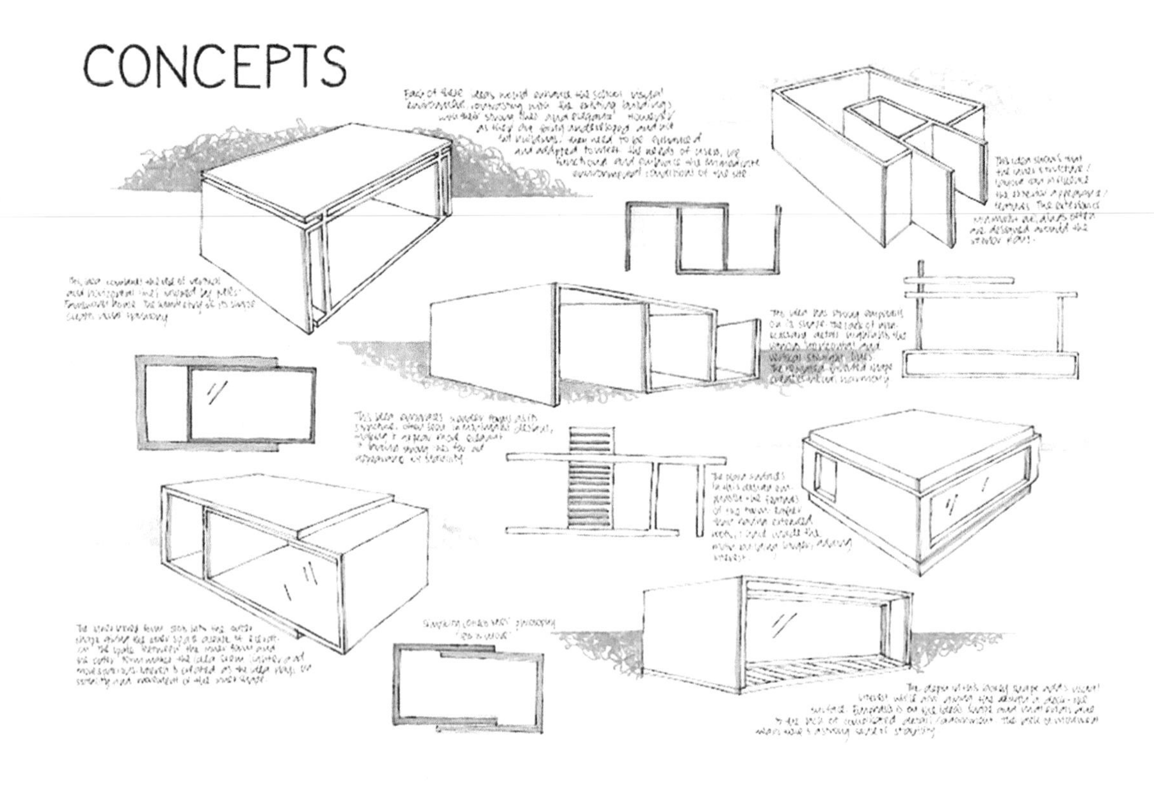

Sheldon Carr — Onslow College

The following work produced by Sheldon Carr is expressive. He explores various forms using the angular idea. Sheldon has generated these ideas from studies of Gunther Domenig, Ben Van Berkel, Tadao Ando, Toyo Ito, Michael Rojkind, Daniel Libeskind, Zaha Hadid and Kiyonobu Nakagame.

ISBN: 9780170233279

Sheldon Carr — Onslow College

ISBN: 9780170233279

Christian Burgos — St Peter's College

The following work produced by Christian Burgos demonstrates extensive exploration of the shape and form for the vehicle design. He does not just sketch his ideas from one viewpoint but is constantly turning the object around.

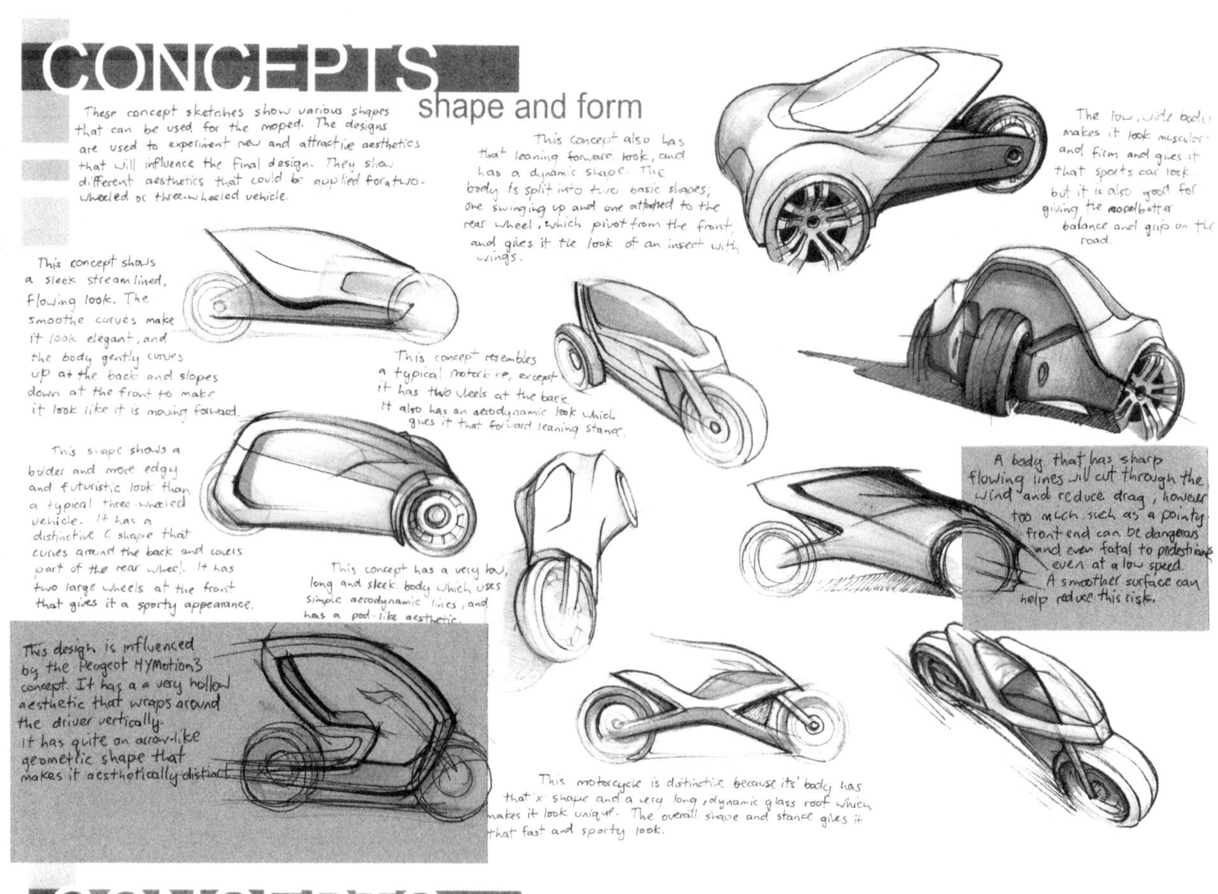

CONCEPTS
shape and form

This idea is quite futuristic and bold, with the smoothe lines and the covered rear wheel. It has a unique split of the glass and the main body. It shows that the split of the surface can create a unique yet sleek look which could be used for further development.

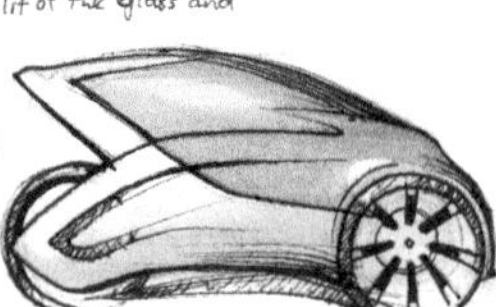

The body of the moped blends and stretches to the rear wheel, instead of a typical moped with plain steel poles that hold the wheel. The smoothe shape makes the moped look fresh and bold and makes it stand out from a typical moped.

The concept also has the dynamic, flowing aesthetic, but the front resembles a car to give it a car look which gives it more character. It has large arches that wrap around the front two wheels which emphasises the wheels.

This concept doesn't blend the body with the wheels, but it makes the wheels stand out more since they are almost completely exposed. It looks more unique since it moves away from the 'sports car' look.

Some sporty wheels would greatly increase the visual appeal of the moped. This design shows sports car style wheels and how it can change the whole appearance.

This design has only two wheels, which is basically a motorbike, except it has a body that covers and wraps around the drivers' body. The roof is detachable, and opens and closes vertically which acts as the door when it is attached. It has a simple, dynamic shape that makes it look sleek and aesthetically appealing.

This concept has a car 'roadster' look with a removable glass roof. It can open upwards, pivoting from the front. It has small proportions that save space, but provides a comfortable amount of interior space. It also uses the sleek, flowing proportions which make it attractive like a car.

ISBN: 9780170233279

Bronwyn Kan — Epsom Girls Grammar School

The following work produced by Bronwyn Kan explores ideas for the lighting design relating to folds, concertinas and repetition.

Ergonomics

Consideration of the human form

Ergonomics is the consideration of the human form when designing spaces or products. The example demonstrates a torch being held and used from different viewpoints. This helps to explain the function (such as grip) and scale of the product.

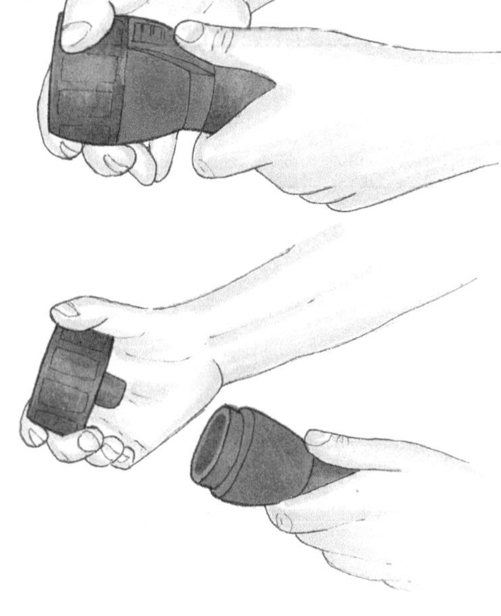

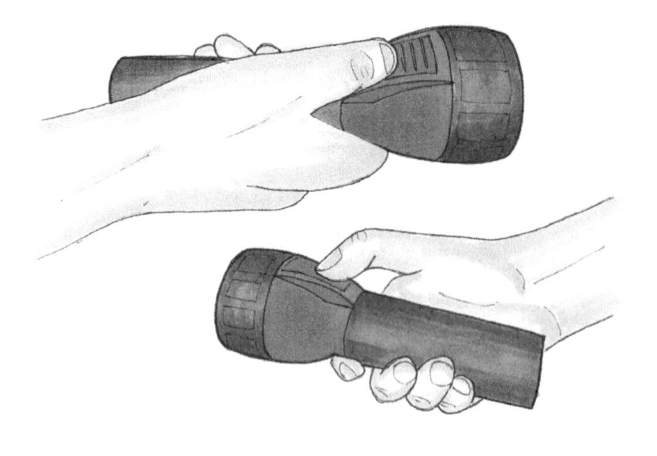

Christian Burgos — St Peter's College

The following work produced by Christian Burgos demonstrates scaled figures positioned around the exterior of the building to indicate scale.

ISBN: 9780170233279

Braeden Scally — St Peter's College

The following work produced by Braeden Scally demonstrates the use of a human form in the interior and exterior of the building using the stairwell, seating and public spaces.

Christian Burgos — St Peter's College

The following work produced by Christian Burgos demonstrates the integration of a human form seated in the vehicle and also considers how the user will enter and exit the vehicle.

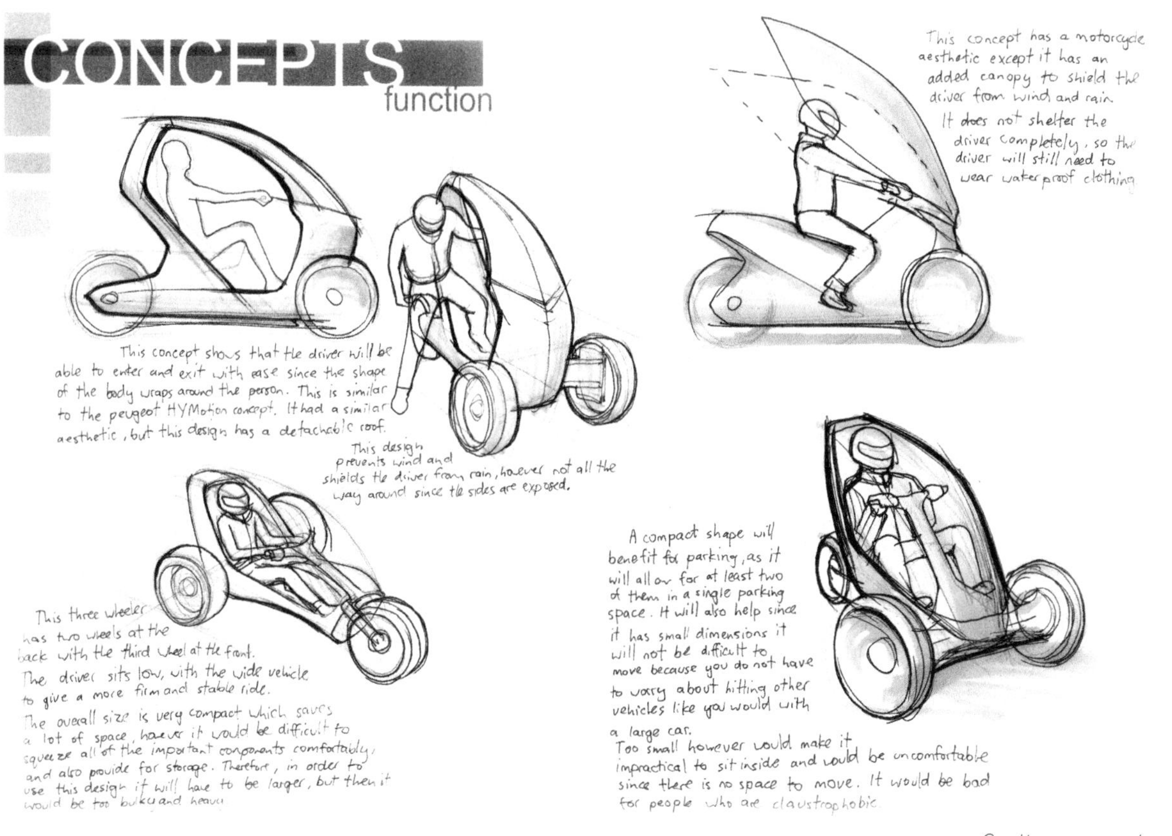

Continues on next page

ISBN: 9780170233279

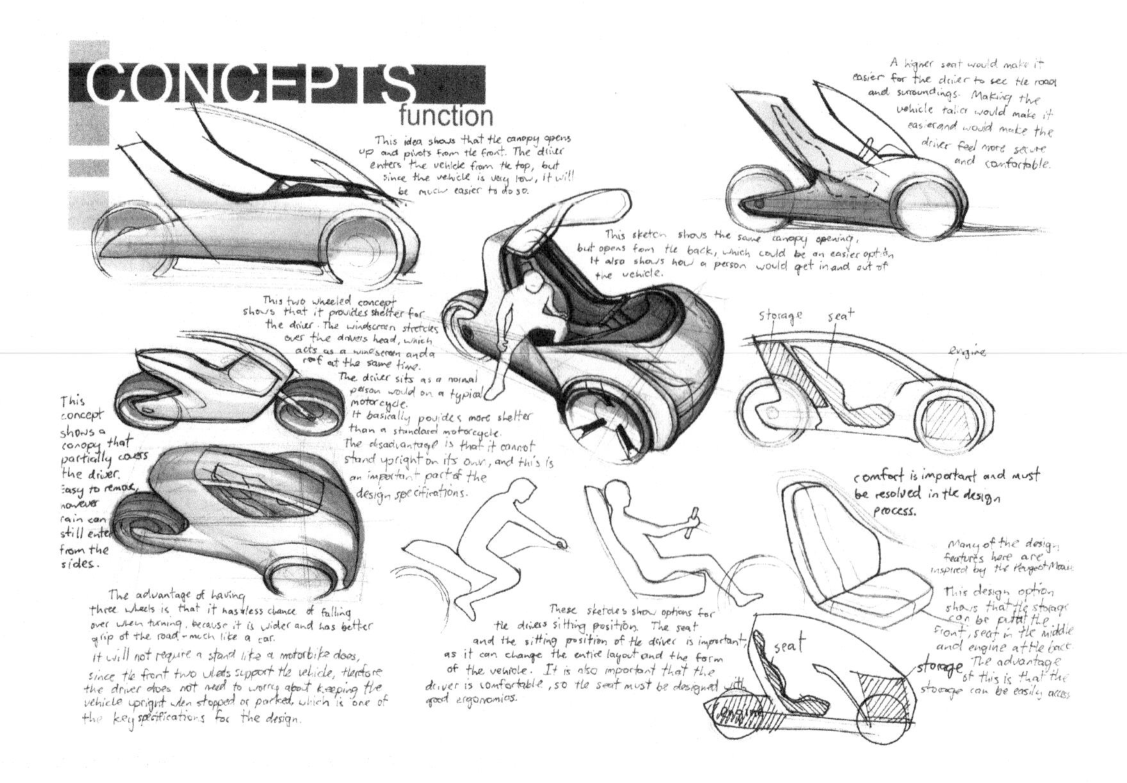

Thinking sketches

Consideration of function, aesthetics and features

Thinking through your ideas visually and demonstrating how the space or product functions and exploring alternatives is essential when developing design ideas.

Braeden Scally — St Peter's College

The following work produced by Braeden Scally demonstrates the thinking through of alternative seating options — how the seat would appear in the space, how the seating would function and how the component parts fit together.

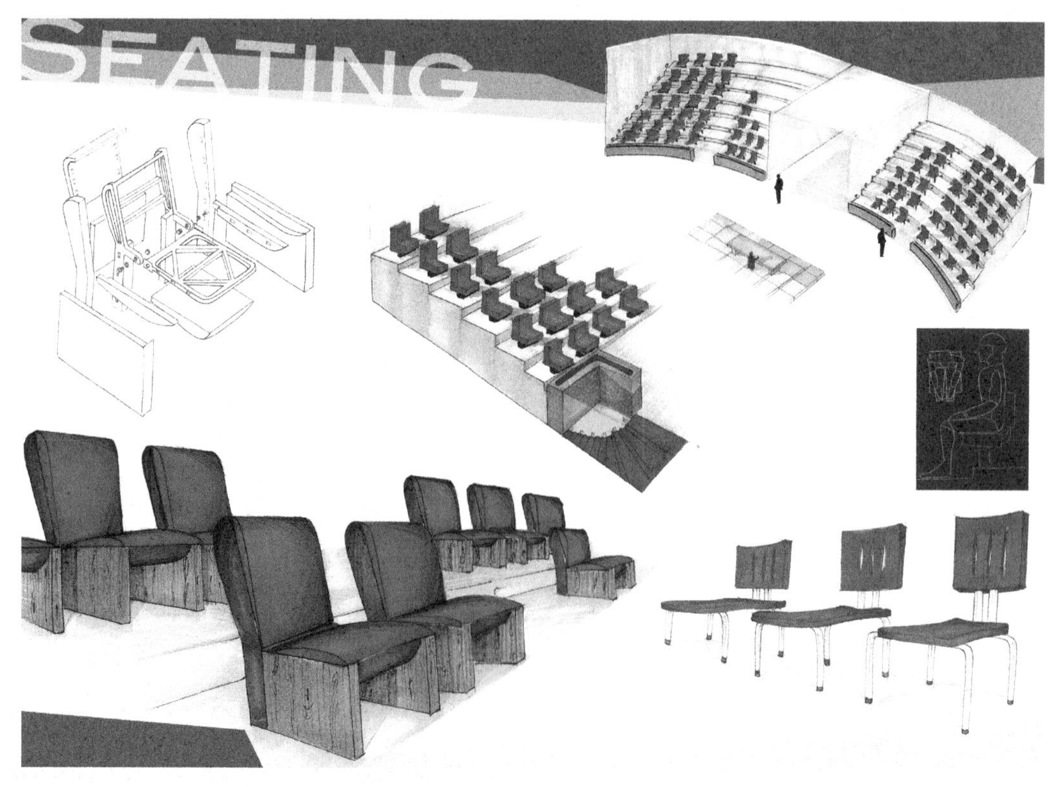

ISBN: 9780170233279

Christian Burgos — St Peter's College

The following work produced by Christian Burgos demonstrates the structure of the building addition and how this will affect the interior and exterior, and the construction and shape of the decking.

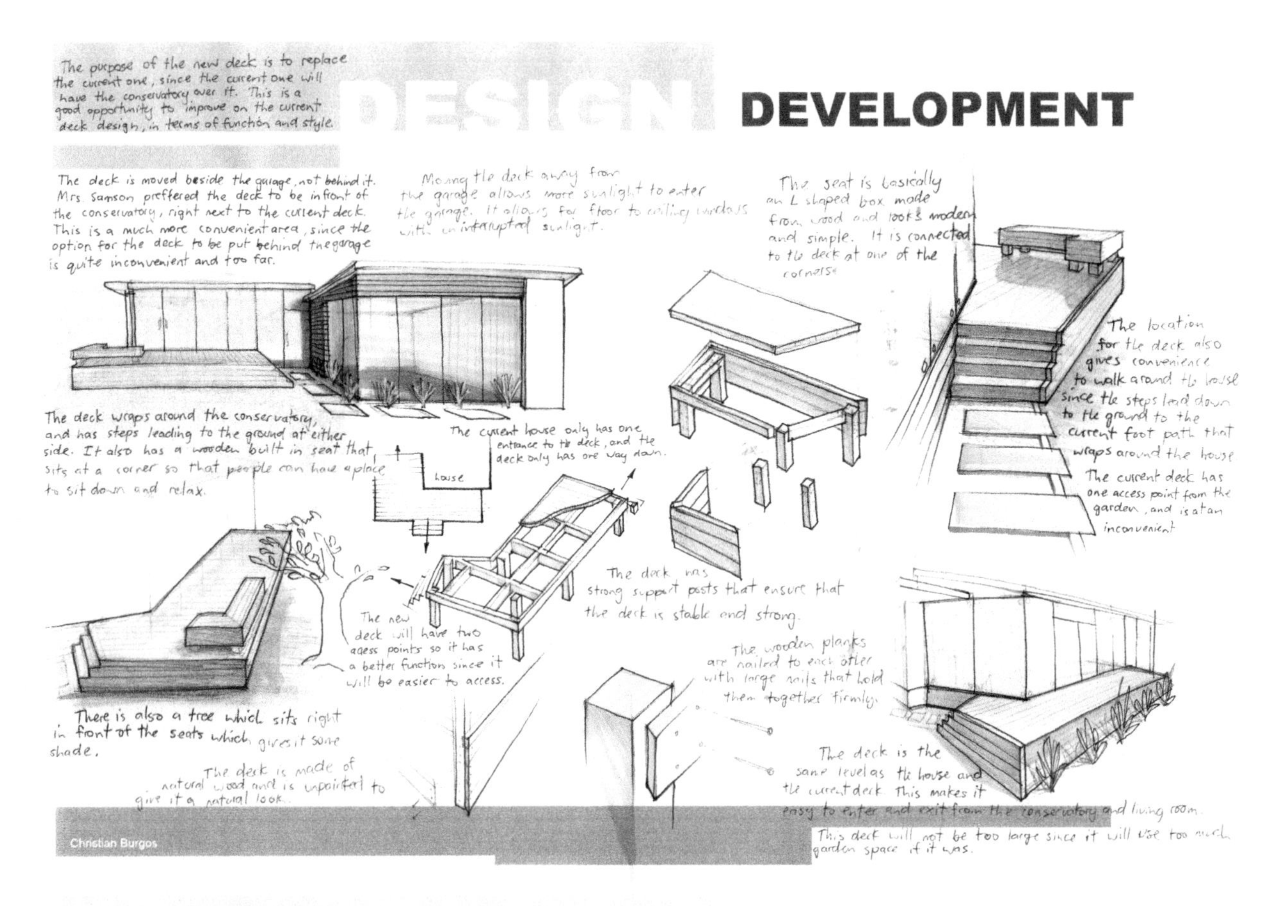

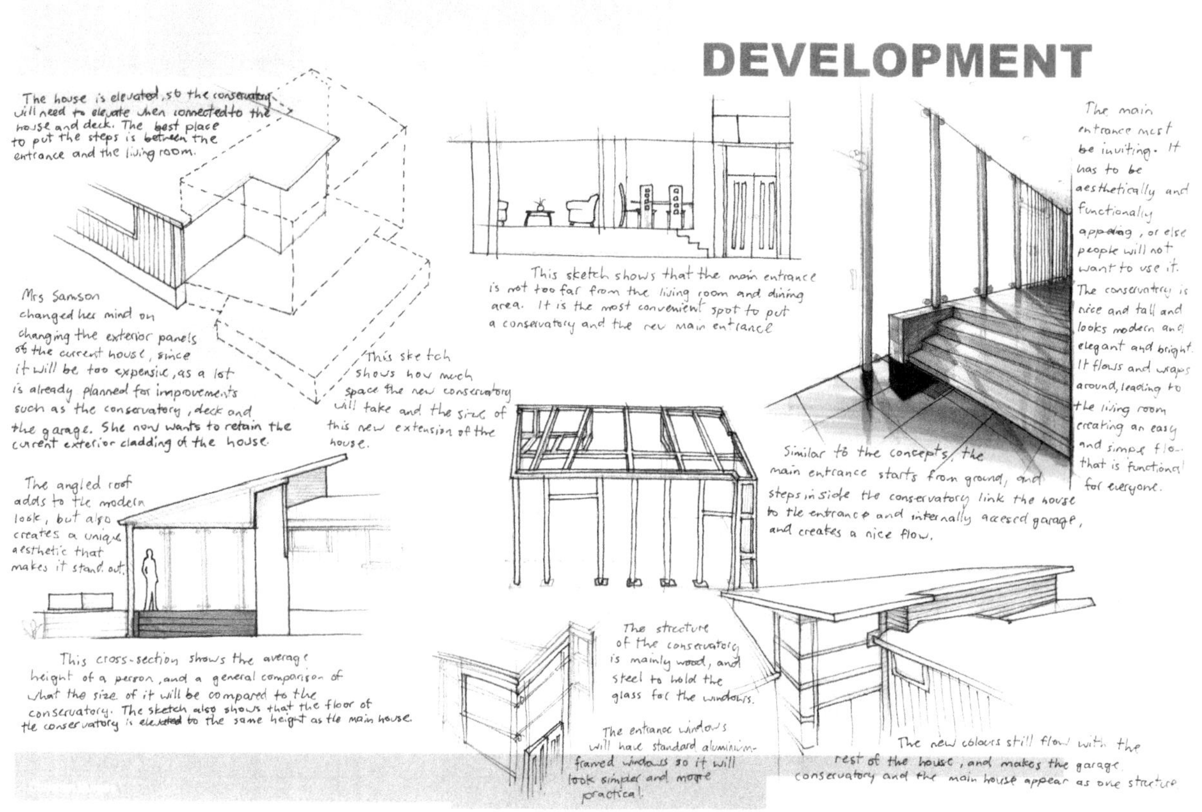

ISBN: 9780170233279

Jamie King — Onslow College

The following work produced by Jamie King demonstrates visual alternatives for the exterior and interior spaces of the design.

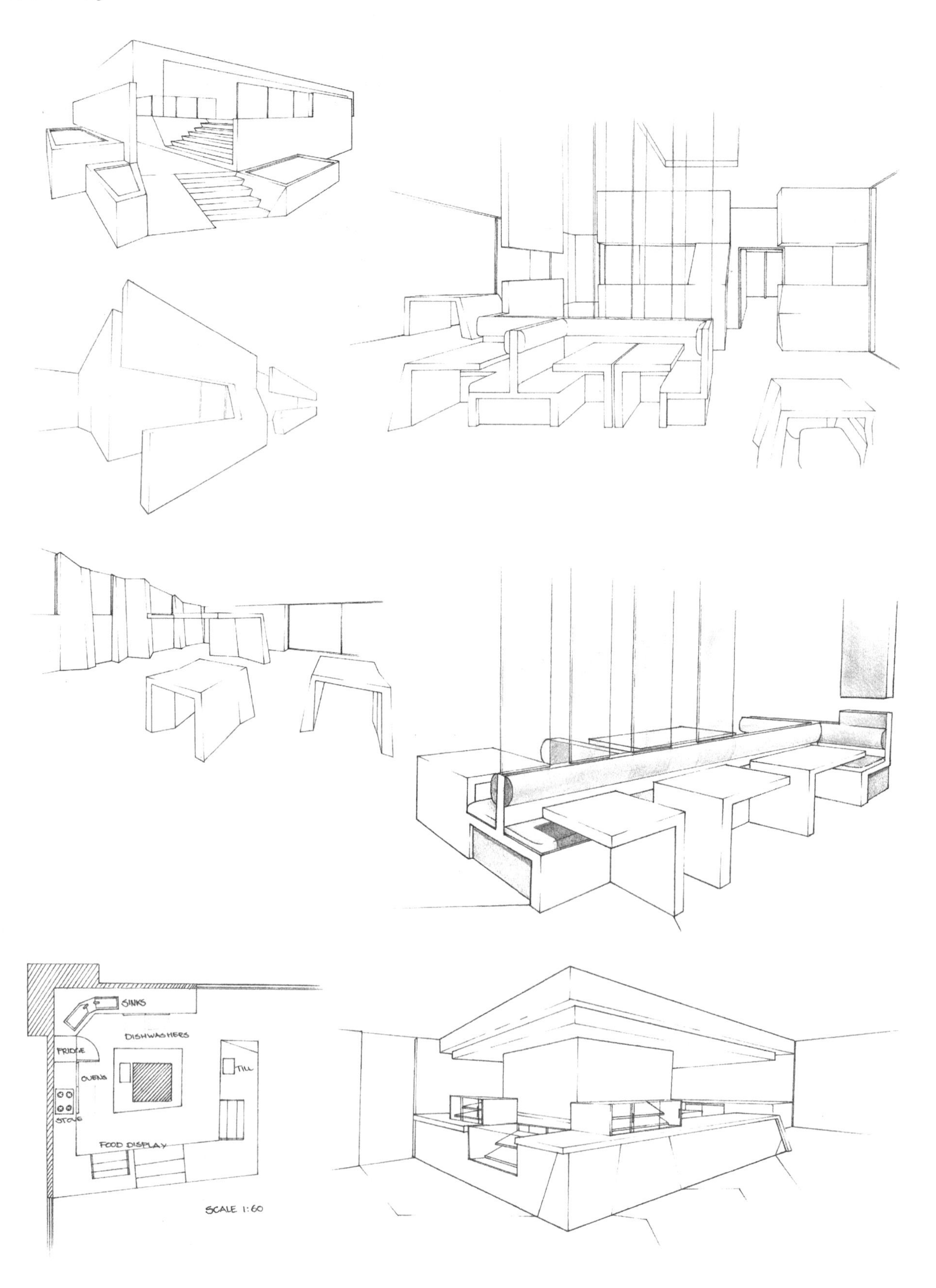

ISBN: 9780170233279

Franklin Mwanza — Onslow College

The following work produced by Franklin Mwanza demonstrates the exterior and interior spaces of the design, including alternatives for the shape of the exterior and layout of the interior.

ISBN: 9780170233279

Christian Burgos — St Peter's College

The following work produced by Christian Burgos demonstrates visually his thinking about how the vehicle will be used, how the user will be positioned, alternatives for the canopy and how the functional components will be encased within the frame. He has used research to inform his ideas. Christian has reviewed and refined his design ideas.

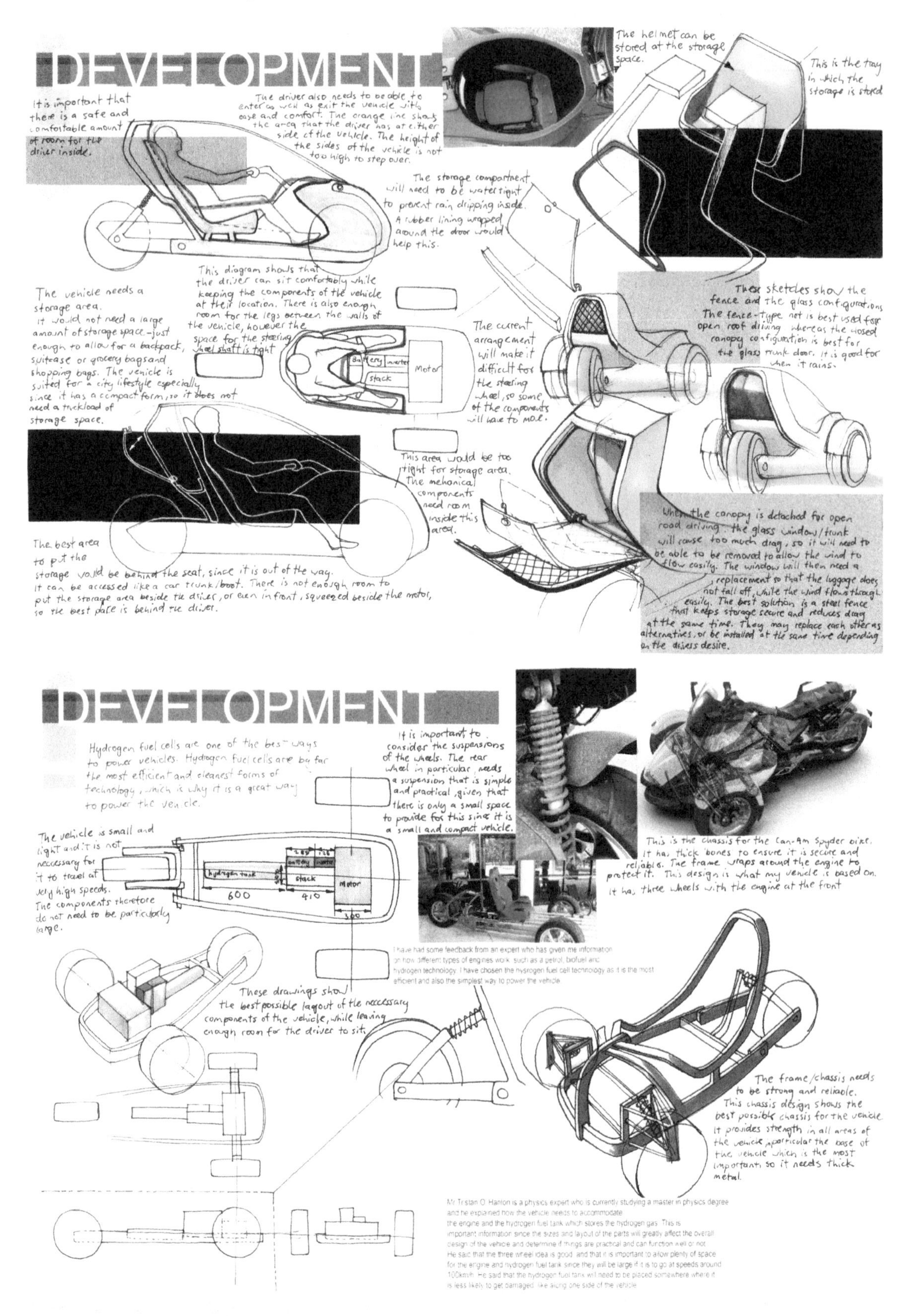

ISBN: 9780170233279

Consideration of function and components

To apply technical knowledge to demonstrate the following you will need to communicate your ideas using exploded, sectional or sequential sketches.

Spatial Design: building materials and details, processes, sustainability and environmental considerations such as climate and space and light.

Product Design: materials, joining, fitting, assembly, fasteners, finish, sustainability and environmental considerations.

EXPLODED ISOMETRIC SKETCH OF A LAMP — PRODUCT DESIGN

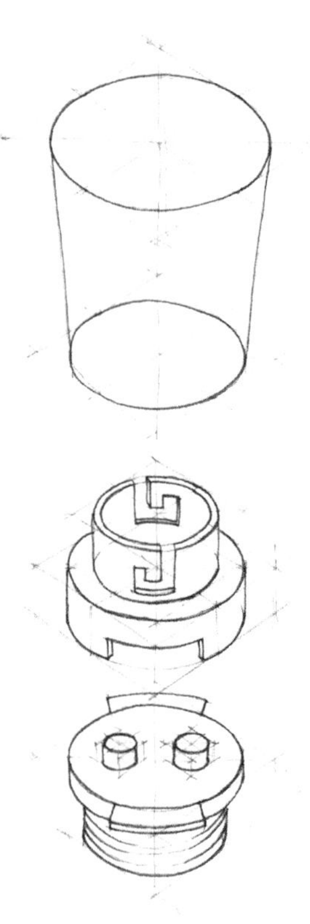

ORTHOGRAPHIC SKETCH WITH SECTIONAL VIEW OF A LAMP FITTING

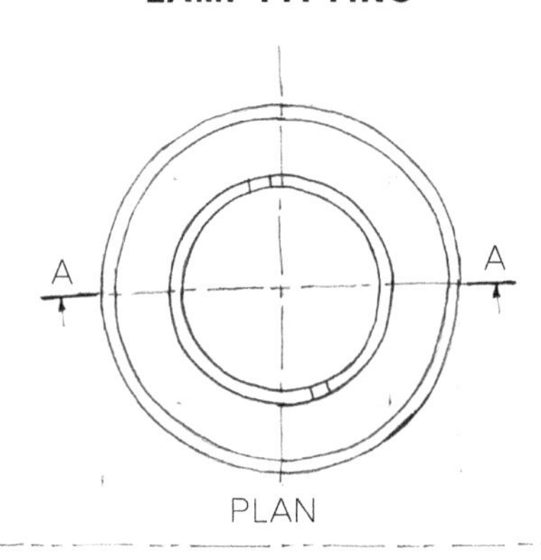

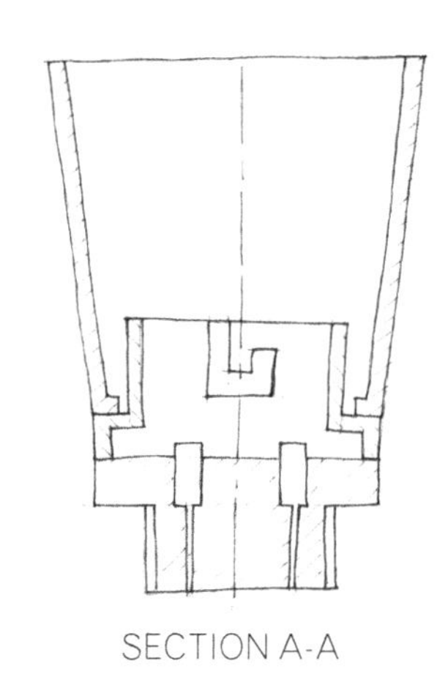

SECTIONAL DETAILS — SPATIAL DESIGN

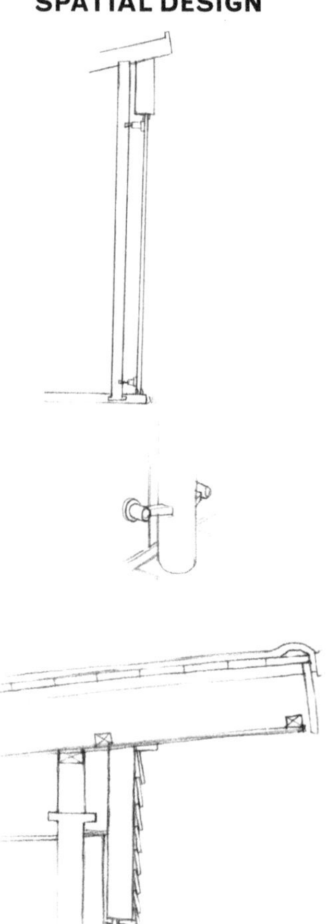

EXPLODED 3D SKETCH OF AN ALARM CLOCK — PRODUCT DESIGN

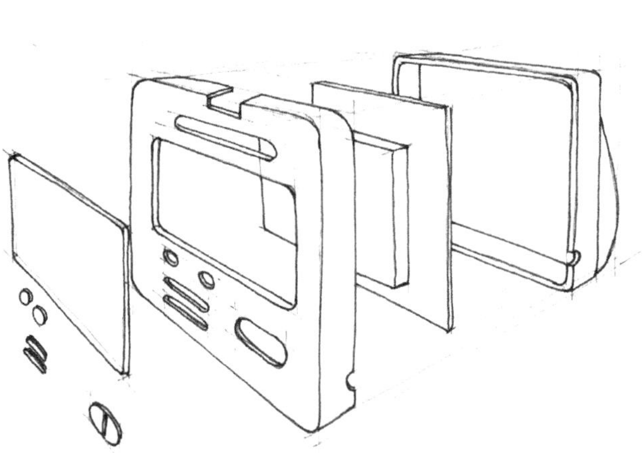

EXPLODED SKETCH OF DECKING — SPATIAL DESIGN

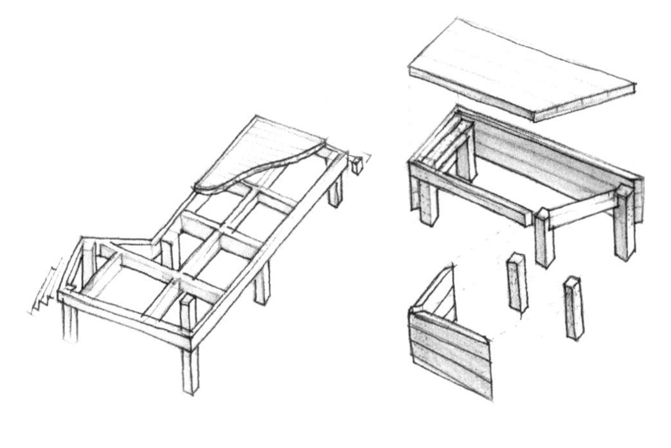

ISBN: 9780170233279

Exploded

Exploded sketching is when parts or components of a product have been pulled apart. Any type of 3D sketching method can be used to present an exploded sketch such as isometric, oblique or perspective.

The exemplar sketch of the torch has been completed in the 3D sketching method isometric. As you can see it explains all the main parts of the torch and how they screw together. Construction lines guide how the product fits together.

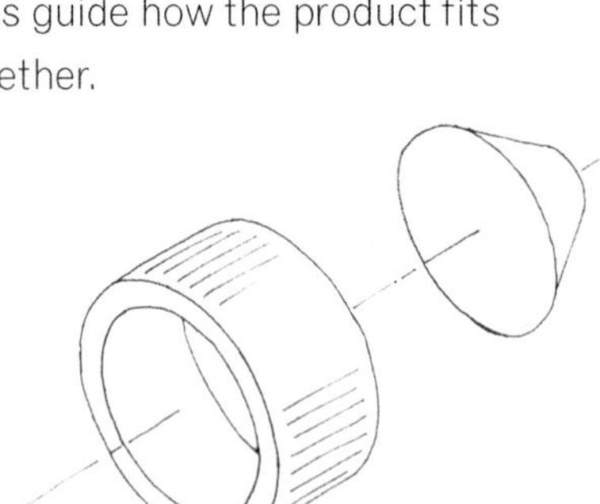

EXPLODED ISOMETRIC SKETCH

Sectional

A sectional sketch is also a view in which you can explain the assembly of components or how something functions on the inside. This type of sketch usually requires at least two views. In one view a section line indicates where the object will be cut and the other view, which is projected from the first view, will explain what will appear on the inside of the product or building.

The exemplar sketch of the barbecue explains the complexity of the interior. This is a 2D orthographic sketch presented with two views.

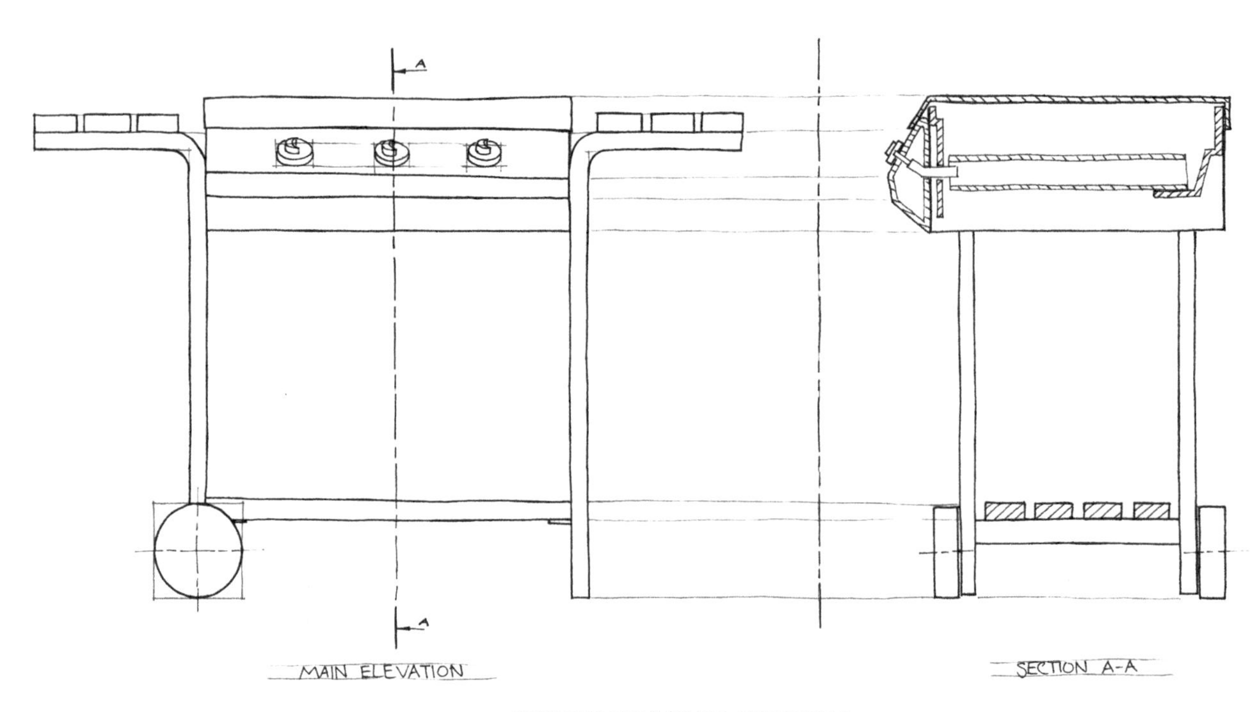

ORTHOGRAPHIC SKETCH

ISBN: 9780170233279

Sequential sketching

Sequential sketching is the step-by-step process to visually explain how a product may function. The 3D sketches of the clasp below communicate how the two parts join together.

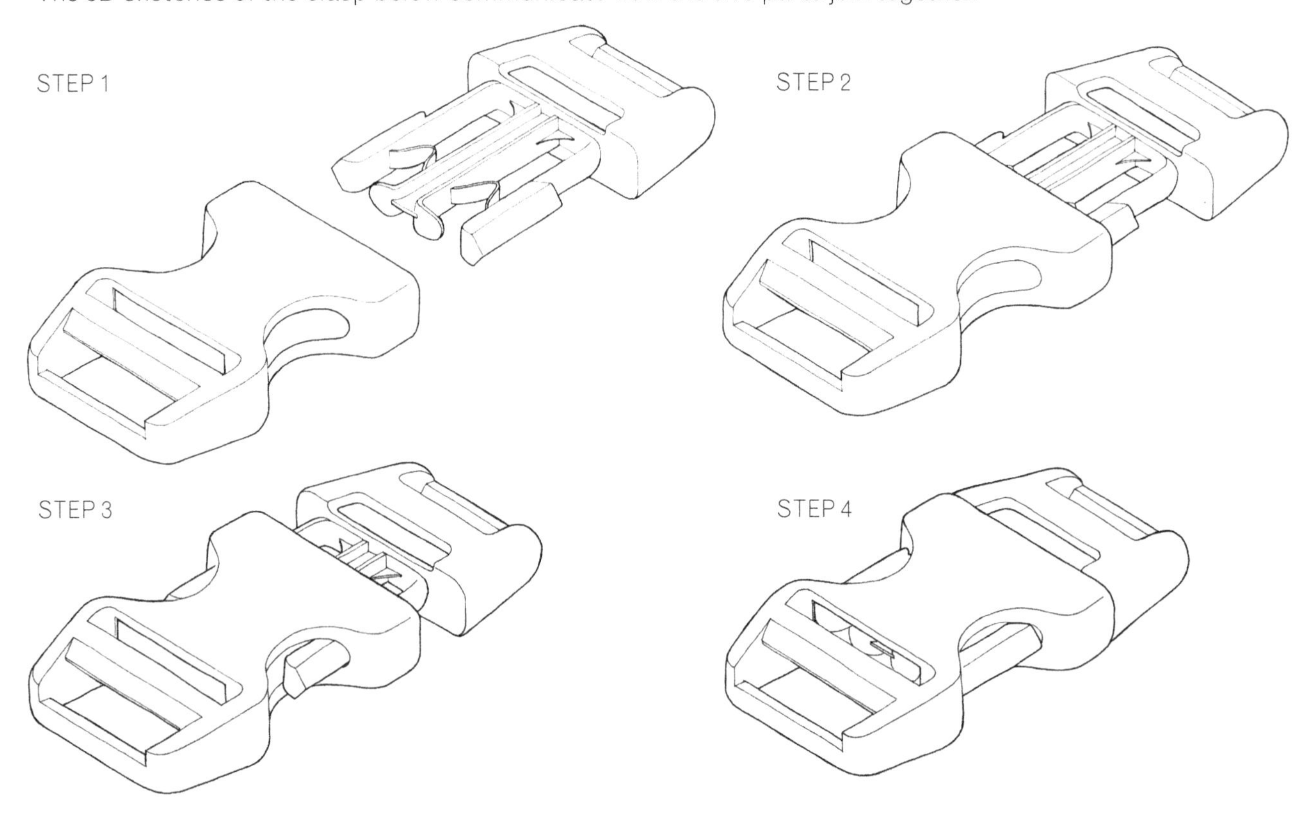

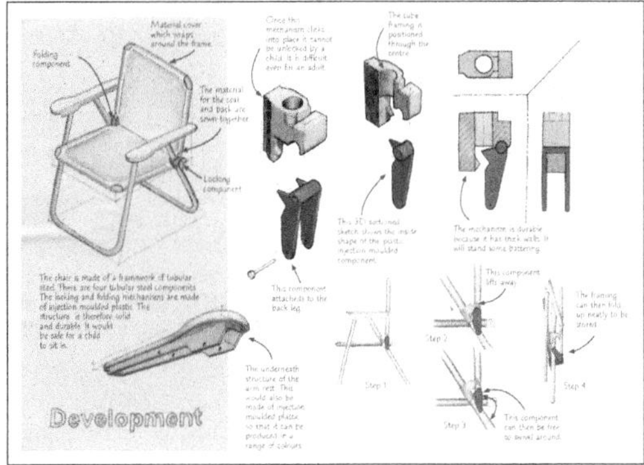

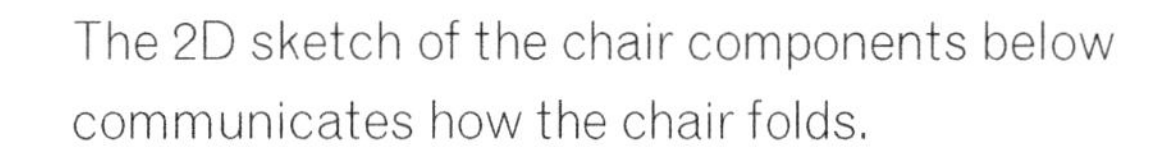

The 2D sketch of the chair components below communicates how the chair folds.

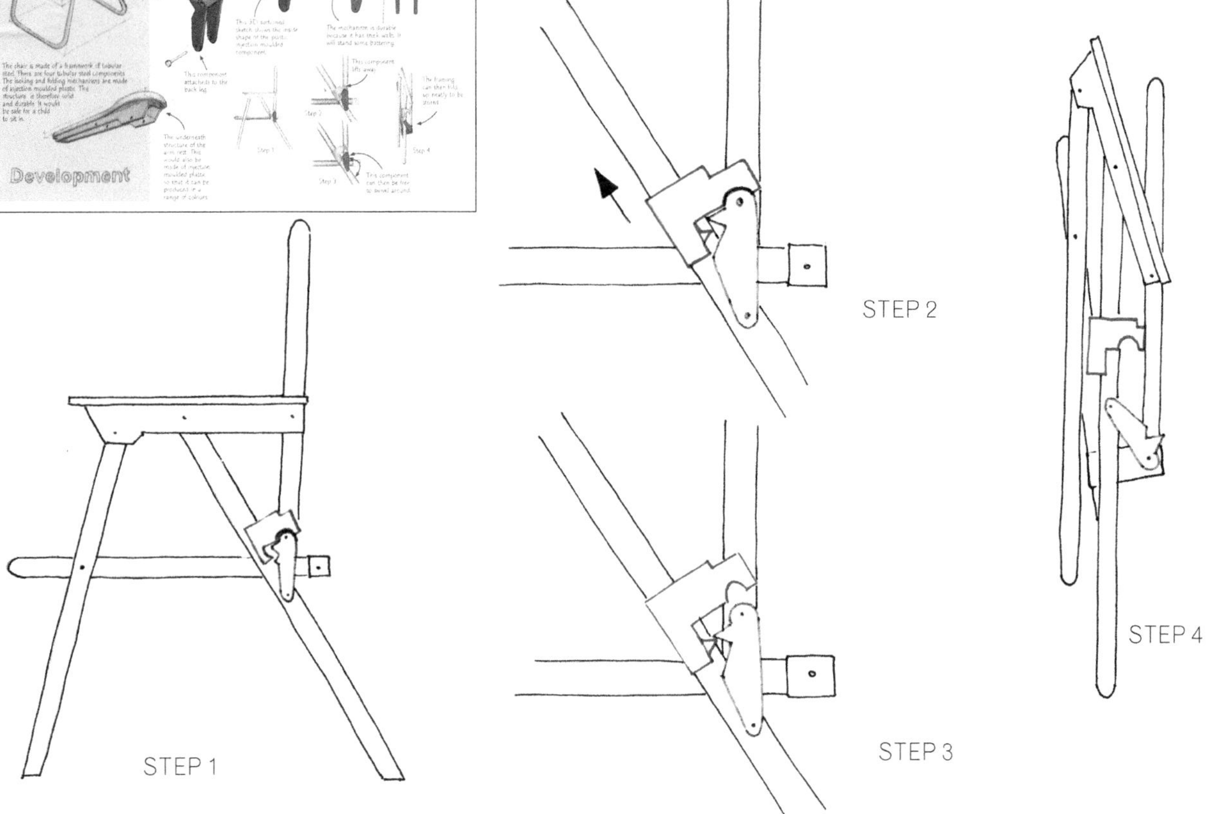

The example pages of development for the chair demonstrates the use of exploded sketches, sequential sketches, three-dimensional and two-dimensional sectional sketches to fully explain how the components fit together.

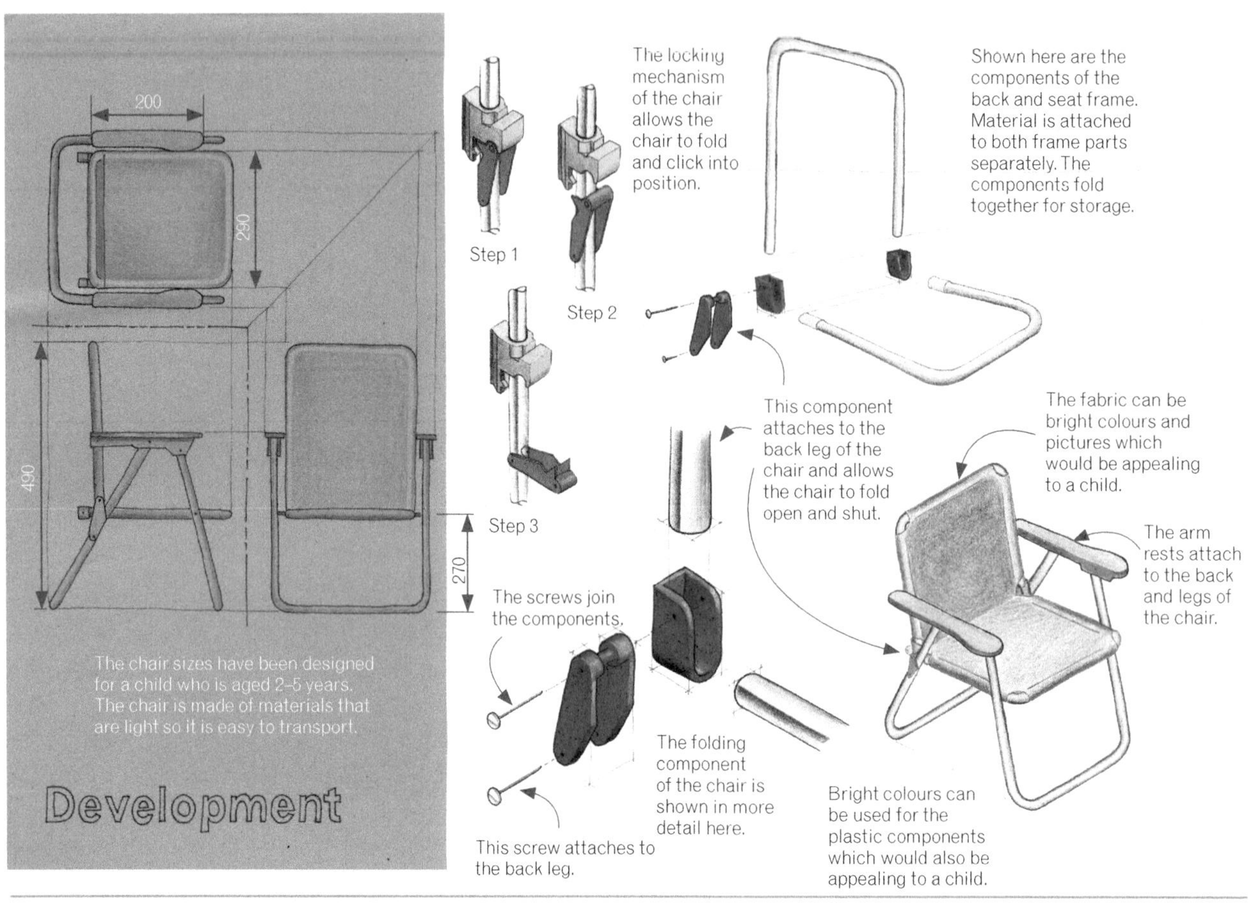

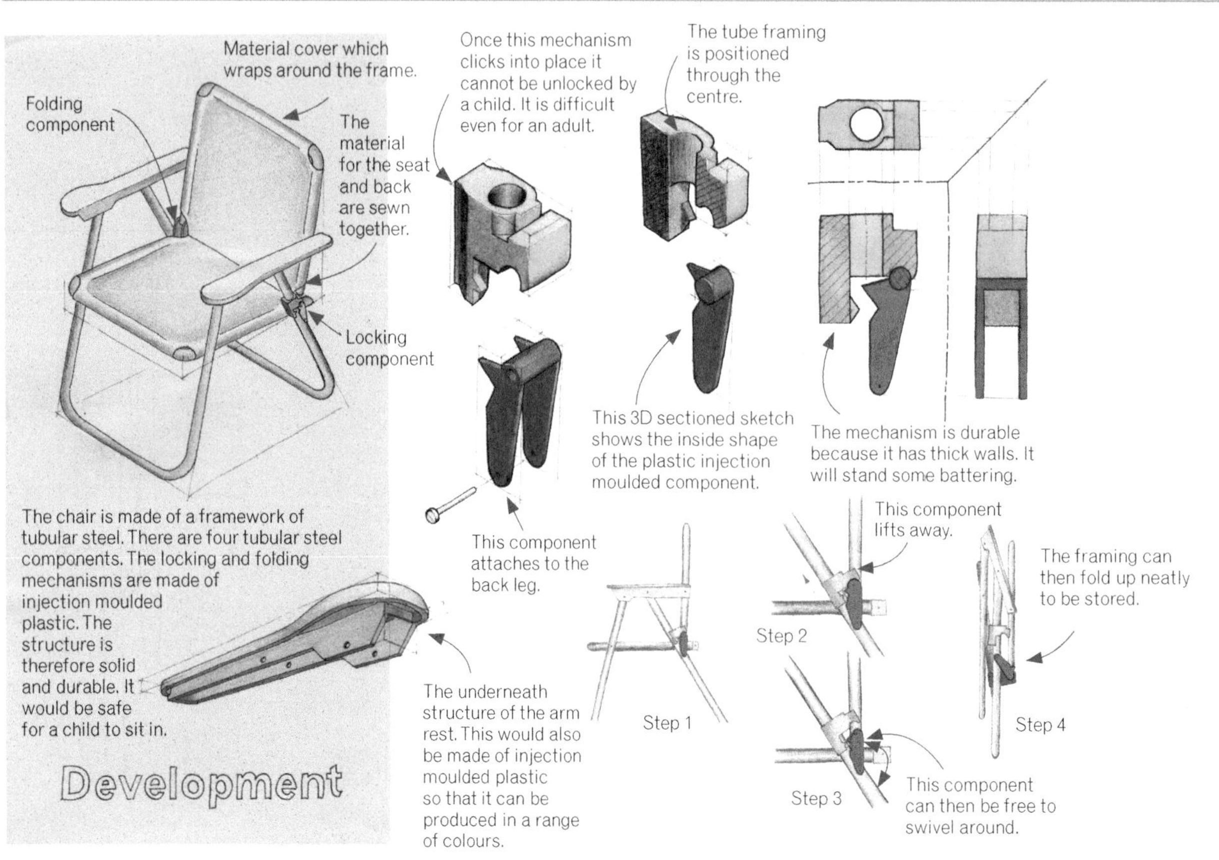

ISBN: 9780170233279

Christian Burgos — St Peter's College

The following work produced by Christian Burgos demonstrates the use of an exploded pictorial to explain the components of the frame of his design idea for a vehicle.

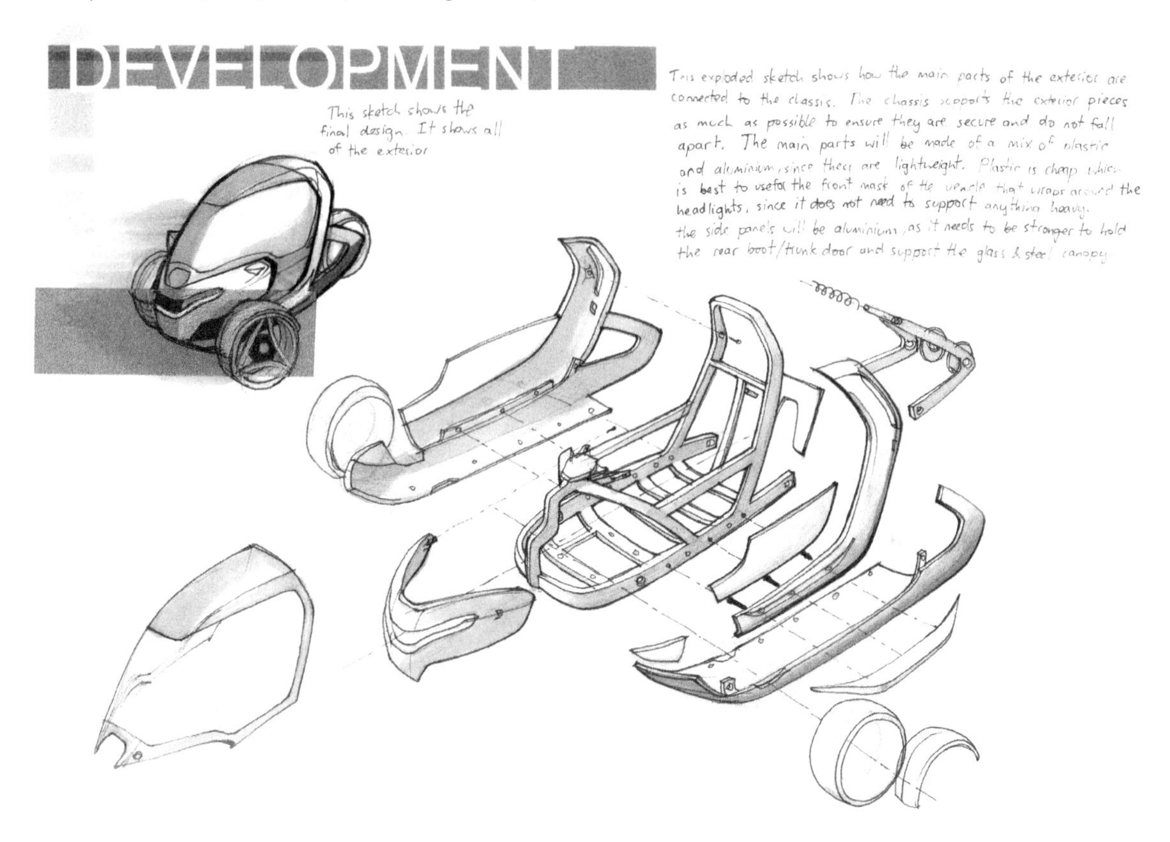

Christian demonstrates the use of a two-dimensional sectional sketch and pictorials to explain the frame of the structure. He gained advice from a knowledgeable source to produce this information.

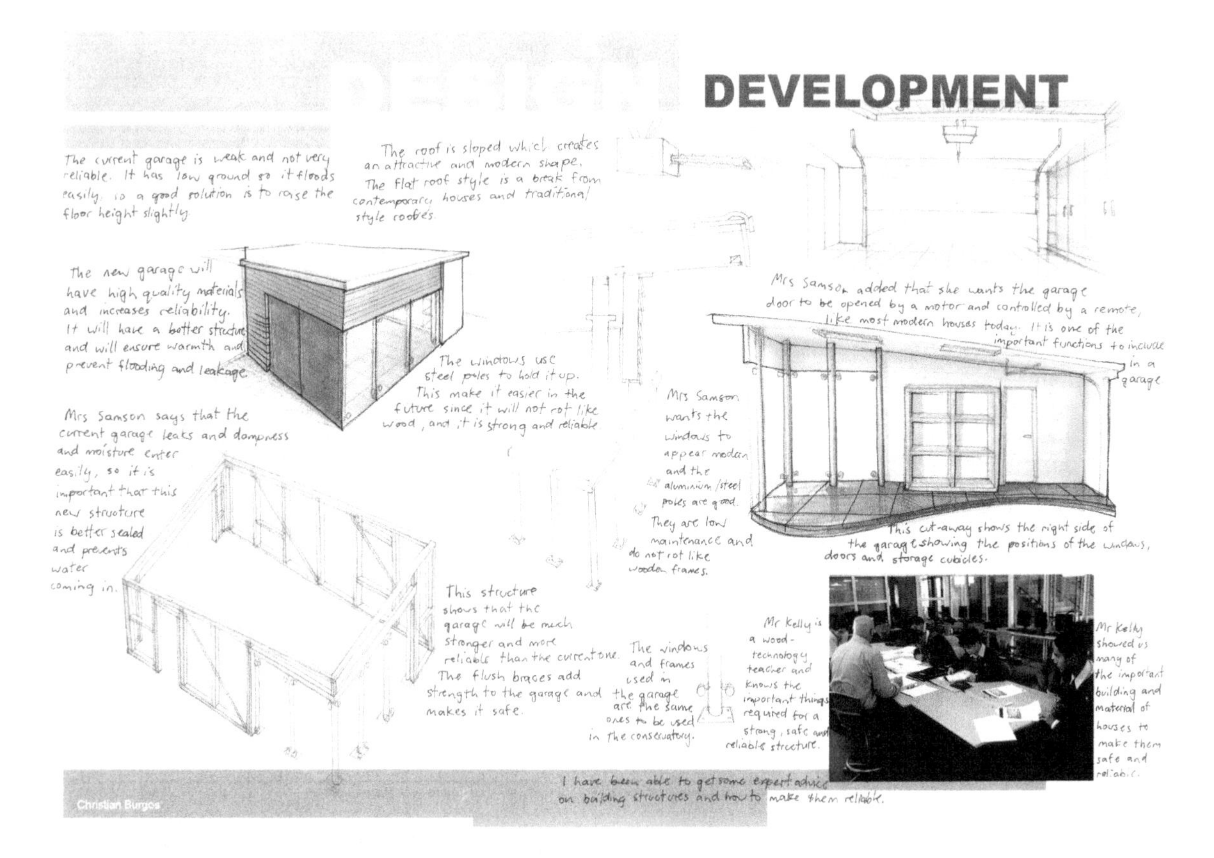

ISBN: 9780170233279

Karl Francisco — St Peter's College

The following work produced by Karl Francisco demonstrates the use of three-dimensional and two-dimensional sectional sketches and an exploded sketch to explain the components of the shelter design.

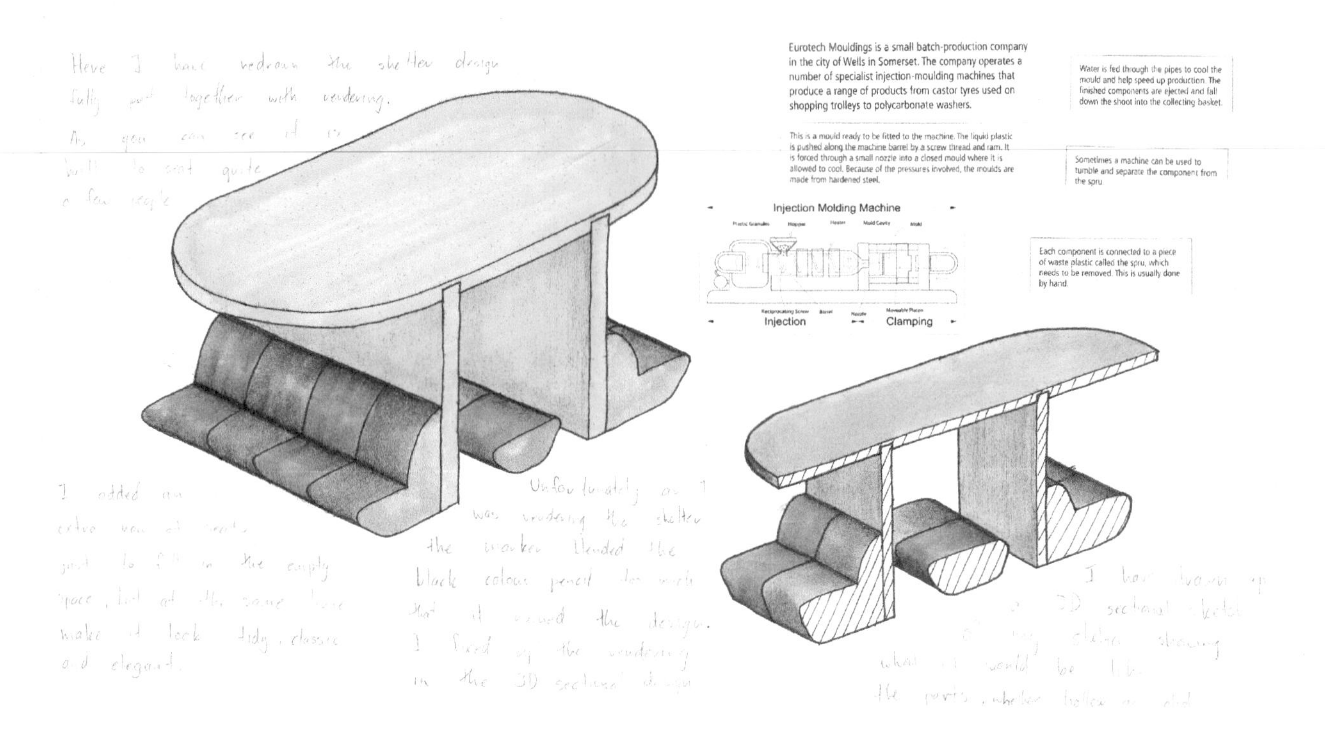

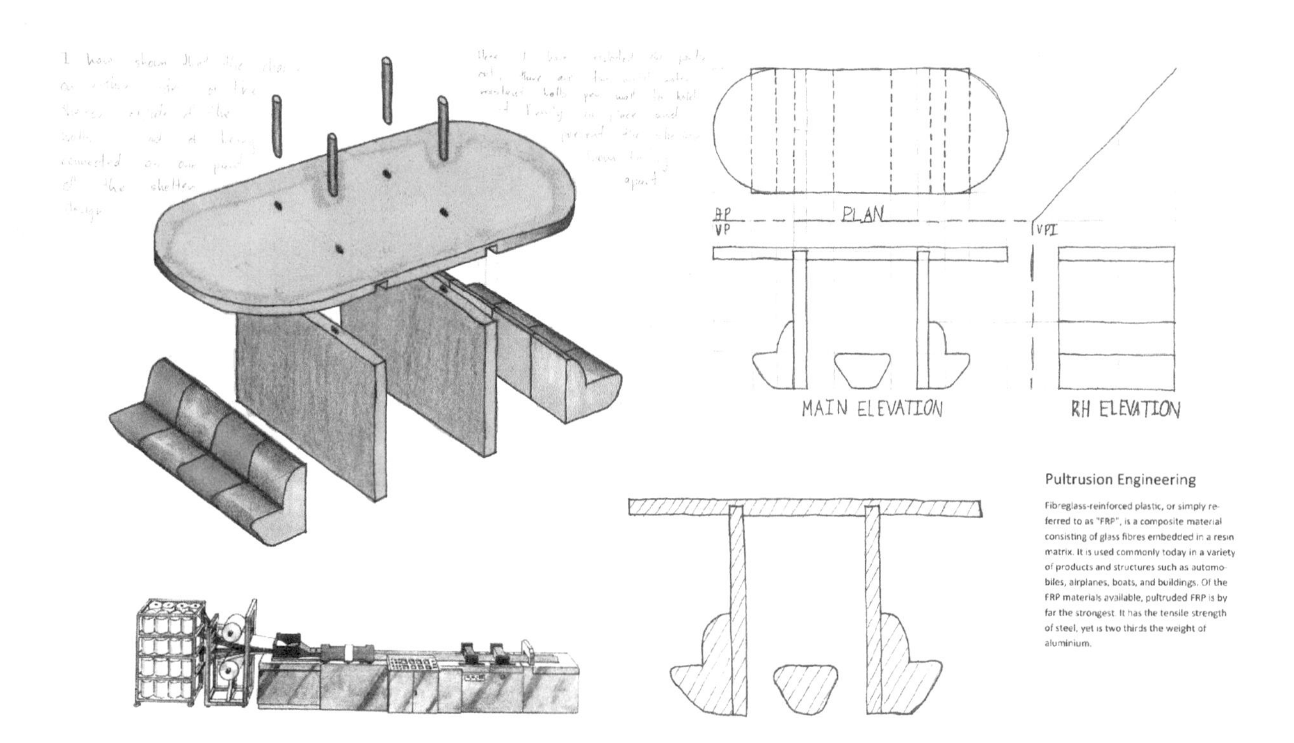

ISBN: 9780170233279

Developing a product or spatial design involves making design judgements based on the positive or negative aspects of the aesthetic and functional features of the design in response to a brief.

Aesthetic (what the product looks like / factors related to appearance) features of a design could be style, form, shape, proportion, pattern, symmetry, contrast, materials and colour.

Functional (how the product works / factors related to use) features of a design could be ergonomics, operation, construction, durability, fitness for purpose, materials, environmental considerations and user friendliness.

Throughout the development of your ideas you will need to review and refine your design ideas, incorporating product or spatial design knowledge. You will also need to make design judgements on the aesthetic and functional features of the design. These judgements will inform the progression of your design ideas. The technical knowledge and how this applies to your design idea can be communicated by integrating research into the development of your ideas as previously shown. Design judgements can be supported by qualitative and quantitative data through research. The judgements are the decisions you, as the designer, make and the opinions you have which may reflect a designer's values or views.

Throughout the development of design ideas remember to:

- Refine and review your ideas
- Critique the positive or negative aspects of aesthetics and function
- Relate your decisions or judgements to the requirements of the brief
- Consider the aesthetic elements - style, form, shape, proportion, pattern, symmetry, contrast and colour
- Consider the functional elements - ergonomics, operation, construction, durability, fitness for purpose and user friendliness
- Consider the changes made in the development of ideas, how the changes have altered the quality of your ideas and what potential changes could improve your design
- Include judgements of your own views, values or perspectives
- Support design judgements with qualitative or quantitative information through continuing research
  Qualitative – relating to or involving comparisons based on qualities
  Quantitative – relating to considerations of amount or size.

Integration of research

Integration of research is a good tool to use at any stage of development. Integration of research could be used in the development of your design ideas in the following ways.

Throughout the development of ideas, research could be used to inform:

- shape and form
- function
- technical aspects of the design.

In this text there are many examples of research being used throughout the development to inform design ideas.

ISBN: 9780170233279

Robert Gorrie — St Peter's College

The following work produced by Robert Gorrie demonstrates the use of research of existing seating throughout the development to help inform his own design ideas for shape and form. He has also researched materials and manufacturing processes that may be applicable to his design such as extrusions for the square tube and plywood bending for the seating.

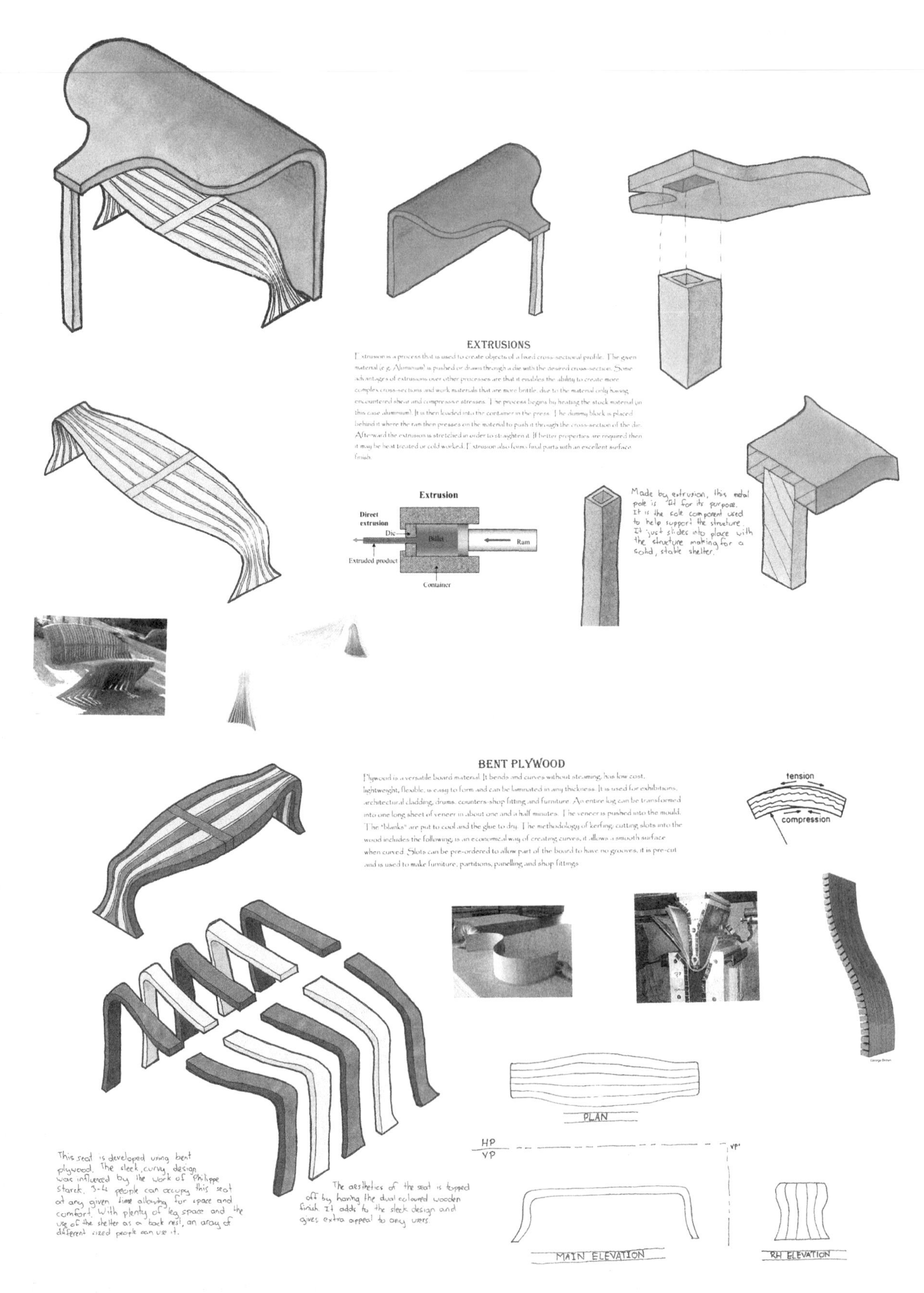

ISBN: 9780170233279

Samuel Matijevich — St Peter's College

The following work produced by Sam Matijevich demonstrates the use of research of existing seating throughout the development to help inform his own design ideas for shape and form. He has also researched materials and manufacturing processes that may be applicable to his design such as casting for the ends of the seating.

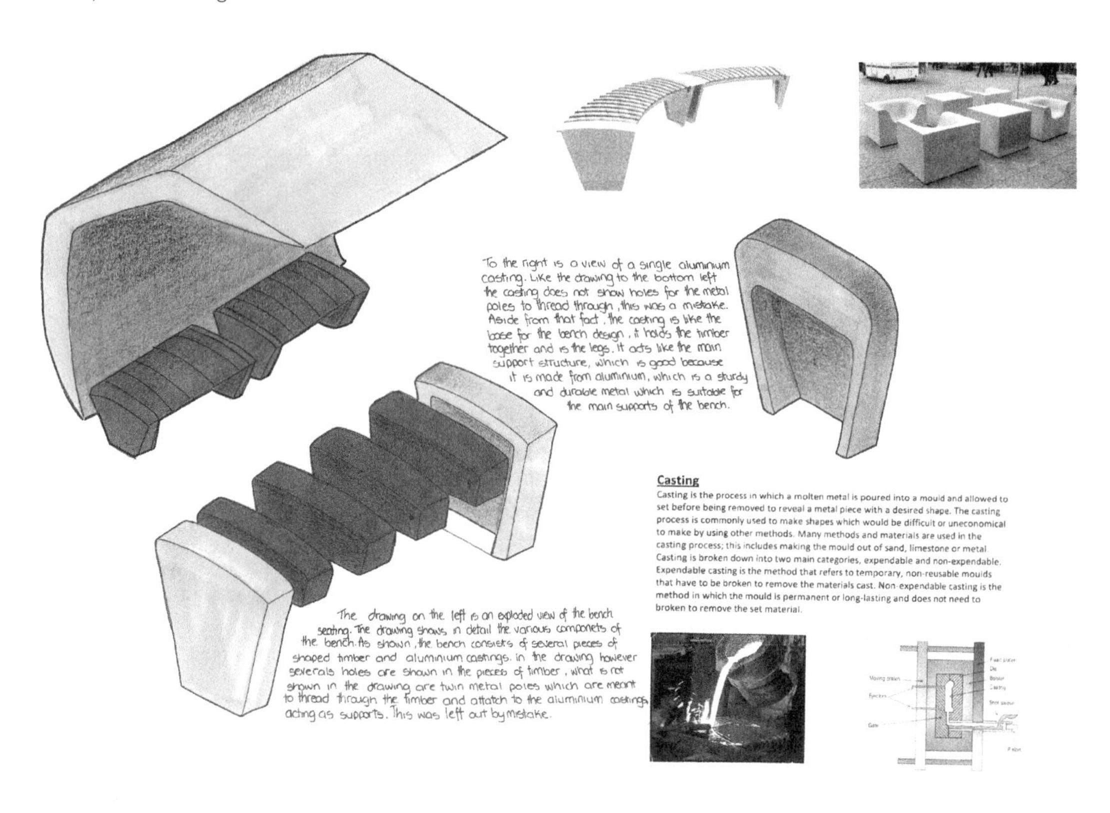

Thanya Chansouk — St Peter's College

Thanya Chansouk has researched materials and manufacturing that could be used, such as press forming for the seating and injection moulding for the shelter component.

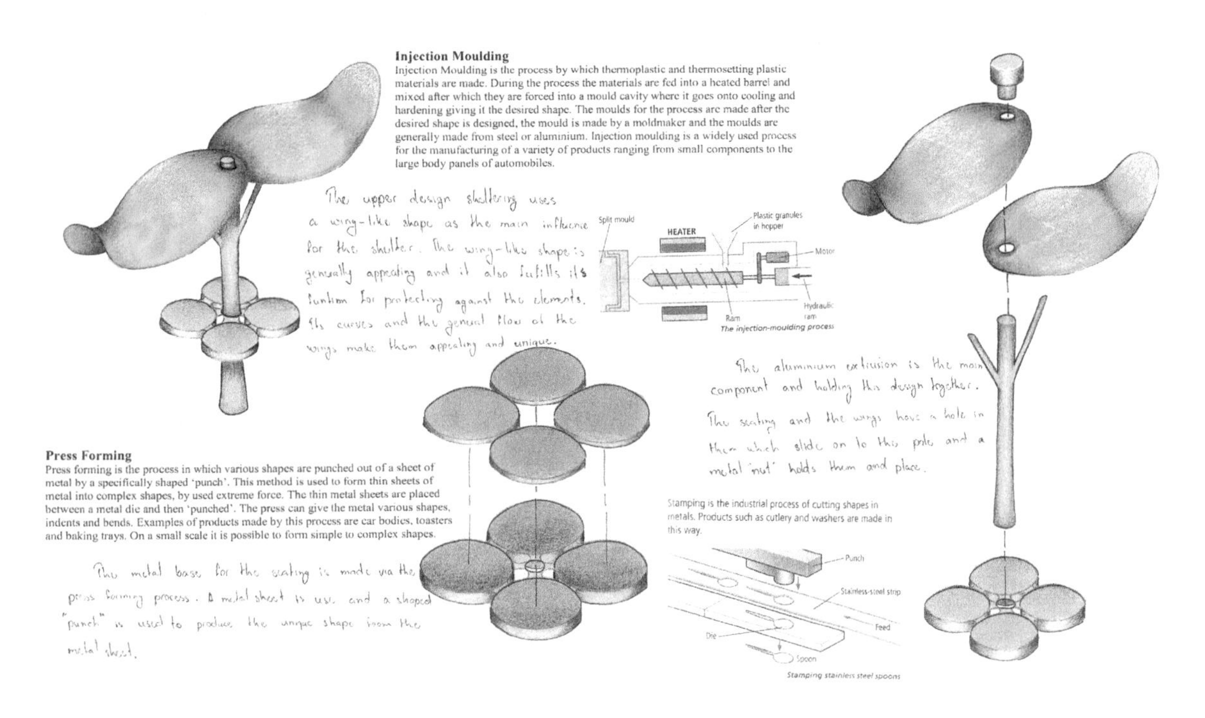

ISBN: 9780170233279

Christian Burgos — St Peter's College

The following work produced by Christian Burgos demonstrates the use of research of existing products throughout the development to help inform his own design ideas for shape, form and details of the interior of the vehicle.

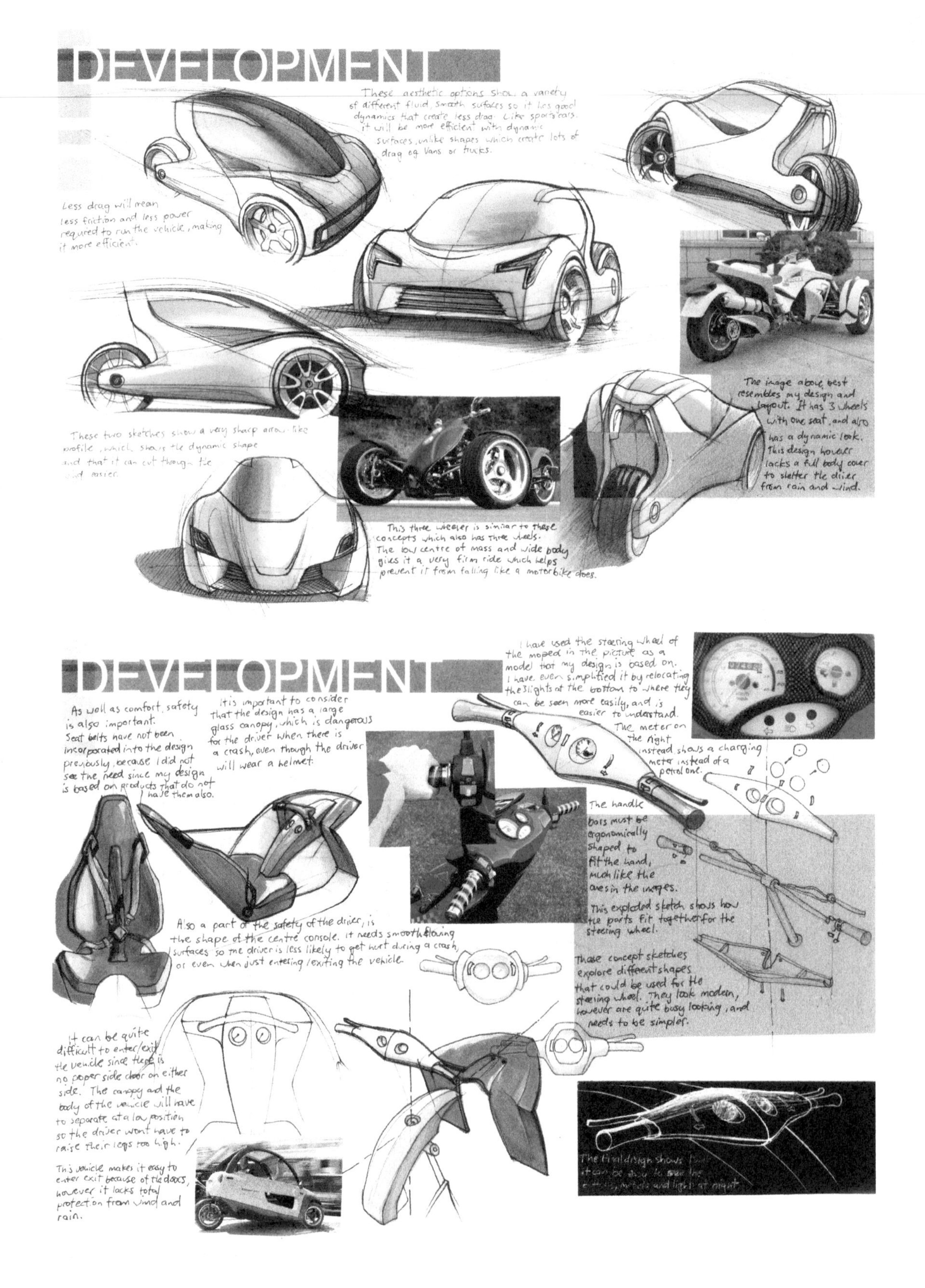

ISBN: 9780170233279

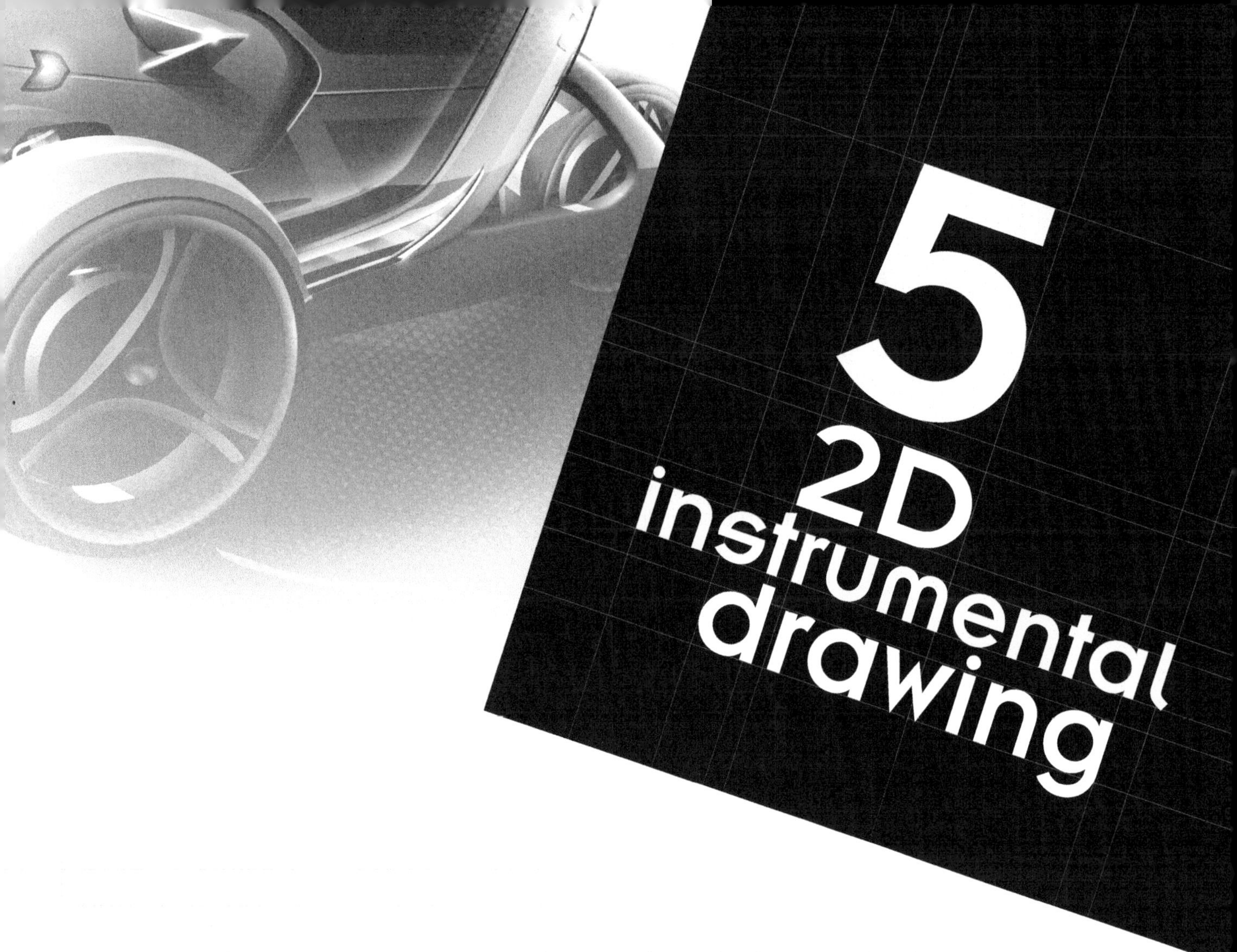

5 2D instrumental drawing

2D Instrumental Drawing is used to construct multi-view orthographic drawings at Level 1 and working drawings at Level 2. Multi-view orthographic drawings and working drawings are used to communicate the technical features or details of a design.

Multi-view orthographic drawings (Level 1) are instrumental drawings that contain two or more projected views (third-angle orthographic projection). Instrumental, multi-view orthographic drawings include the use of line types and projection. Conventions to be used are reference lines, labelling, scale and dimensioning.

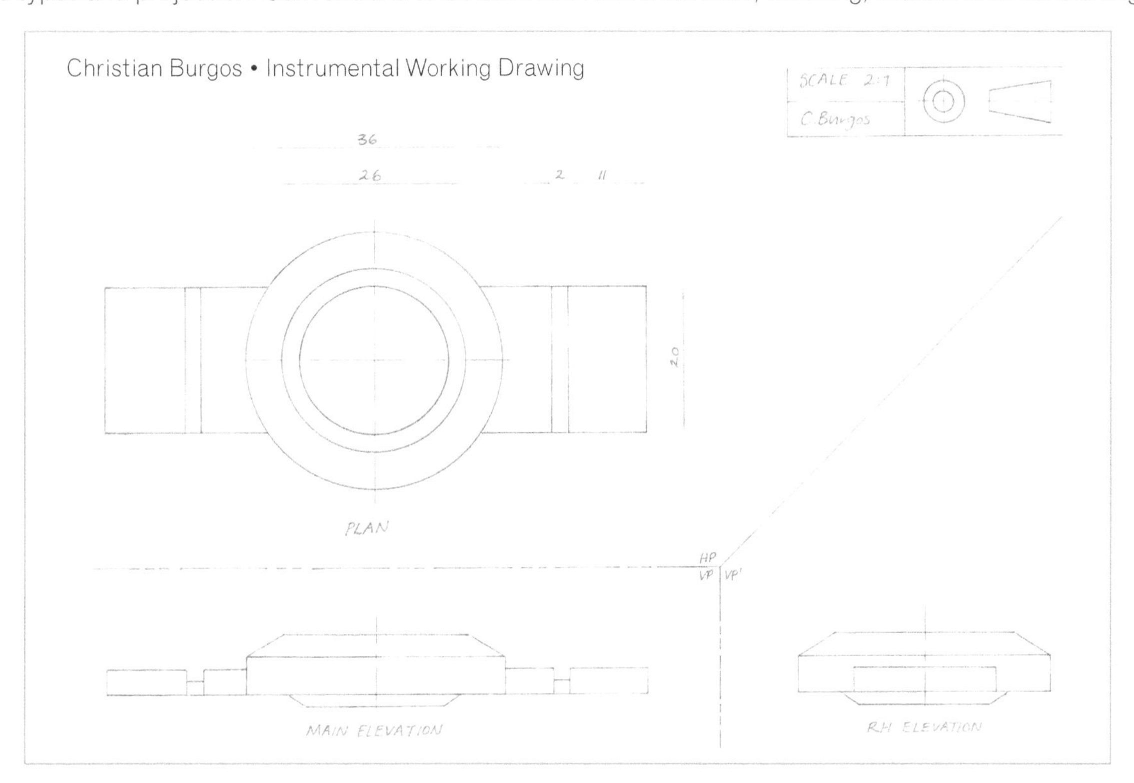

Christian Burgos • Instrumental Working Drawing

ISBN: 9780170233279

Sectioning or surface developments may be used to communicate technical details of the design. Technical features are the dimensions, component shapes, and construction methods necessary to produce the design.

Working drawings (Level 2) are a set of related 2D (orthographic) drawings that could include components, assembly, sectional view, auxiliary view, true shape, surface development and construction details. A set of related drawings are multiple drawings that communicate details of a design's shape and form. The technical details describe the functional and aesthetic qualities of the design. Conventions associated with orthographic drawing define such things as: line types (eg construction lines, outlines, and section lines), drawing and text layout, and dimensioning.

To effectively communicate multi-view orthographic drawings or working drawings you will need to produce accurate and precise 2D instrumental drawings that communicate in-depth details of a design.

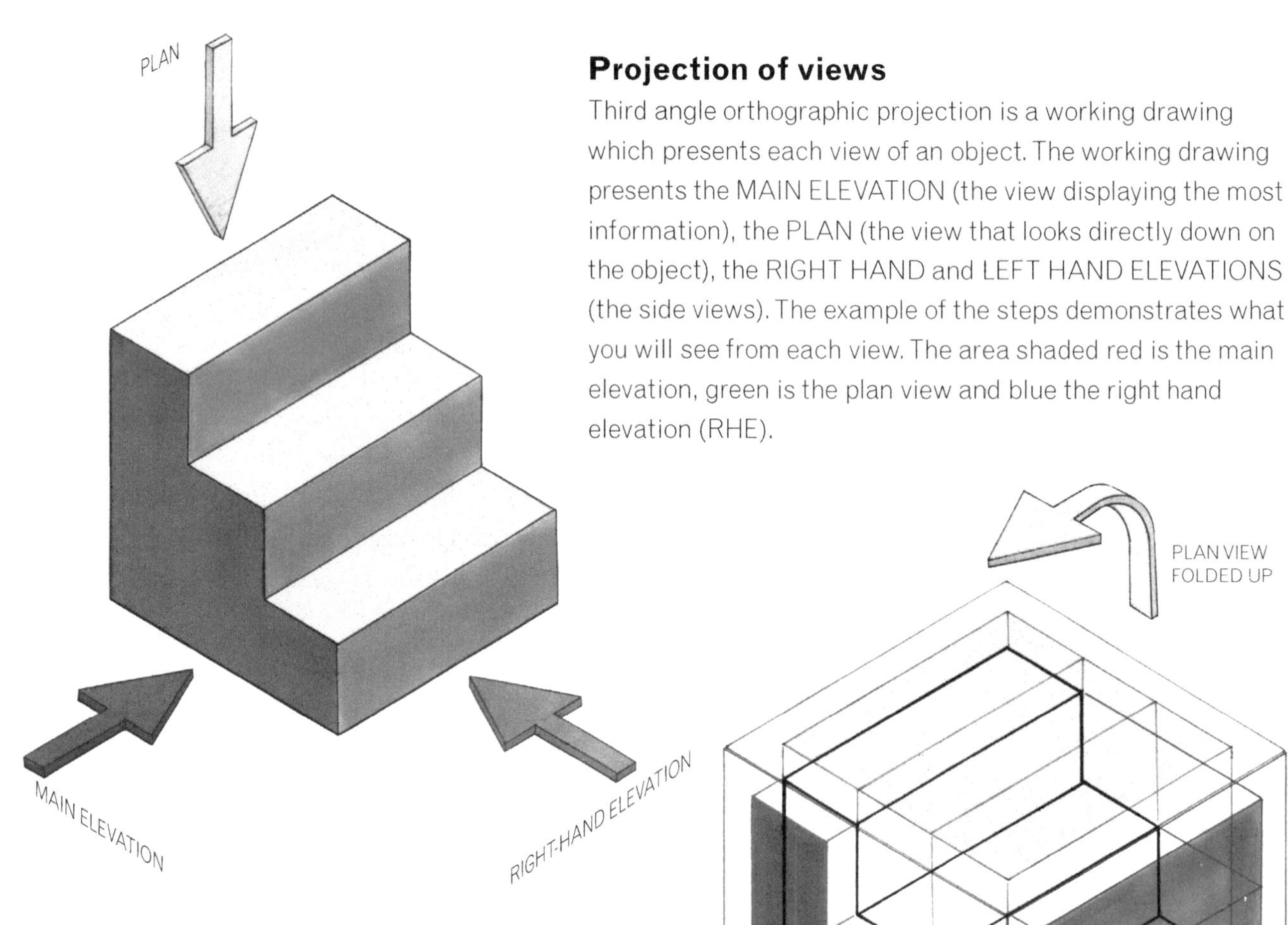

Projection of views

Third angle orthographic projection is a working drawing which presents each view of an object. The working drawing presents the MAIN ELEVATION (the view displaying the most information), the PLAN (the view that looks directly down on the object), the RIGHT HAND and LEFT HAND ELEVATIONS (the side views). The example of the steps demonstrates what you will see from each view. The area shaded red is the main elevation, green is the plan view and blue the right hand elevation (RHE).

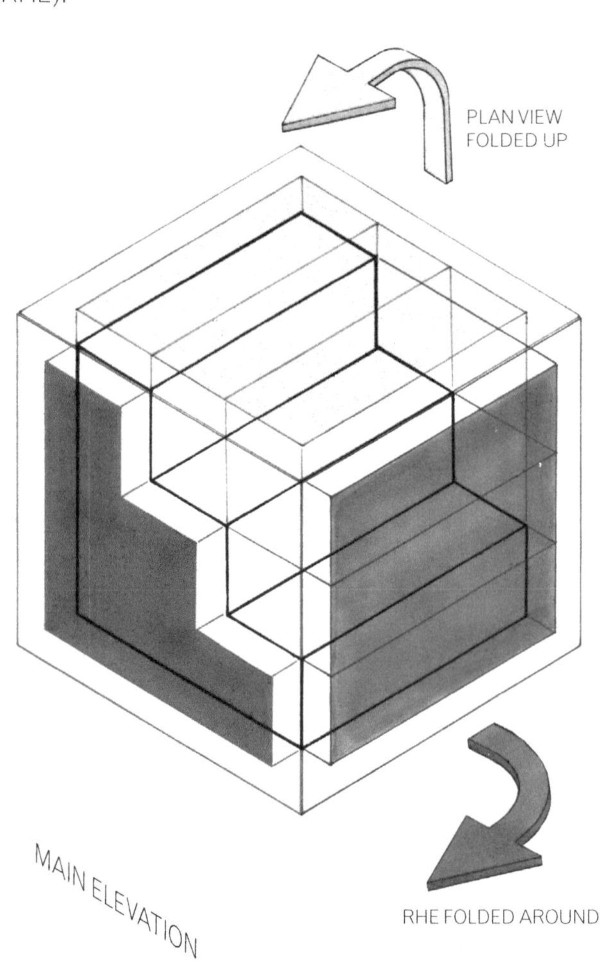

You will notice in the second image that the plan view and the RHE are folded onto the same plane as the main elevation.

ISBN: 9780170233279

The orthographic projection is drawn so that all views are drawn flat on one plane and only show what is on each view. Lines are projected from one view to another so that the views line up with one another. All heights, widths and depths need to line up between the views.

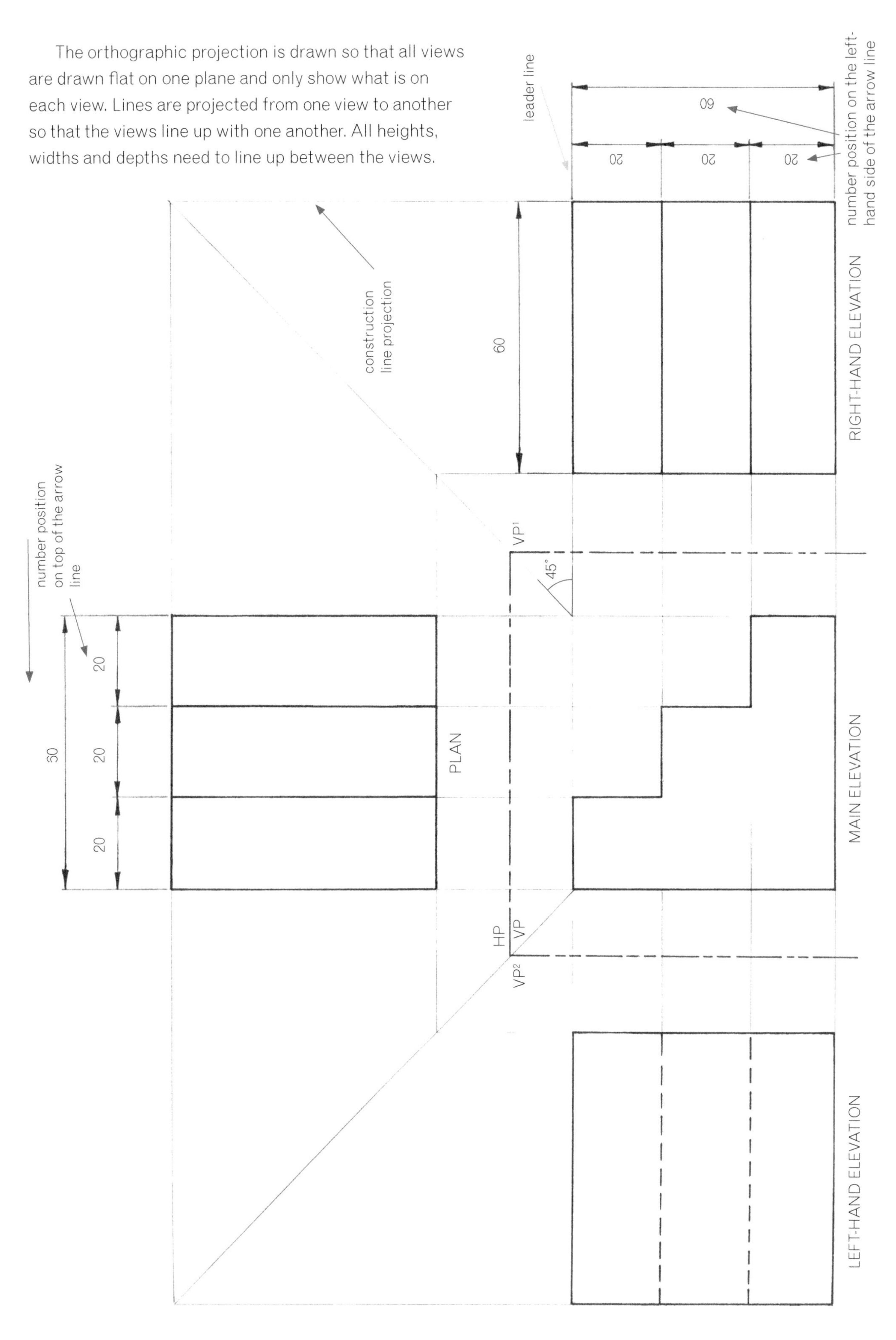

ISBN: 9780170233279

Conventions

Certain standards and conventions are used to produce 2D instrumental drawings. This is so that everyone who reads the drawing can understand it, e.g. the manufacturer.

Line work

Different types of lines are used to represent information on the drawing.

Construction lines – light lines used to initially construct the object.

Outlines – dark lines which are used to outline the important lines of an object.

Reference lines – used for orthographic projection to separate views and project lines and points.

Centre line – indicates the centre of a circle or a cylindrical solid.

Hidden detail – presents parts of an object that cannot be seen from a certain view.

Section line – indicates the plane where an object is being cut through.

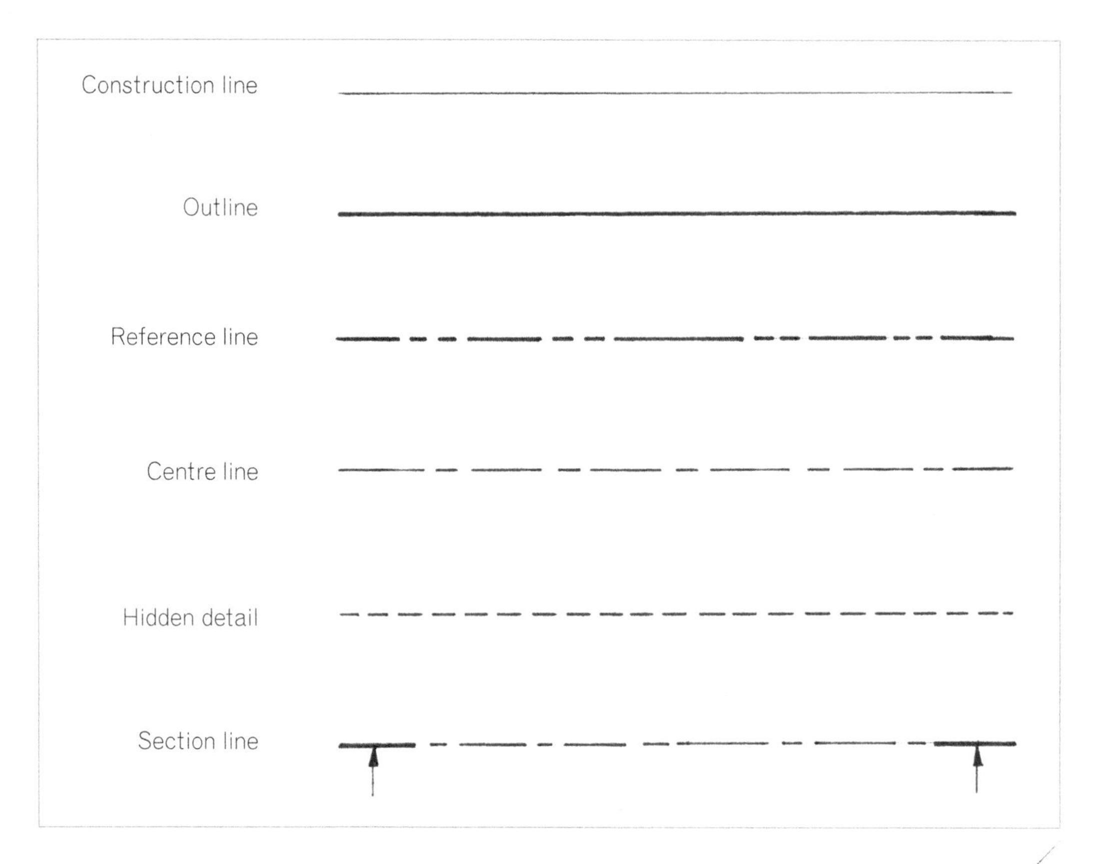

The plan view is projected through a 45° line to the RHE. The top and side edges are projected to find the starting point for the 45° line.

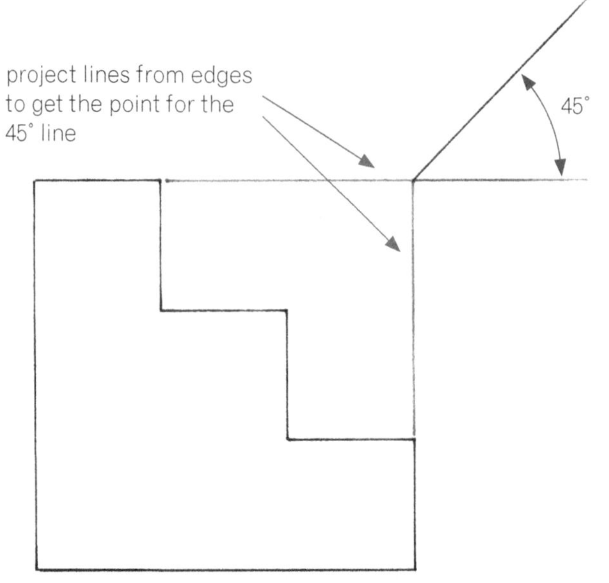

ISBN: 9780170233279

Reference line

The reference line on the orthographic projection acts as a fold line. You could imagine that it is a fold line of the box folding out onto one plane to present all the views as in the example of the work station. A reference line is always a long line and then two short lines. This is repeated as needed.

Labelling of planes

Each view needs to be labelled to indicate which plane it is on. HP stands for horizontal plane, VP stands for vertical plane. On the example of the work station instrumental drawing you will notice that HP is written above the reference line on the plan view. VP is written directly under the same reference line. The rest of the vertical planes are then labelled as shown.

Labelling of views

Each view needs to be labelled with plan, main elevation, RHE or LHE. The labels need to be written between 3mm guidelines in CAPITAL letters.

Title block

To present the information needed on your working drawing use a title block. The title block should include the following information:

1. Name
2. Drawing title
3. Scale
4. Orthographic symbol

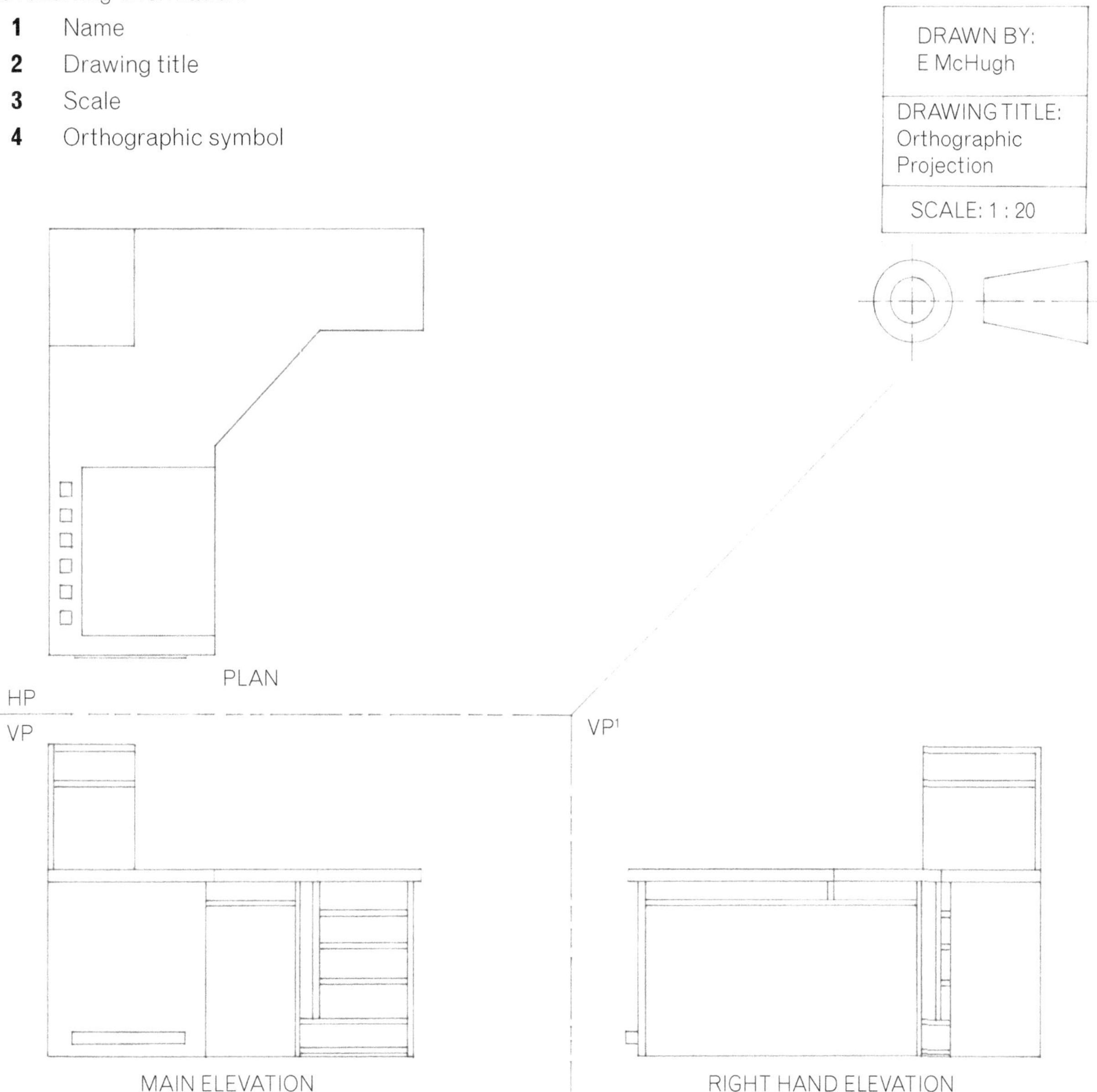

ISBN: 9780170233279

Scale

The scale of the product needs to be documented on the orthographic projection. Scale indicates how much smaller or bigger the drawing is compared to the actual product. If the object is drawn the same size on paper as the actual size of the product, the scale documented is 1:1. If the object is drawn half the size of the actual product, the scale documented is 1:2. If the object is drawn double the size of the actual product, the scale documented is 2:1. You cannot make up your own scale, you will need to work to the recognised scales shown below:

Enlarging the product sizes to fit the page the scales used are: **10:1**, **5:1**, **2:1**

Full size – the actual size of the product the scale is: **1:1**

Reducing the product sizes to fit the page the scales used are: **1:2**, **1:5**, **1:10**, **1:20**, **1:50**, **1:100**

The measurements presented on the drawing are the actual sizes of the product. All measurements should be in millimetres unless otherwise stated. Always indicate the scale on your drawing.

Dimensioning

Leader lines — light lines to indicate where a dimension starts and finishes. The leader lines act as a guide for the arrow line. The leader lines start 1mm away from the object and finish 1mm after the arrow line.

Arrow lines — dark lines with arrowheads to indicate the measurement for the length of an edge, side or arc. Any length should only be dimensioned once. Heights or widths will be shown on two views but should only be dimensioned on one. The arrow line should be positioned 10–12mm away from the object. Each dimension after that should be 10mm away from the previous arrow. Smaller dimensions should be shown closest to the object and larger dimensions on the outside of these. Arrowheads should be 3mm by 1mm in size.

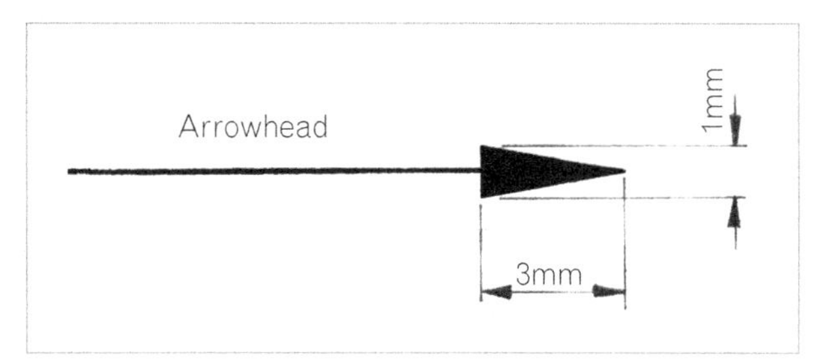

Number positioning — numbers should always be positioned on top of the arrow line or to the left-hand side in the middle of the arrow line. Remember the number on the arrow head should be the real size not the size of the measurement on the drawing.

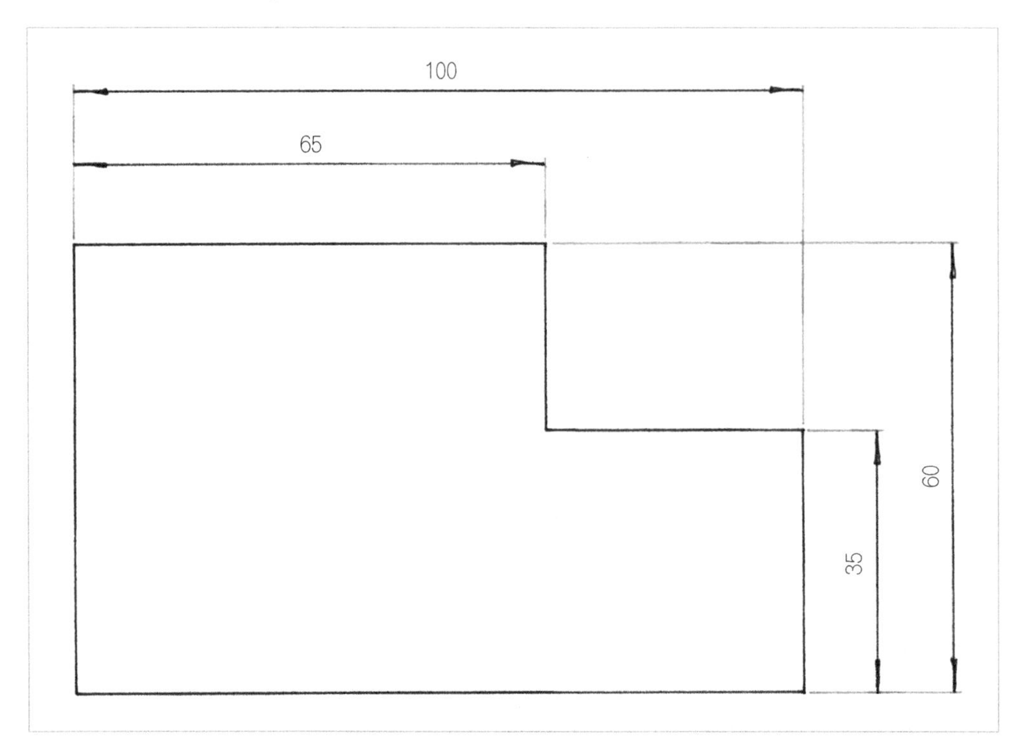

ISBN: 9780170233279

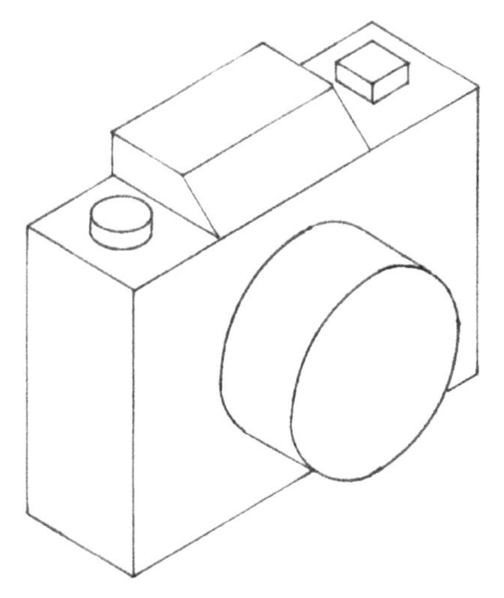

The main elevation of the camera has been shown dimensioned correctly and incorrectly. You will notice how confusing the measurements are to read when the conventions for dimensioning are not followed.

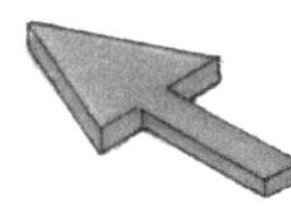

MAIN ELEVATION OF CAMERA

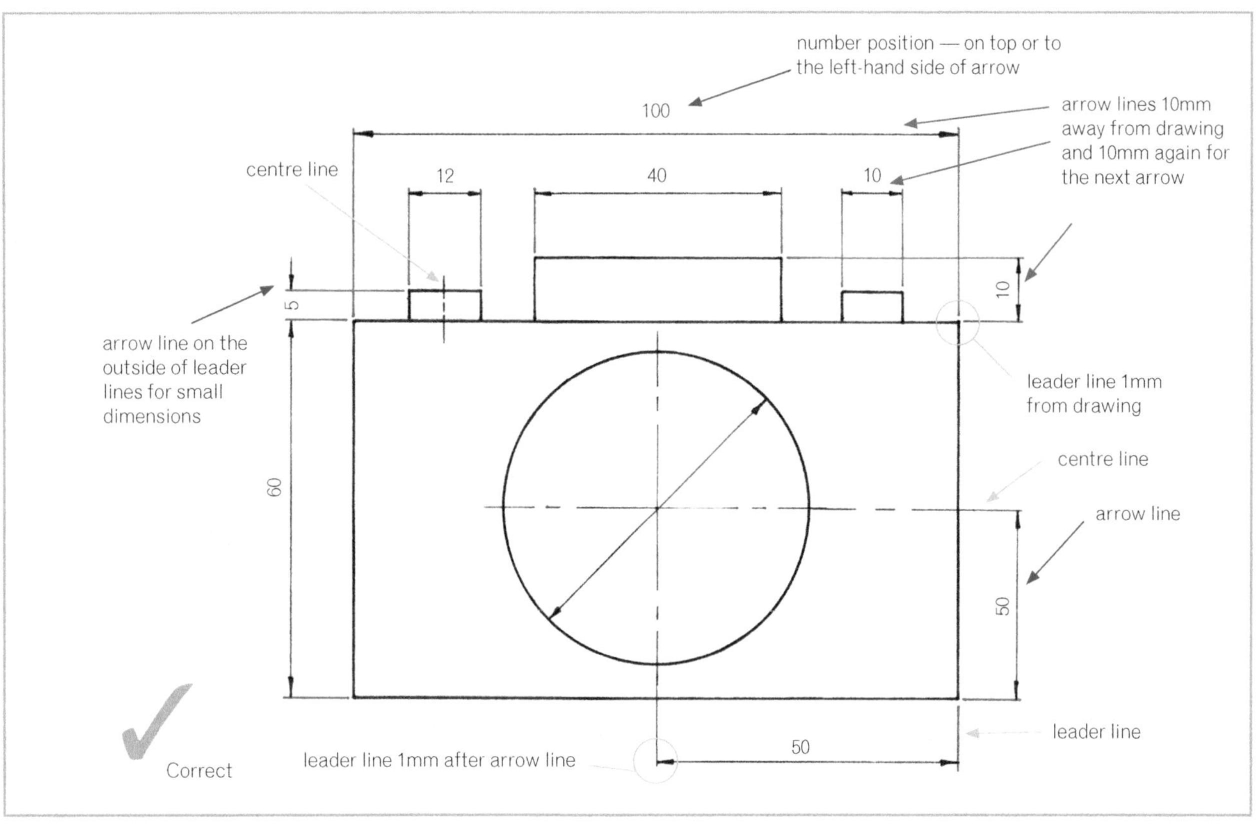

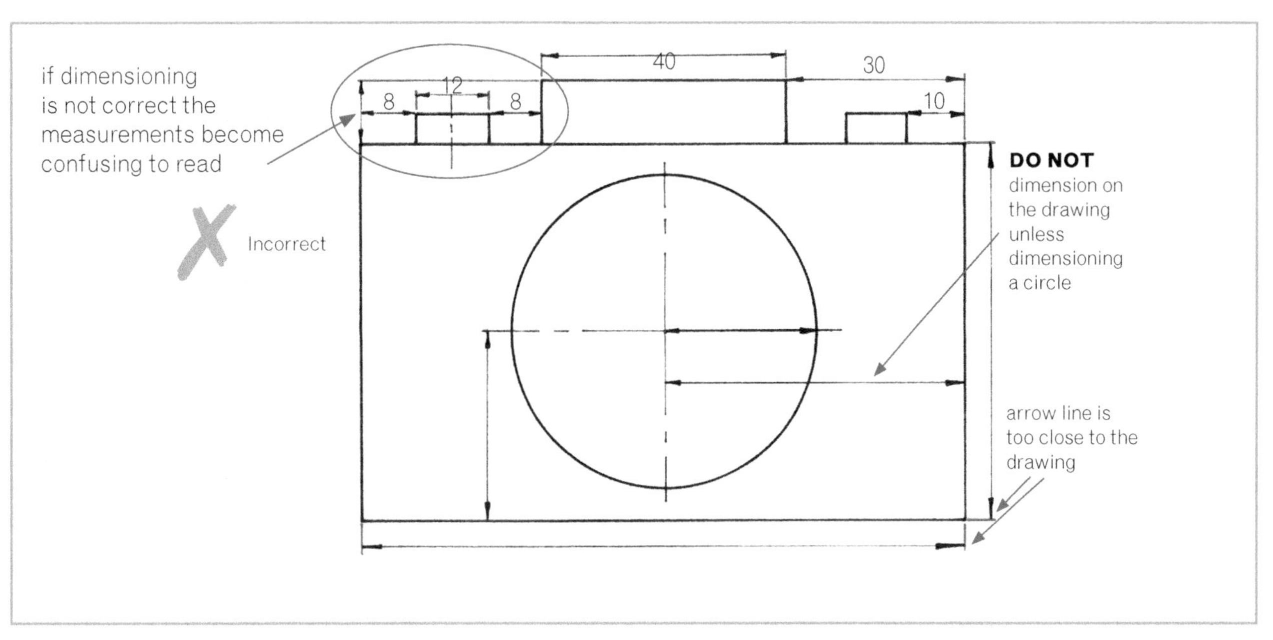

ISBN: 9780170233279

The work station instrumental drawing presents the correct conventions for dimensioning.

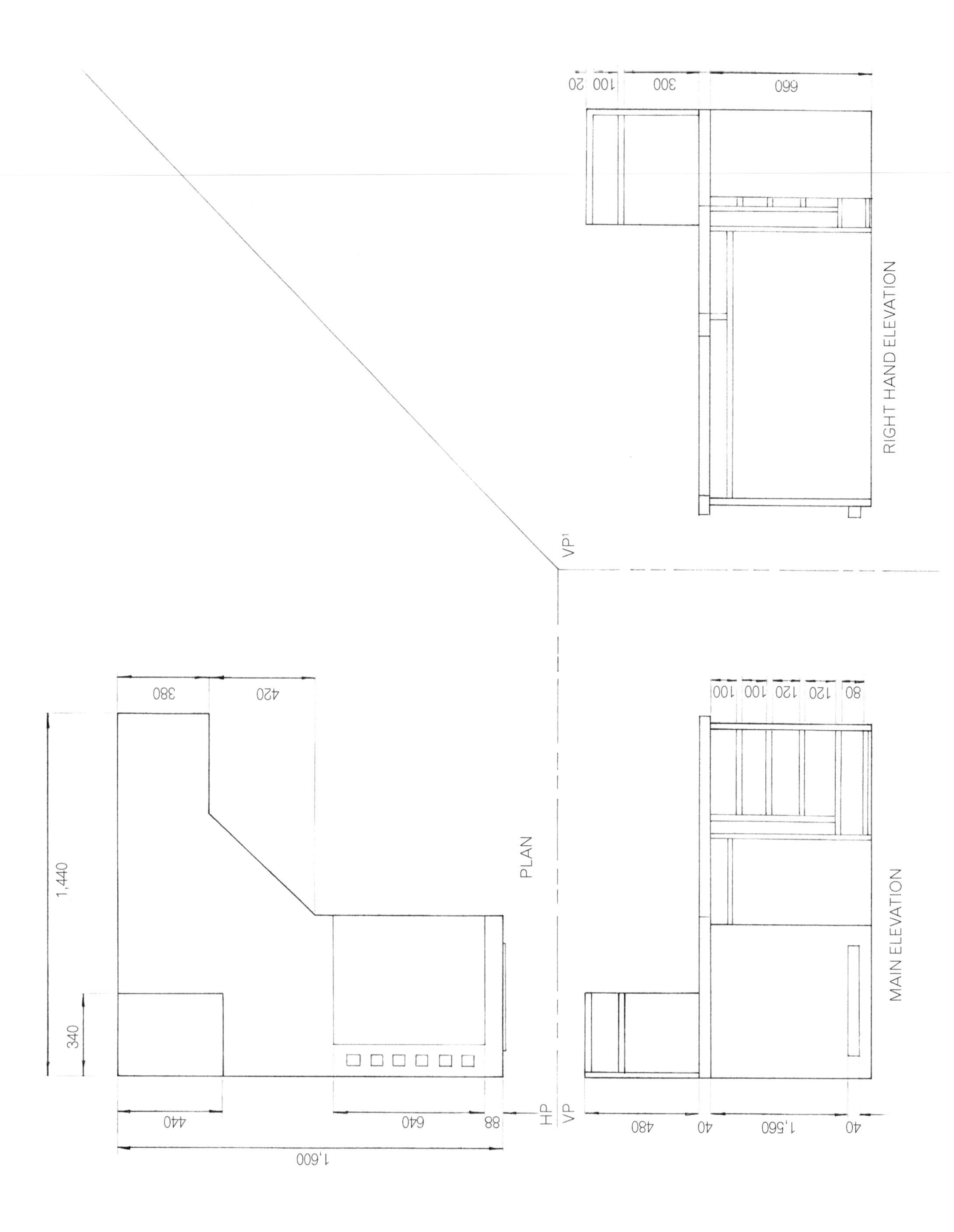

ISBN: 9780170233279

Dimensioning circles, arcs and angles

The following drawings demonstrate the different methods that can be used to dimension circles and arcs. A circle or an arc can be dimensioned on the actual drawing whereas all other dimensions should be positioned 10mm away from the view. Remember to include a centre line when dimensioning an arc, circle or cylinder.

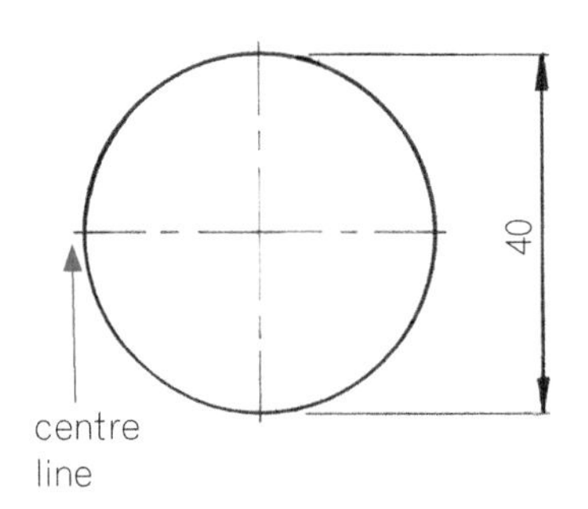

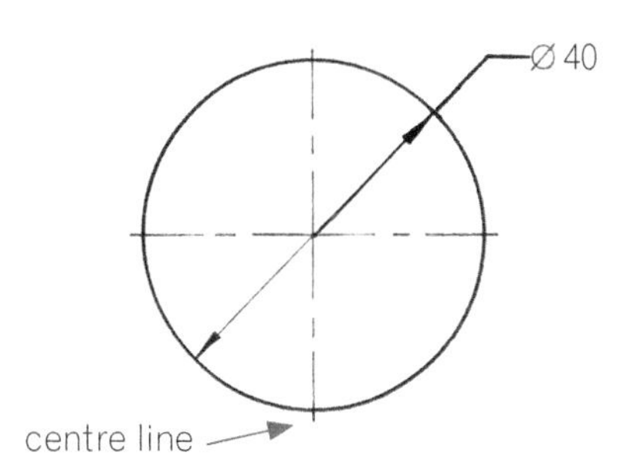

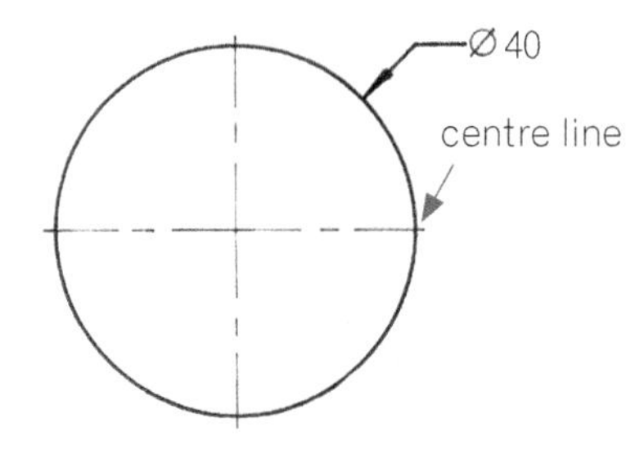

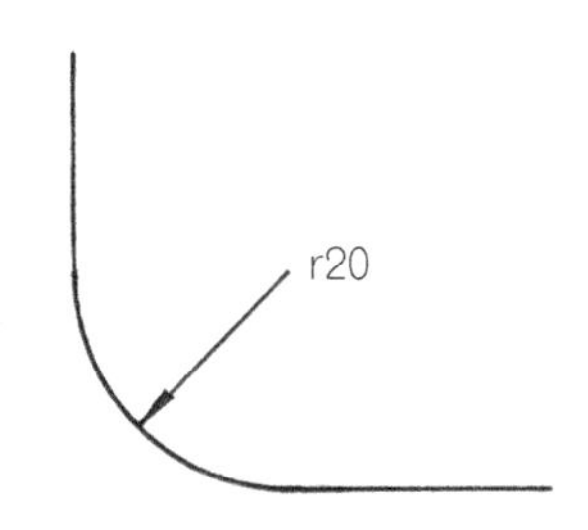

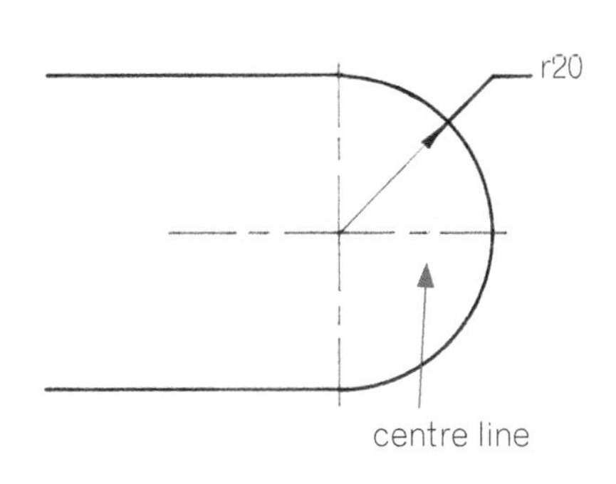

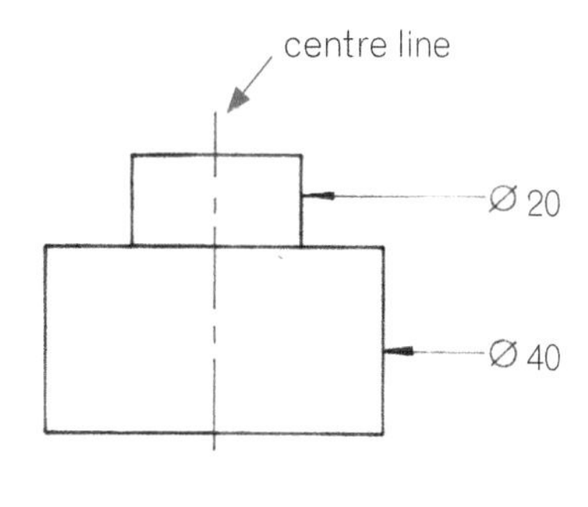

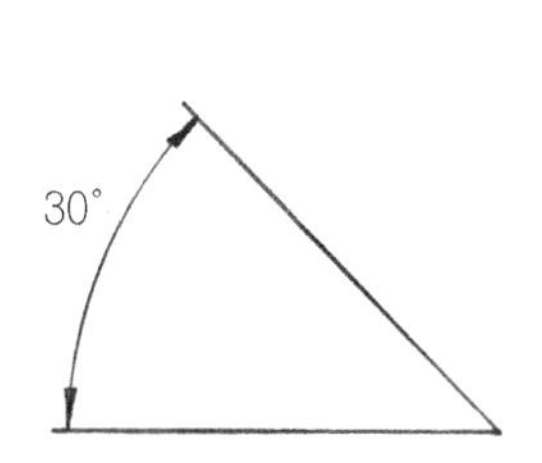

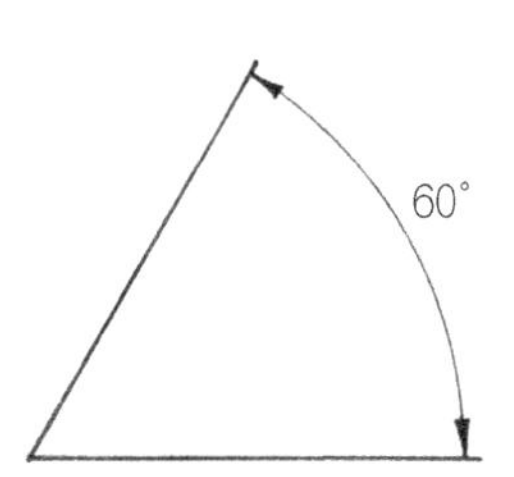

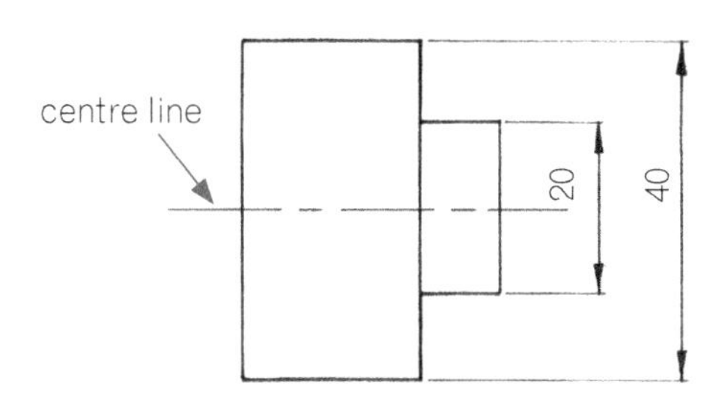

DIMENSIONING OF ANGLES

ISBN: 9780170233279

The orthographic projection of the toy truck demonstrates more complex projection and therefore the dimensioning needs to be more comprehensive. You need to think about the layout of the drawing and leave enough space around all the views for the dimensioning. You will notice that some of the measurements on the drawing are too small to draw an arrow line in the space between the leader lines. In this situation draw arrows on either side of the leader line and position the number in the gap between the arrows or above one of the arrows.

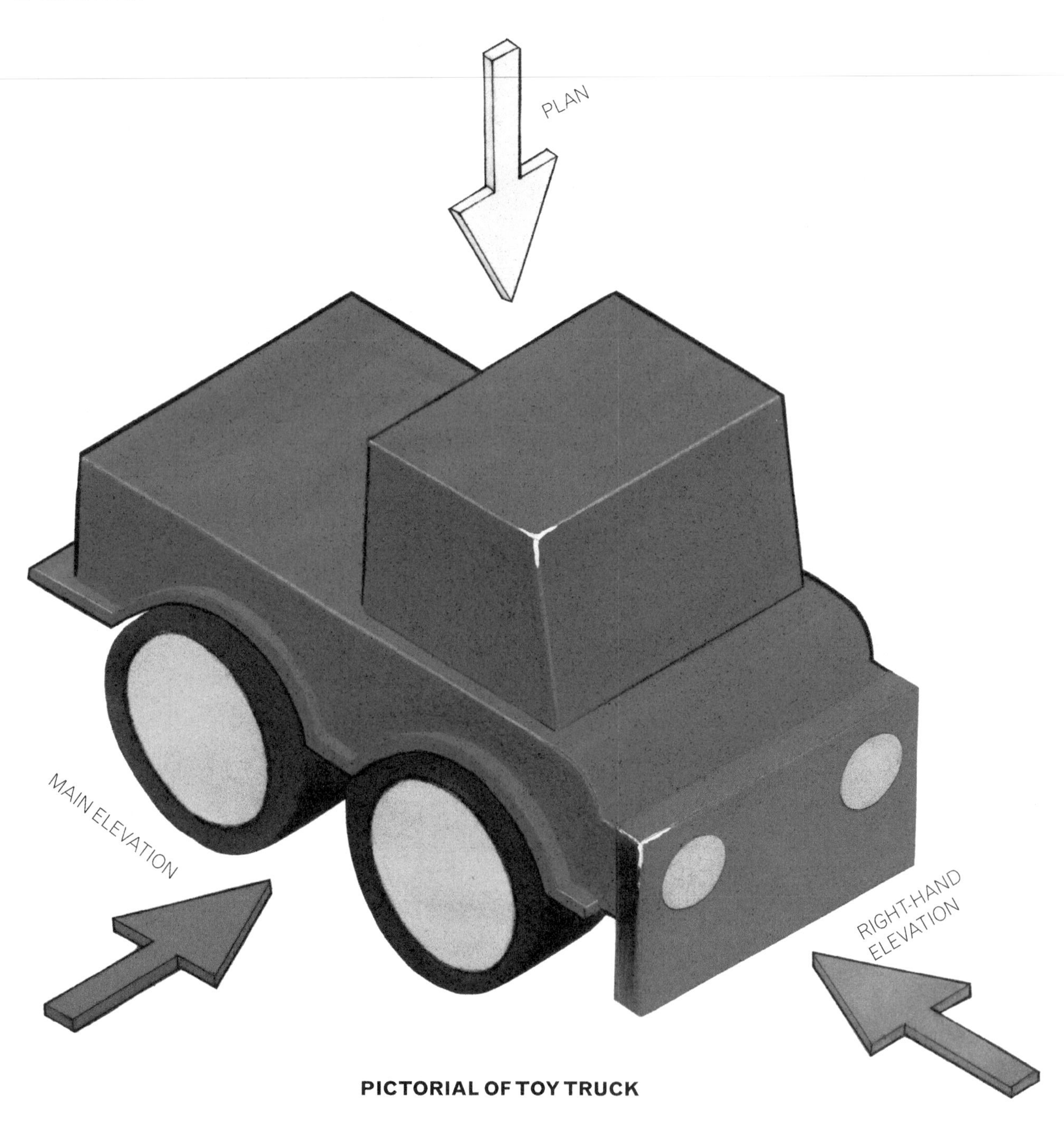

PICTORIAL OF TOY TRUCK

ISBN: 9780170233279

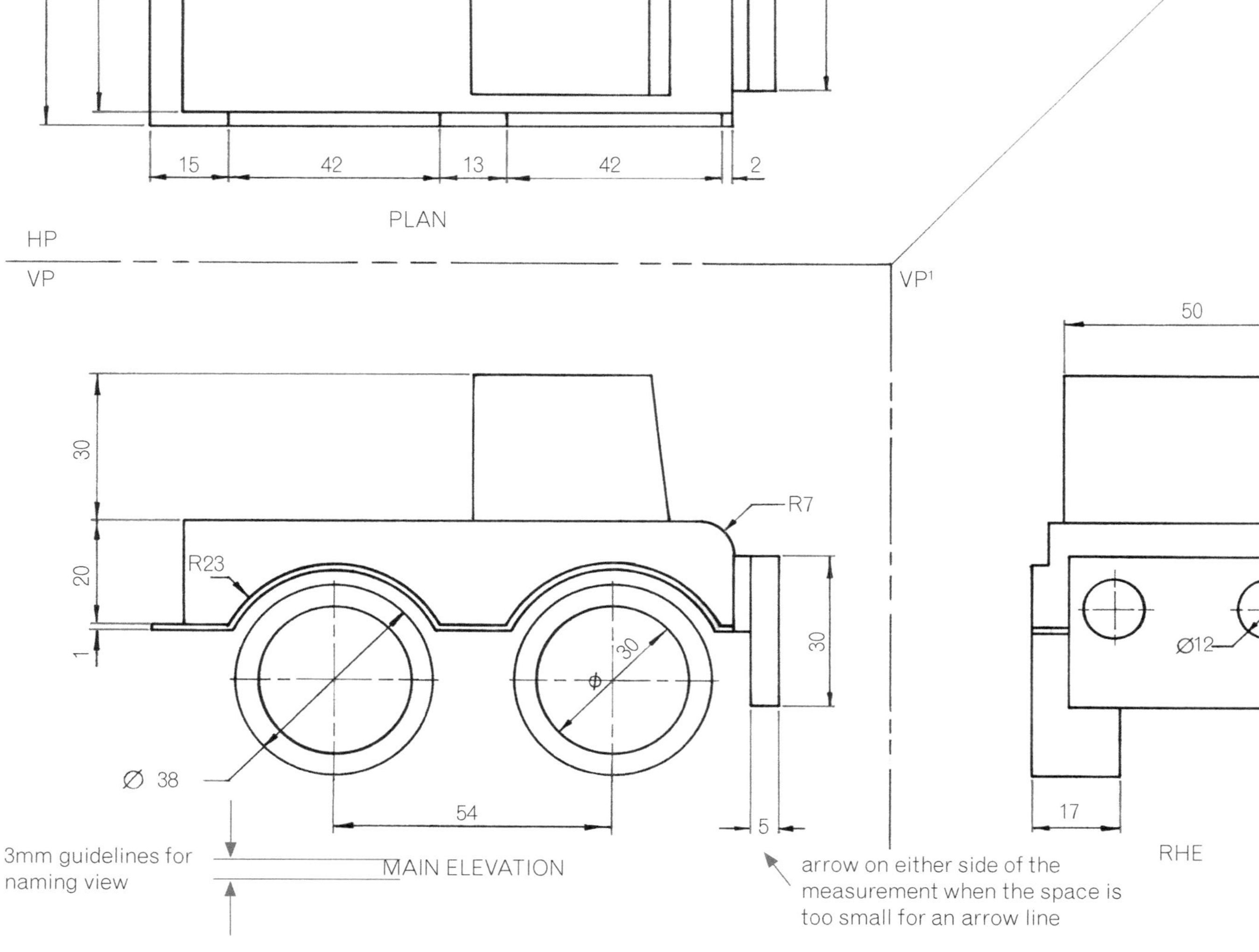

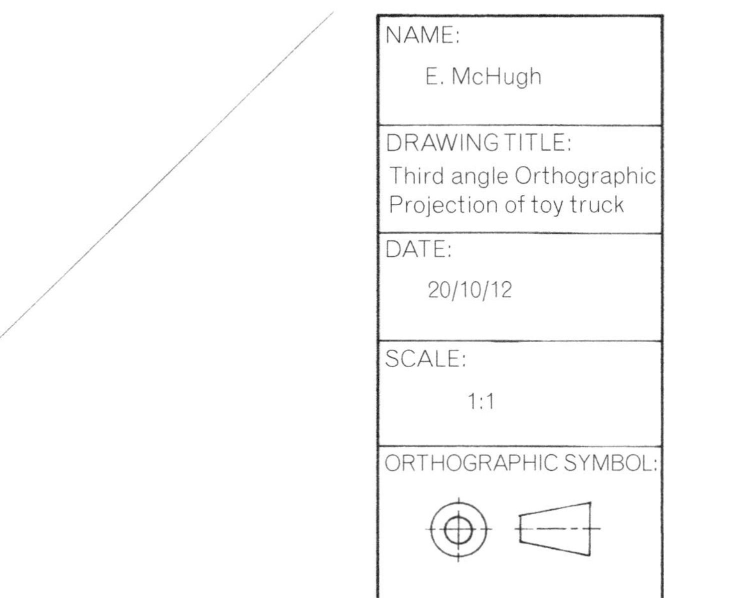

ISBN: 9780170233279

Working drawings are used to finalise and explain the specific sizes of a product that has been designed. Working drawings need to be accurate as they are used to manufacture the product. At the end of a design brief you may use one of the following drawing techniques to explain your design idea in detail.

Assembly

Assembly working drawings explain how the components of a product fit together.

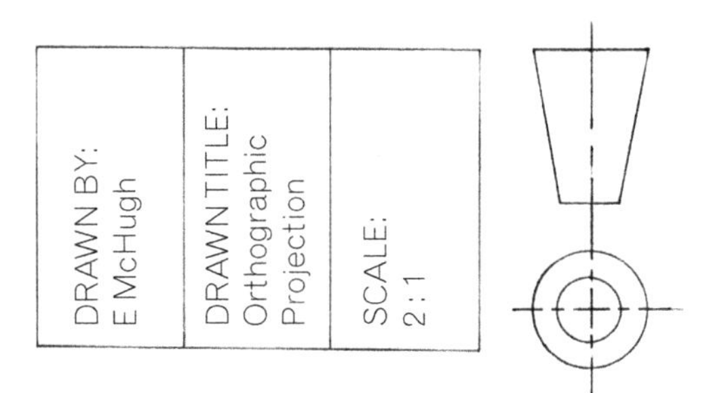

This orthographic projection presents the parts of a bottle opener assembled. The drawing shows how the parts of the bottle opener fit together. Although the overall sizes of the product are shown, the bottle opener could not be manufactured from this drawing as there is not enough information about the individual components given. As the bottle opener would appear too small on the page if it is drawn full size the product is drawn at a scale 2:1 so that the information on the drawing can be presented clearly.

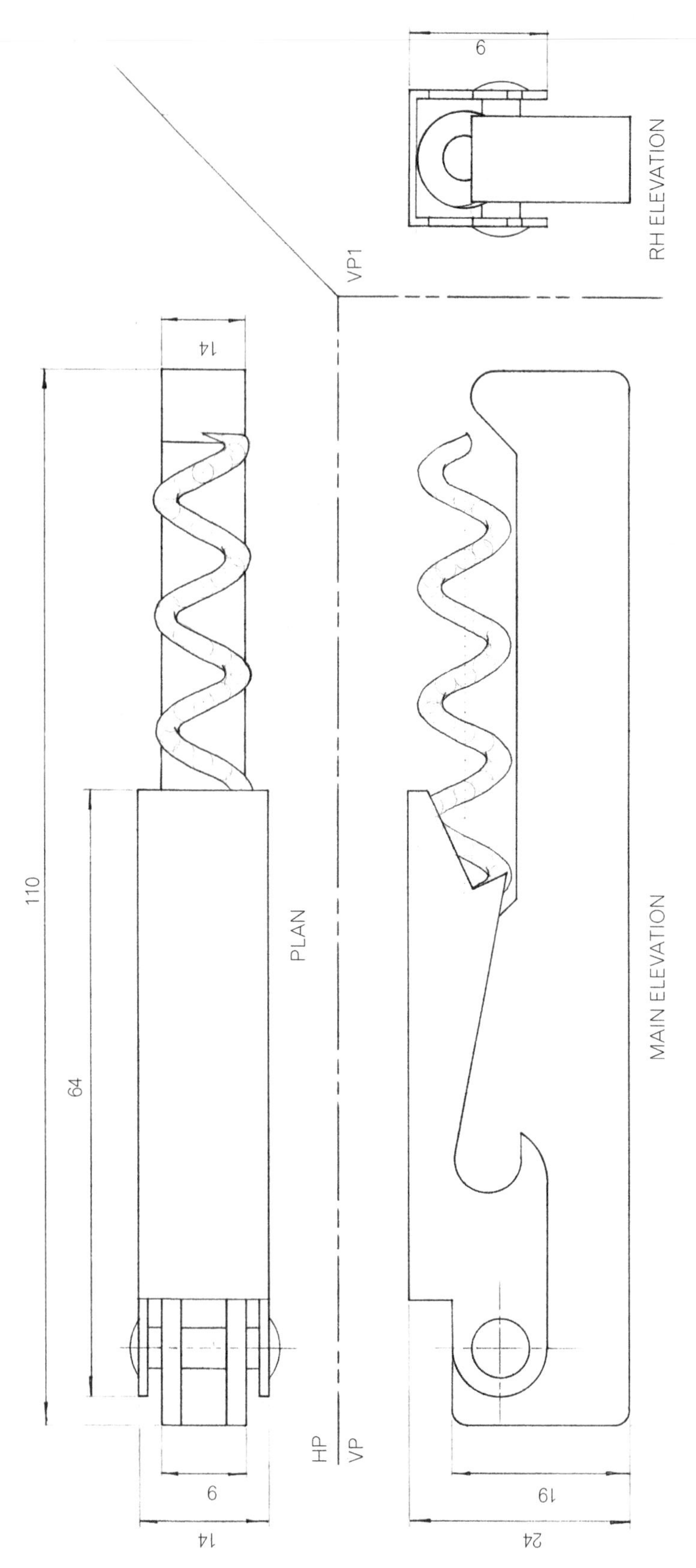

ISBN: 9780170233279

Component details

The components of a product can be presented with separate orthographic projection drawings. This enables you to present the product or object in more detail. The size of each part can then be documented also.

The following shows each component of the bottle opener drawn separately.

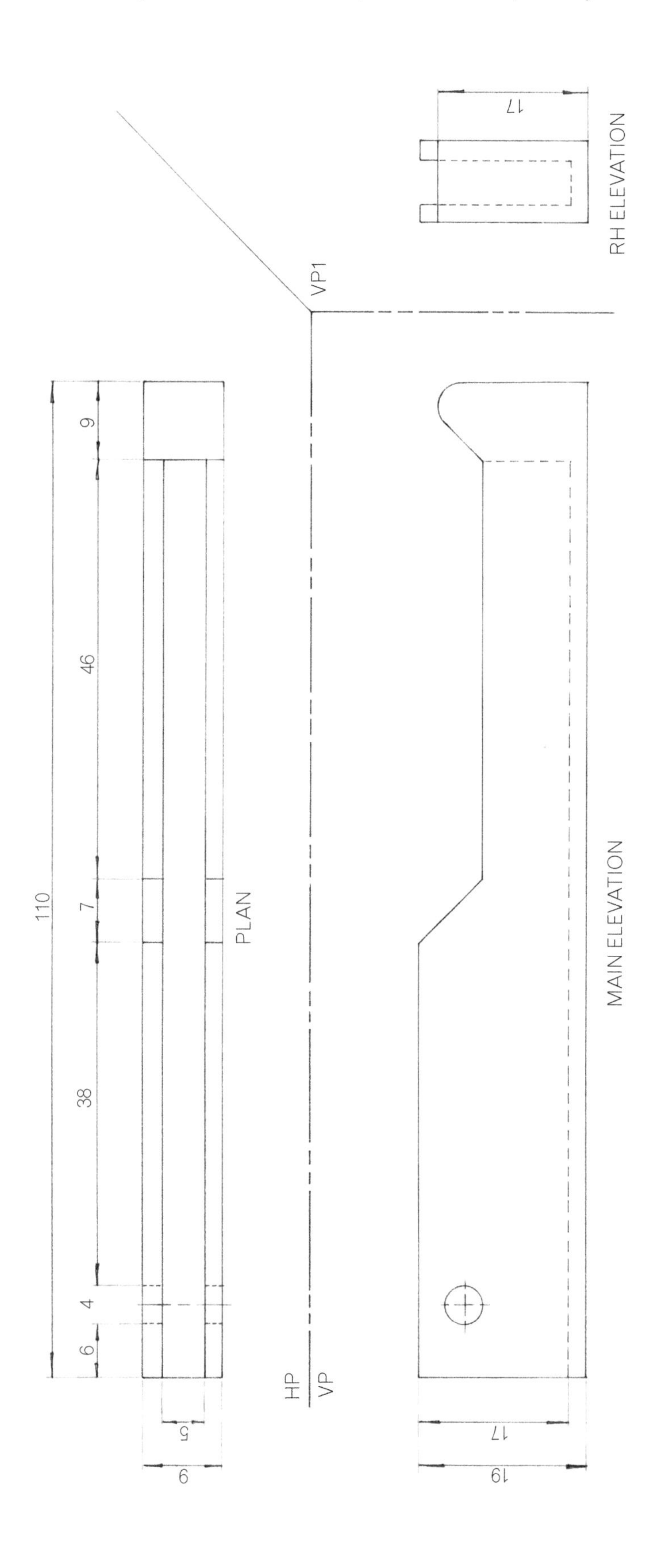

PLASTIC COMPONENT

ISBN: 9780170233279

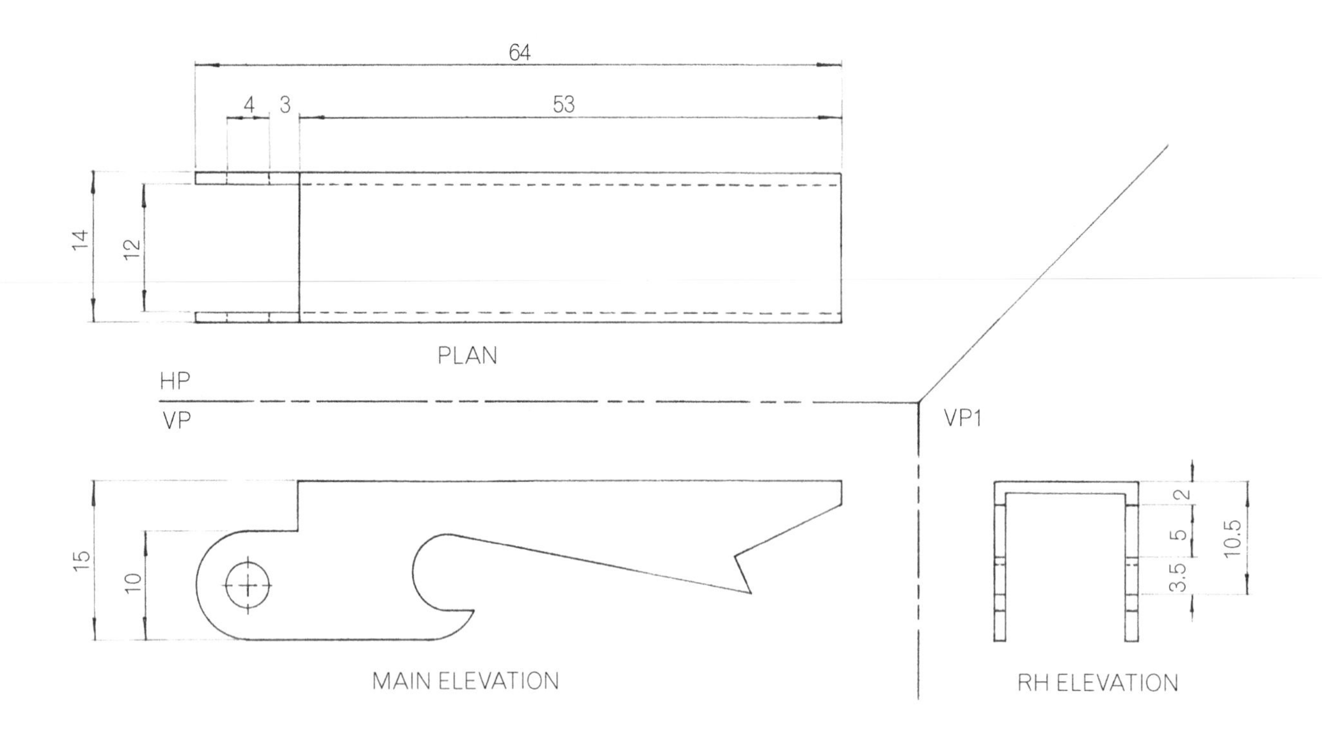

METAL COMPONENT

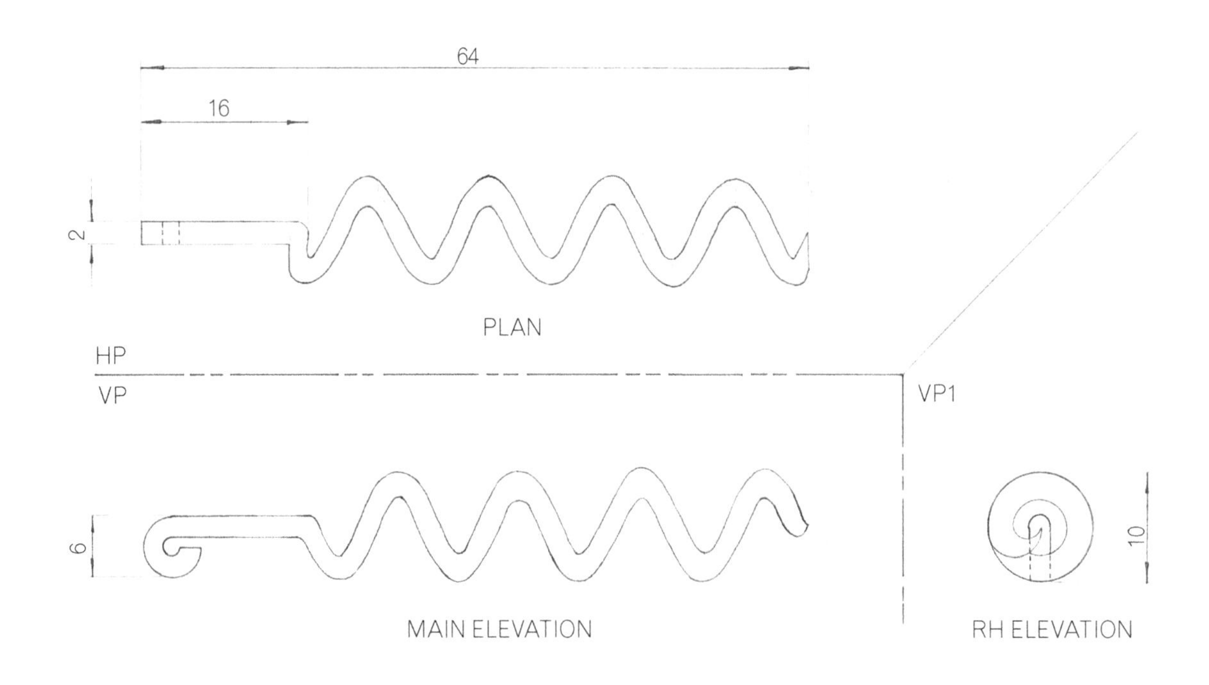

CORKSCREW COMPONENT

ISBN: 9780170233279

Thanya Chansouk — St Peter's College

The following work produced by Thanya Chansouk demonstrates a third angle orthographic projection with hidden detail of a shelter.

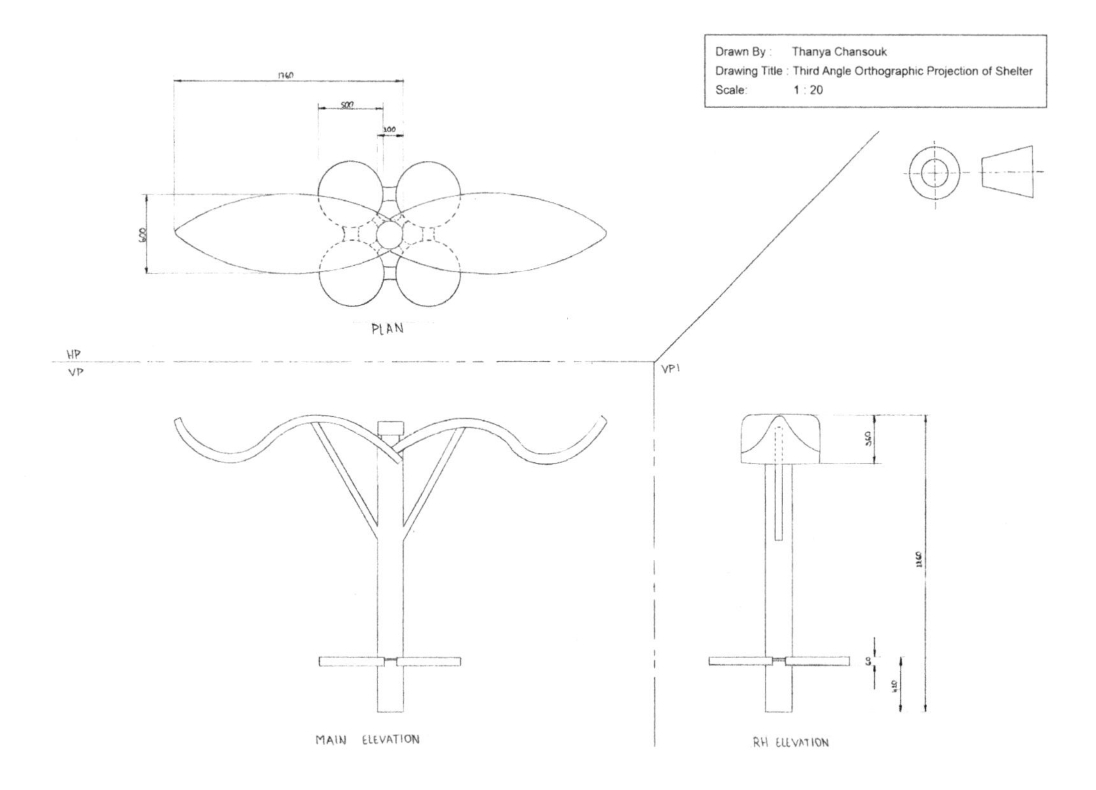

The following orthographic projection shows the seating for the shelter as a separate component and therefore in more detail.

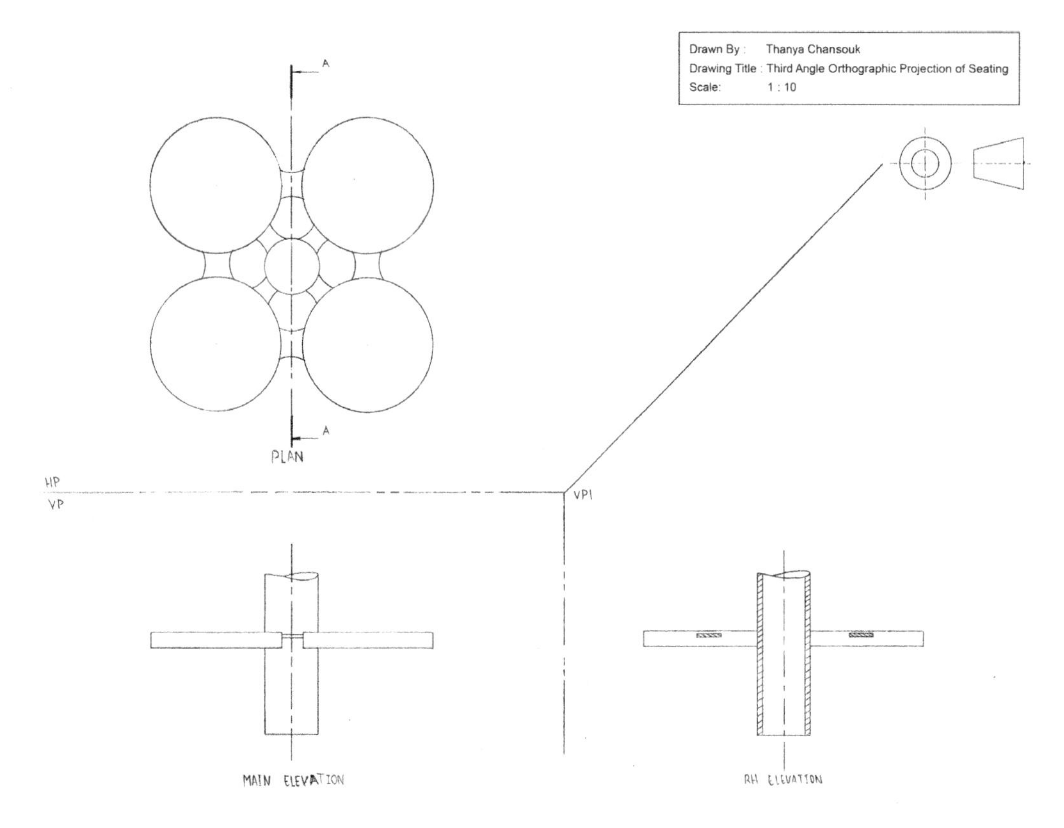

A sectional view shows a part of an object after a portion has been removed. It will describe the information on the inside of an object. There are standards and conventions that are also used for sectioning so that the drawing can be easily read.

Full section

The example demonstrates a full section. The object is cut through a plane. To indicate what material has been cut the area is hatched. Only the material that is actually cut through is to be hatched. Hatching is angled lines that are drawn 2-5mm apart. The pictorial shows what has actually been cut. The centre circle is not hatched as there is no material in this space.

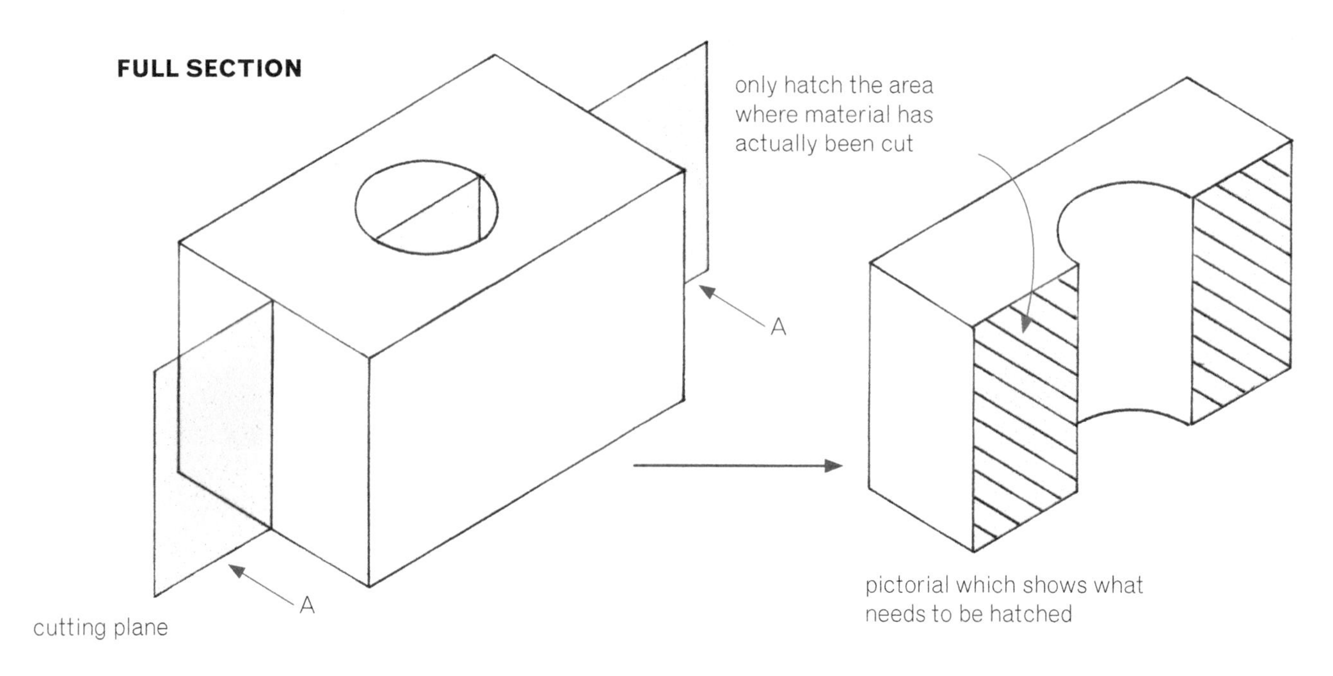

The view that is being hatched is called Section A-A. The cutting plane is indicated on a separate view on the same drawing with a section line. If there is more than one sectioned area the next section line would be named Section B-B. The arrows at either end of the section line point towards the part that you will see.

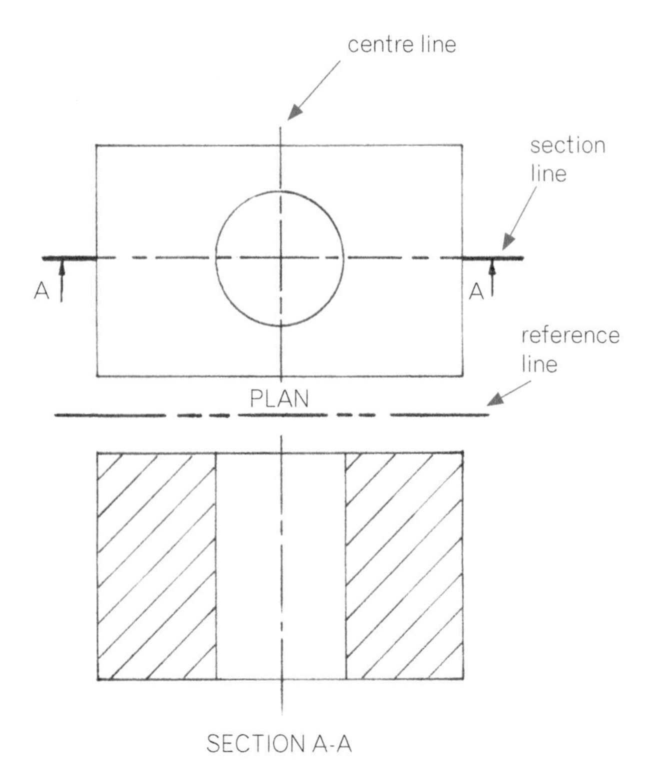

ISBN: 9780170233279

Hatching

So that the hatching is evenly spaced scribe a guideline on your set square. You can then line this guideline up with the previous line on the hatched area. The hatching lines need to be accurate and drawn up to the edges of the part.

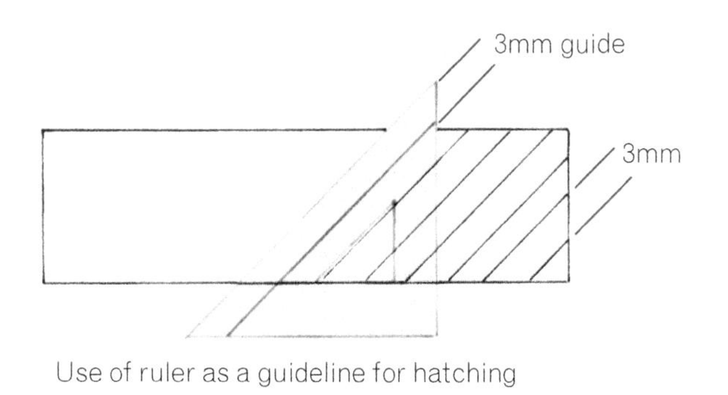

Use of ruler as a guideline for hatching

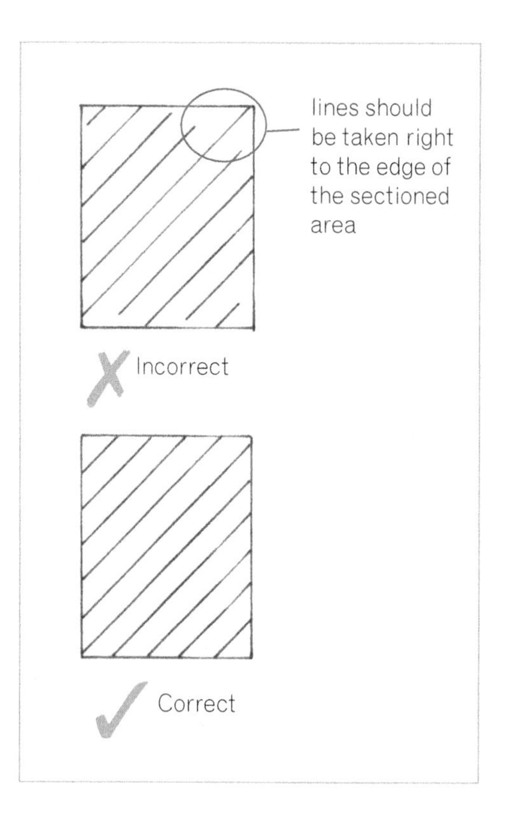

Adjacent parts

Separate parts need to be hatched in different directions to communicate this to the reader. In this example 45° lines are used in opposite directions. If there is a third part this needs to be hatched on a different angle again. Make sure you take the hatch lines right to the edge of the area as in the example.

There are exceptions to the rule that all material that is cut through is hatched. When the section line passes through the centre of webs, shafts, bolts, rivets, keys and pins these are not hatched.

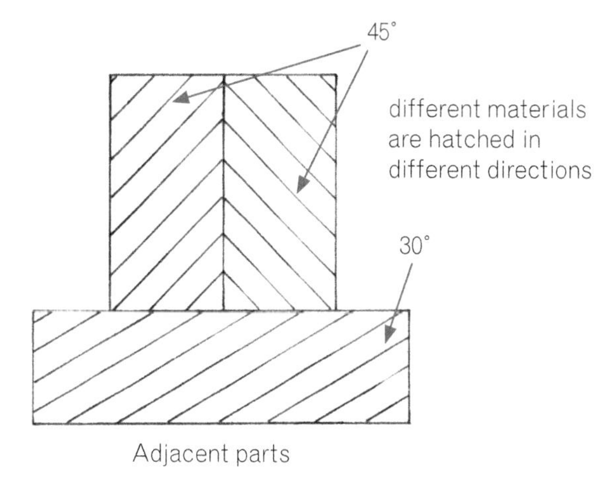

Adjacent parts

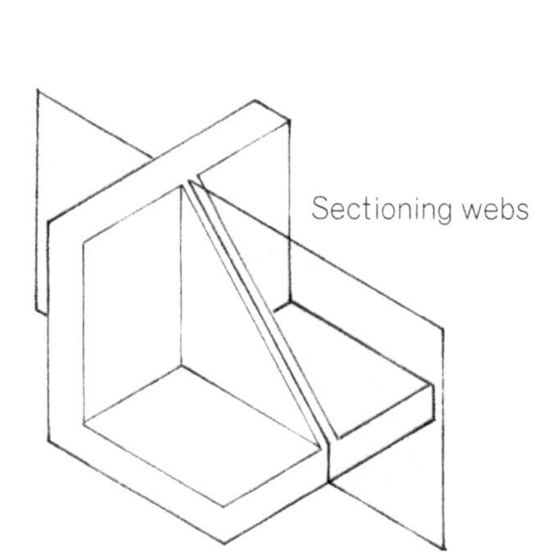

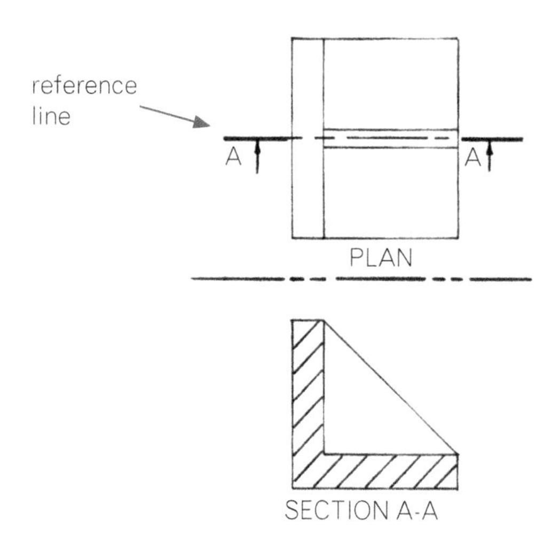

ISBN: 9780170233279

Half section

A half section can be used on objects which are symmetrical about a centre line. As shown, the plane takes away a quarter of the object. The advantage of this view is that you can see the exterior information and interior details on the same view. The section line stays the same in the plan view but the naming of the elevation is half section A-A.

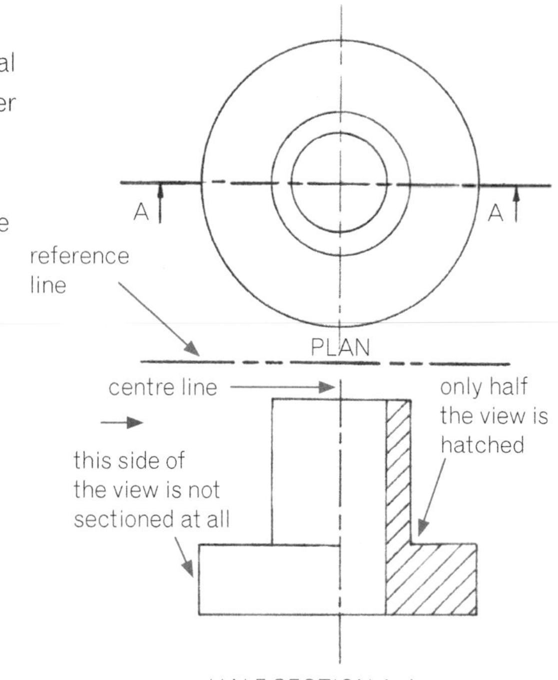

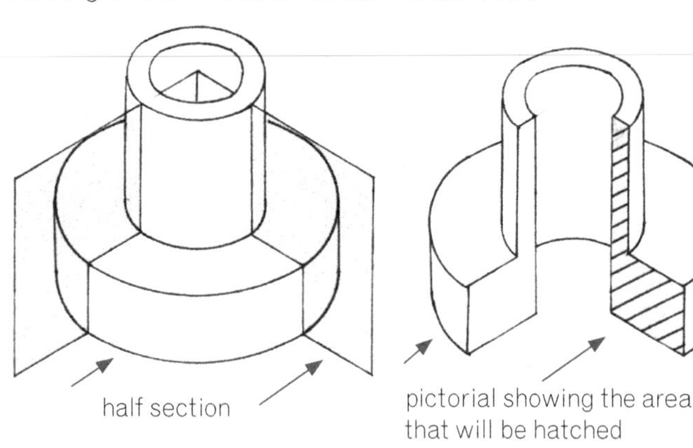

Aligned section

If important information is not located along one plane, the section line can be bent round to the same line to show the detailing. In the example the points have been rotated around onto the same plane. You will notice that the section line runs through the part before the points have been rotated.

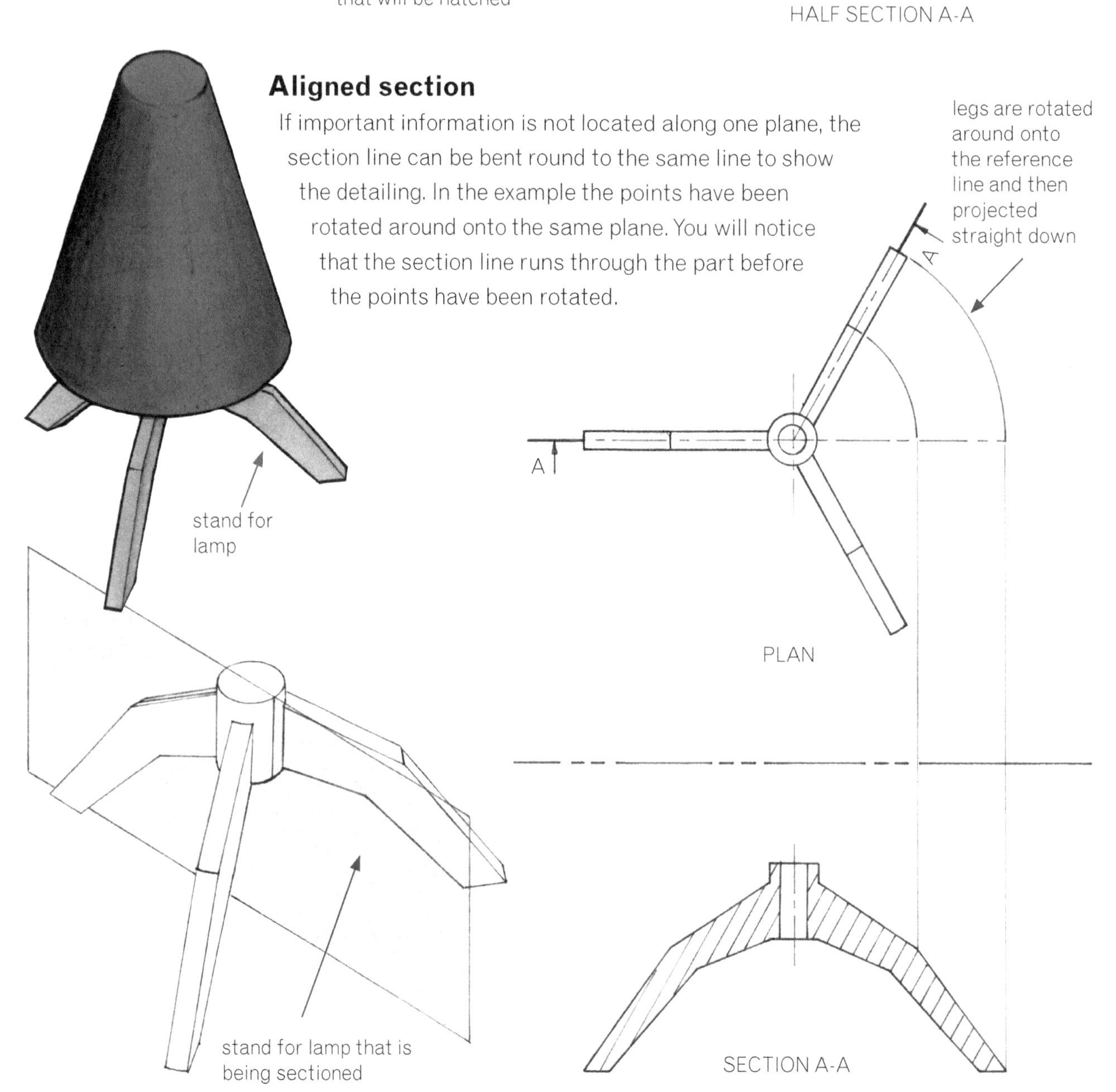

ISBN: 9780170233279

The same applies where the detail such as holes are located on the same radius from the centre.

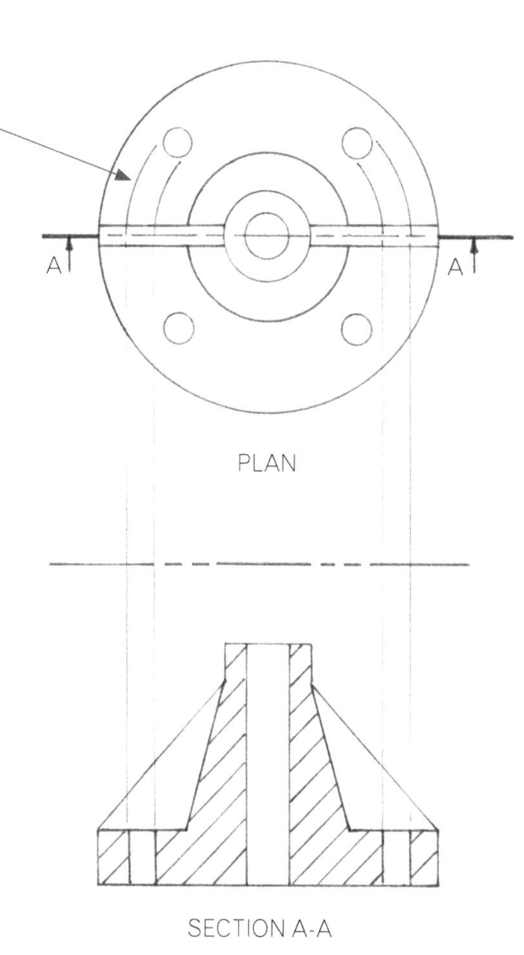

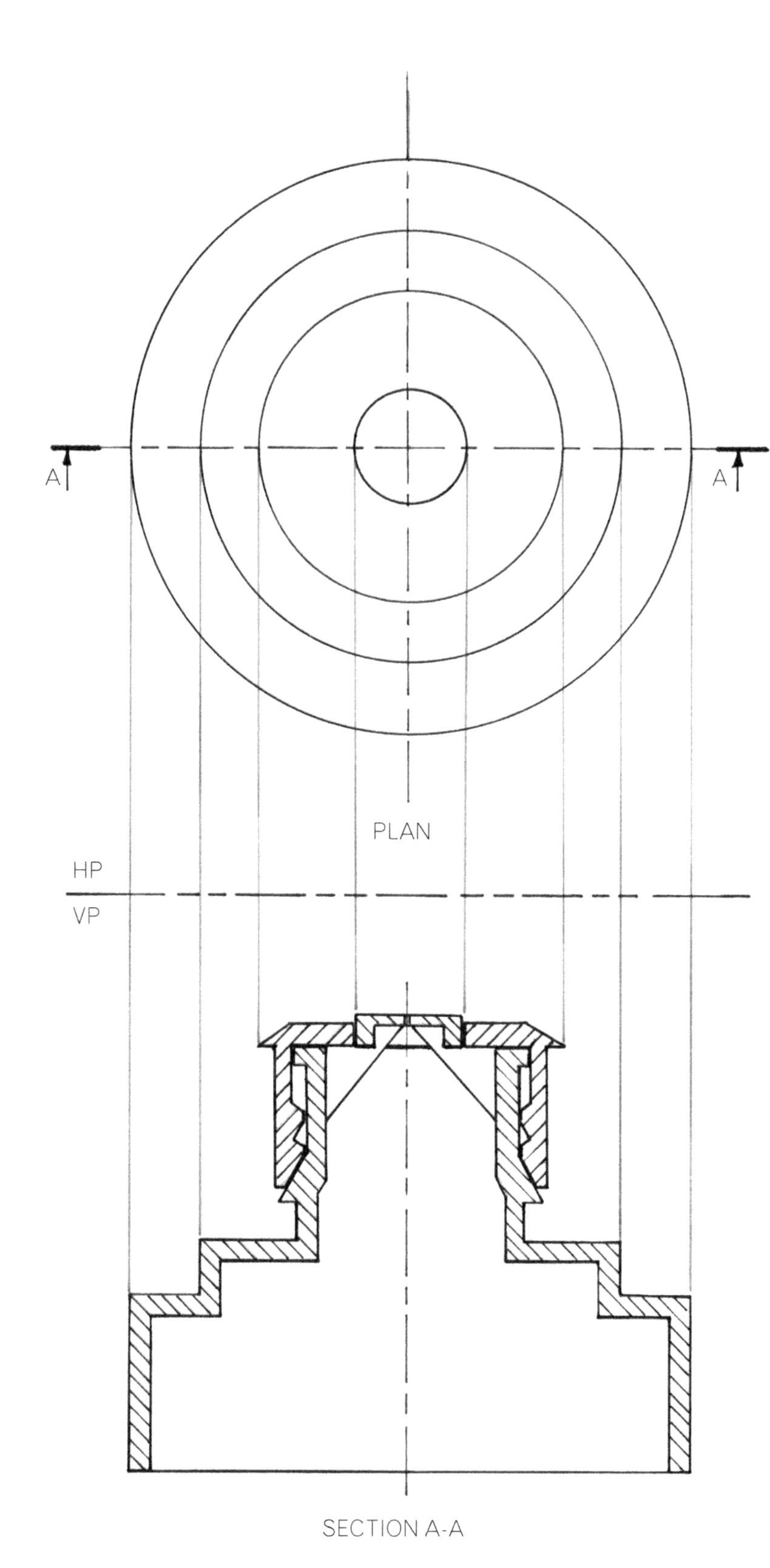

The drink bottle top shows a sectioned view with two different parts. You will notice that these are sectioned in opposite directions. The projection lines from one view to another have been shown in green.

ISBN: 9780170233279

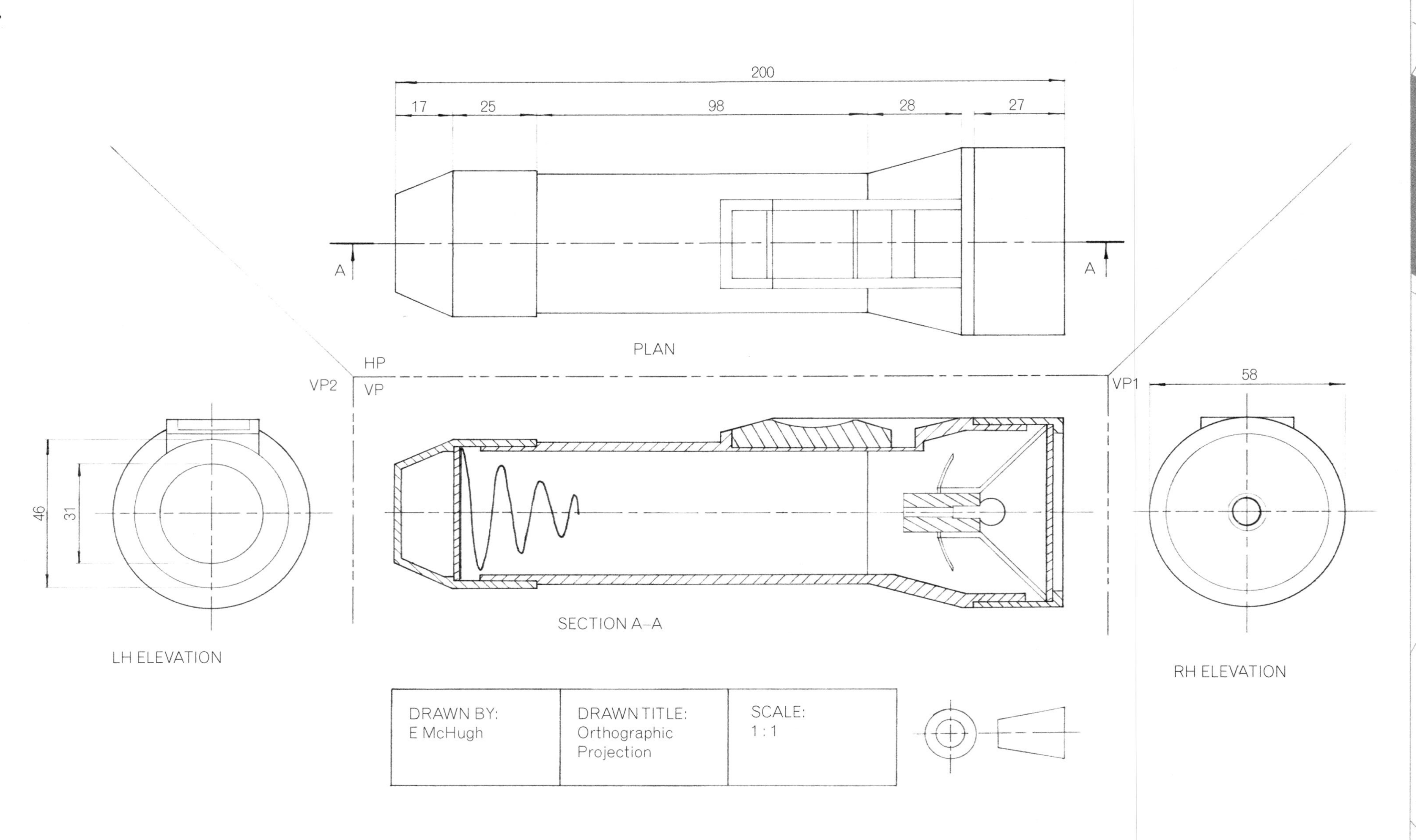

ISBN: 9780170233279

Paihere Tims — St Peter's College

The following work produced by Paihere Tims demonstrates a third angle orthographic projection with a sectional view. The section view shows how the components are joined together.

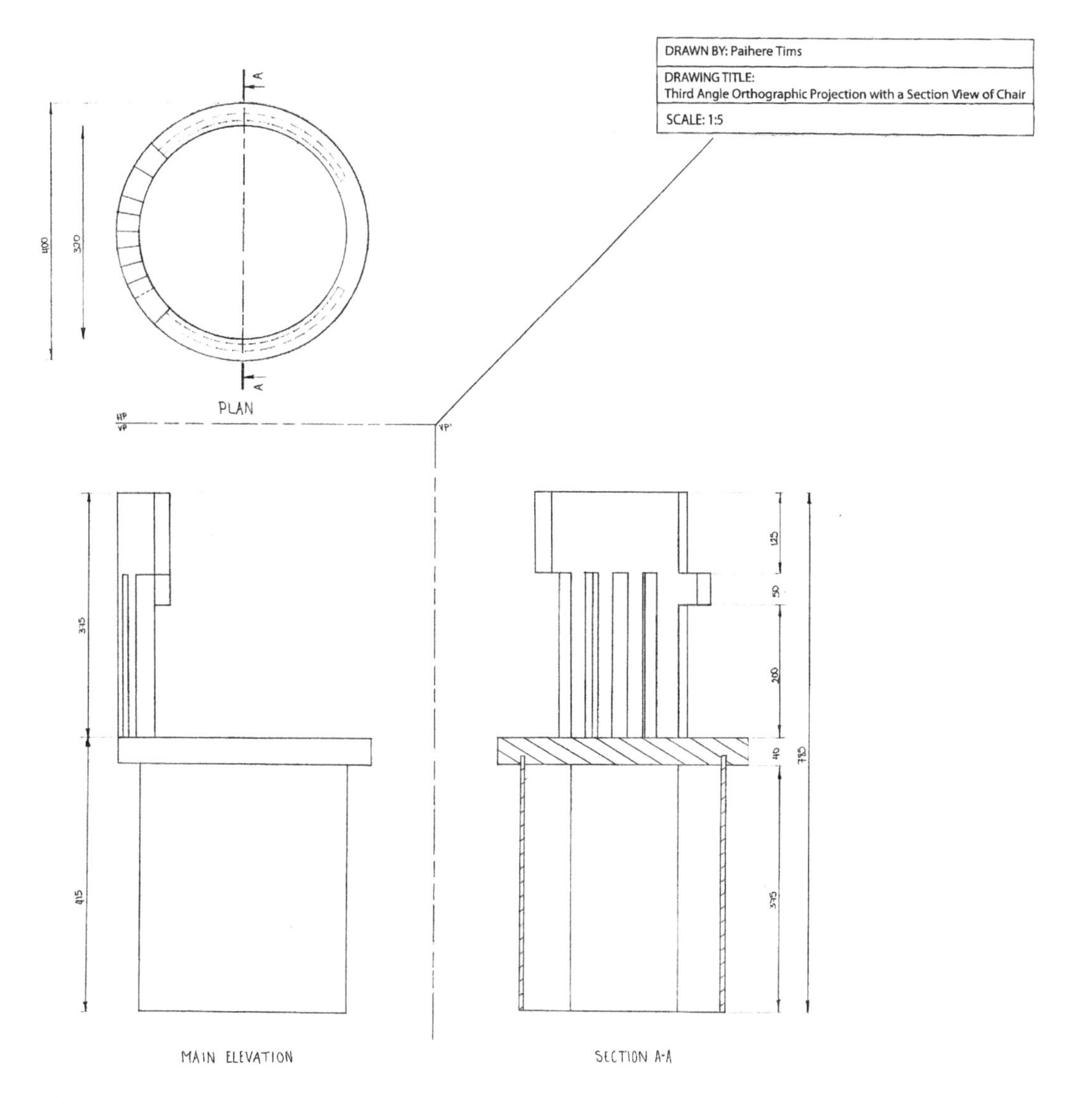

An auxiliary view shows another view of an object, for example an auxiliary view can be used to gain a true face. In the pictorial view of the block the auxiliary view is shown in grey.

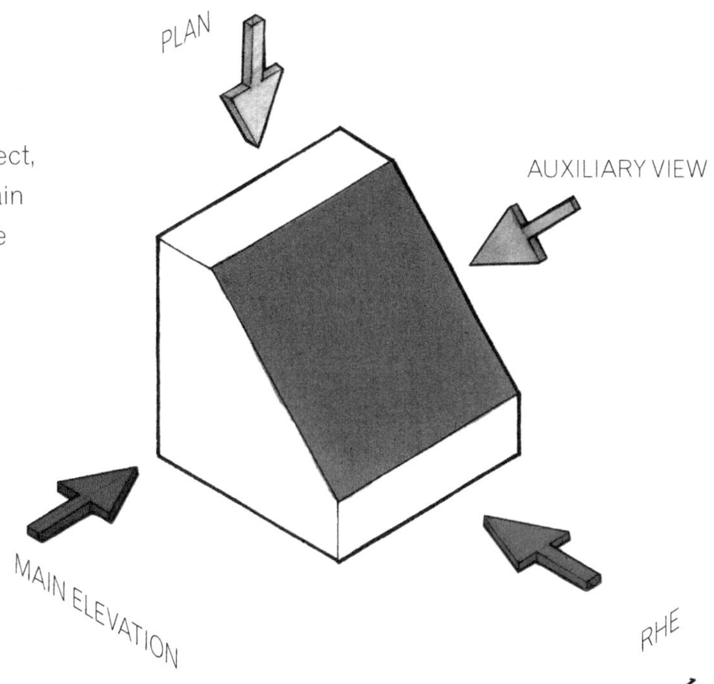

ISBN: 9780170233279

Step 1 • Draw a reference line that is parallel to the slanting surface. It does not matter what distance the reference line is from the slanting surface. To transfer the angle of the slanting surface use two set squares. Line one of the set squares up with the slanted line and position the other set square directly underneath the first set square in this position. Then move the first set square along the set square you are holding in place. Once you have moved the first set square the distance needed, draw a reference line along the edge.

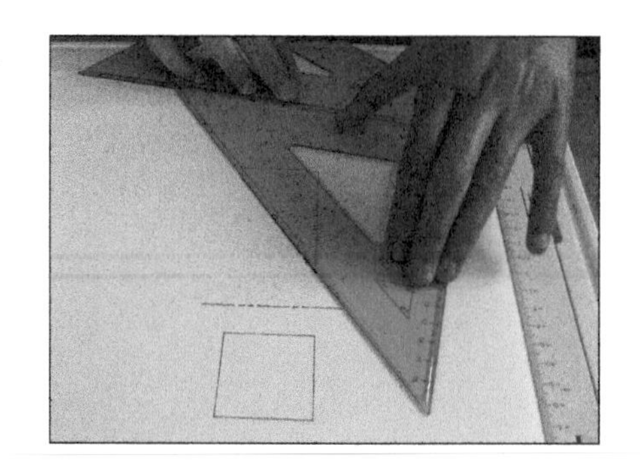

Step 2 • Draw lines at 90° to the slanting surface. These have been shown in green. To draw lines accurately at 90° to the sloping surface use the two set squares again. Line the first up with the slanting surface and move it back from the line. Hold this set square in place. Use the second set square to draw lines 90° away from the slanting surface by positioning it against the first set square.

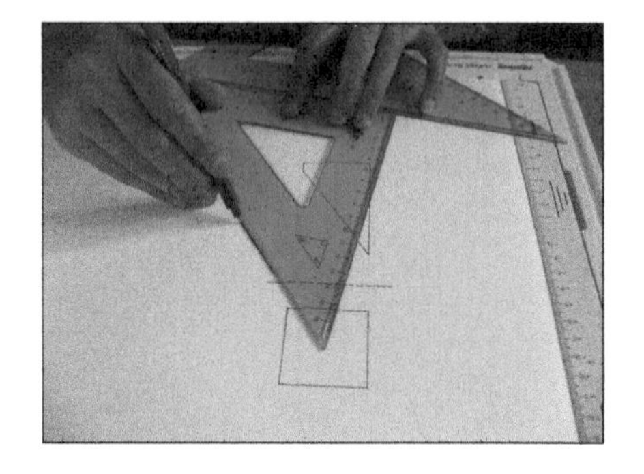

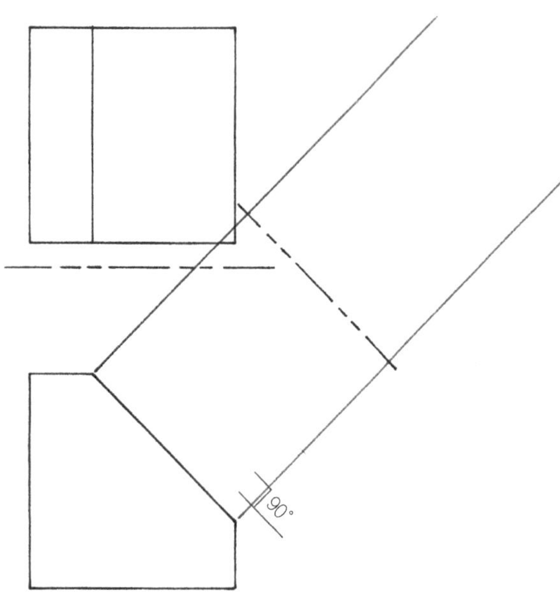

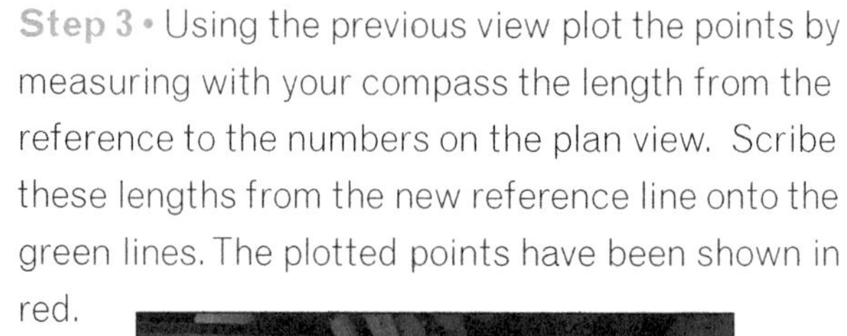

Step 3 • Using the previous view plot the points by measuring with your compass the length from the reference to the numbers on the plan view. Scribe these lengths from the new reference line onto the green lines. The plotted points have been shown in red.

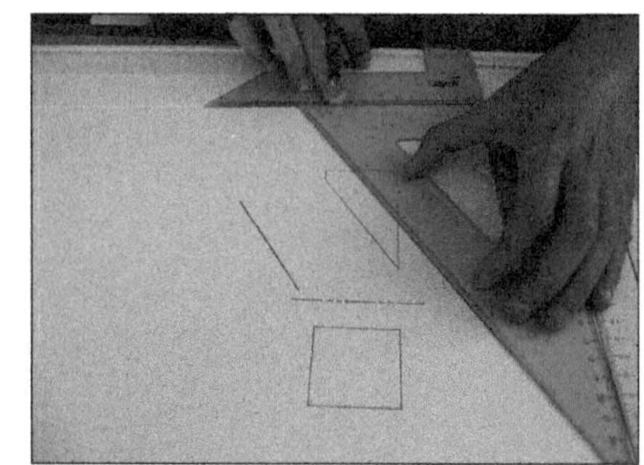

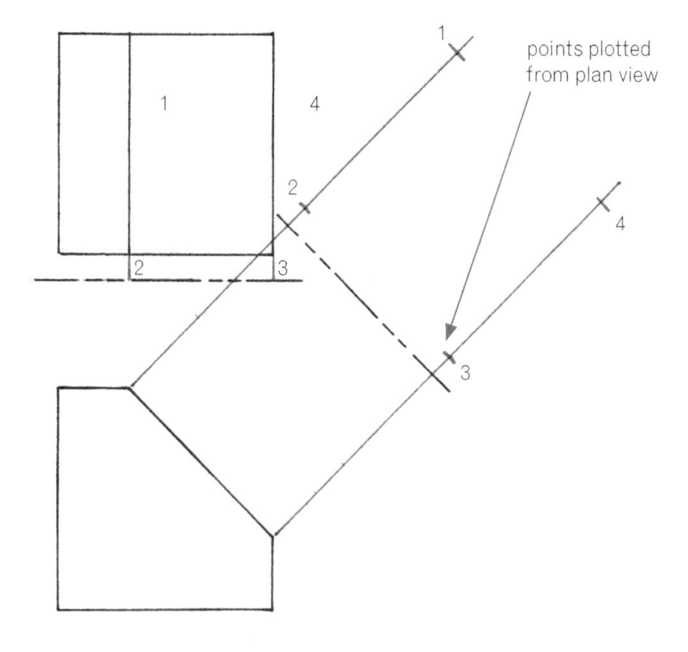

Step 4 • Outline the auxiliary view.

ISBN: 9780170233279

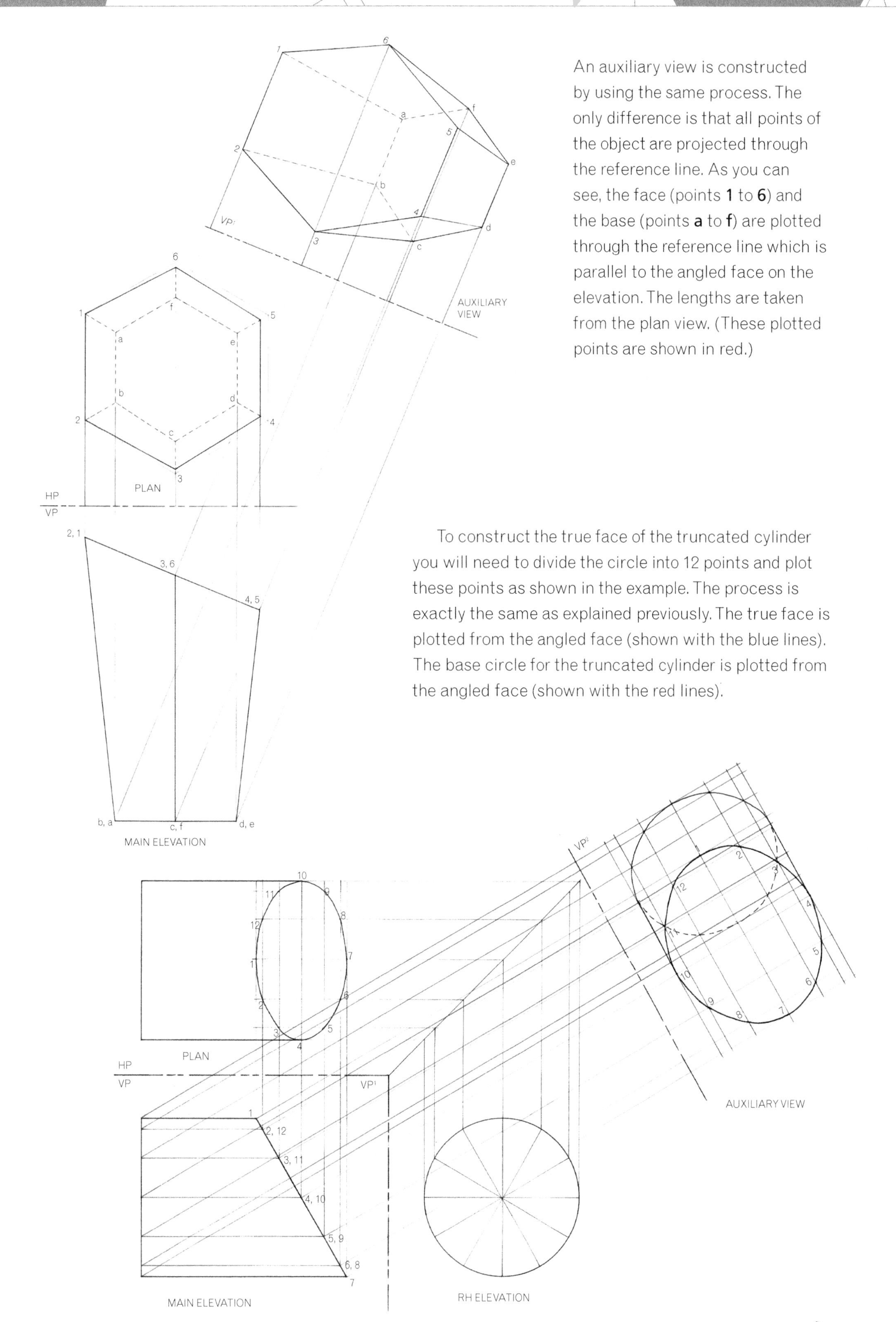

An auxiliary view is constructed by using the same process. The only difference is that all points of the object are projected through the reference line. As you can see, the face (points **1** to **6**) and the base (points **a** to **f**) are plotted through the reference line which is parallel to the angled face on the elevation. The lengths are taken from the plan view. (These plotted points are shown in red.)

To construct the true face of the truncated cylinder you will need to divide the circle into 12 points and plot these points as shown in the example. The process is exactly the same as explained previously. The true face is plotted from the angled face (shown with the blue lines). The base circle for the truncated cylinder is plotted from the angled face (shown with the red lines).

ISBN: 9780170233279

Paihere Tims — St Peter's College

The following work produced by Paihere Tims demonstrates a third angle orthographic projection with an auxiliary view. The auxiliary view gives a view of what the chair looks like in three dimensions.

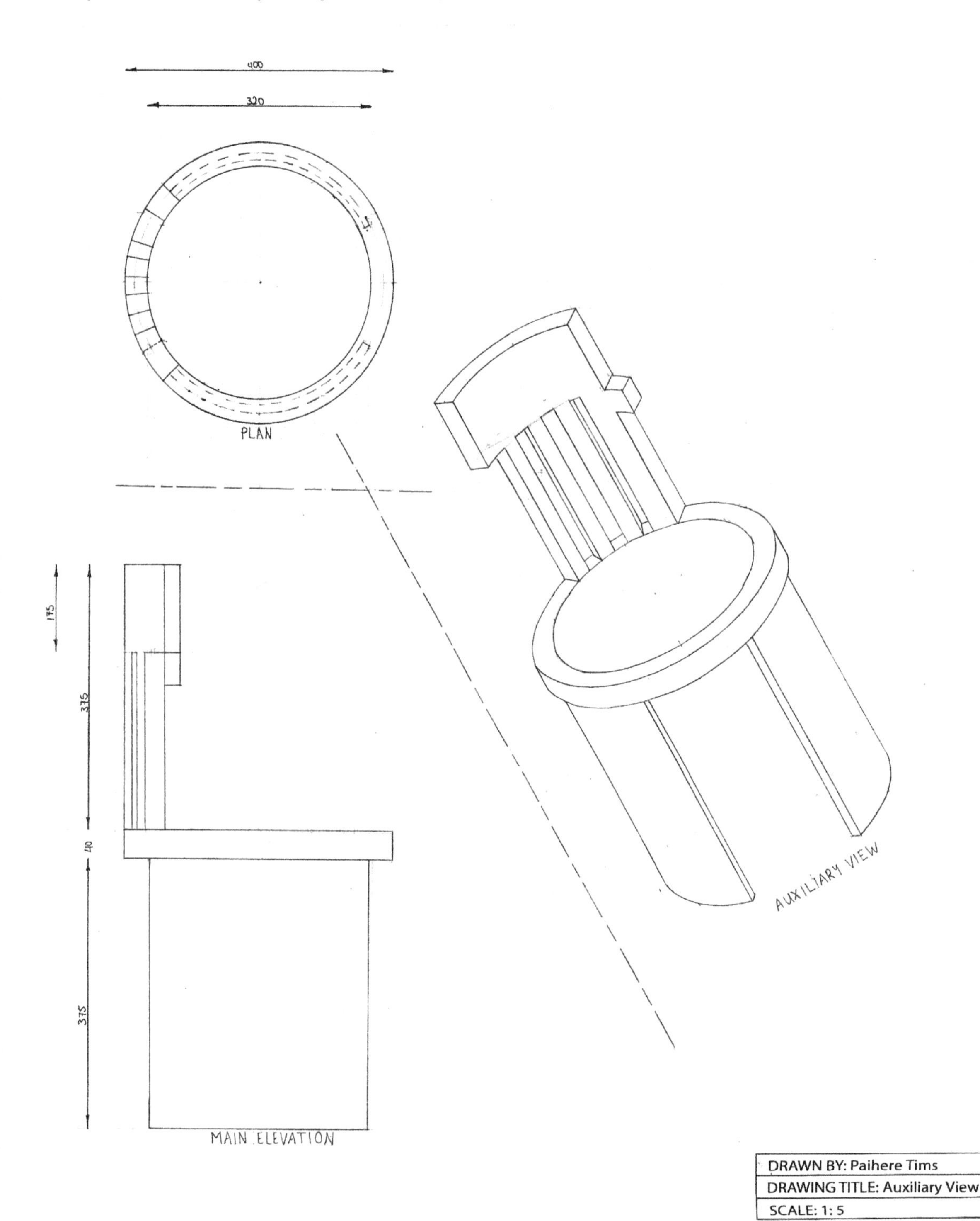

DRAWN BY: Paihere Tims
DRAWING TITLE: Auxiliary View
SCALE: 1: 5

ISBN: 9780170233279

A surface development is a solid (a three dimensional prism or pyramid) that has been drawn so that it is folded out flat on the paper. There are two types of surface developments: parallel line and radial line.

Parallel line surface developments

Parallel line developments have parallel sides. The same process is followed for the construction of a hexagonal prism, cylinder or any other prism.

Step 1 • Construct the plan and elevation for the prism.

Step 2 • Project the top and bottom of the main elevation horizontally.

Step 3 • With your compass take one of the side measurements of the object from the plan. For the cylinder you will first need to divide the plan into 12 sections with your 30°/60° set square.

Step 4 • Plot this length along the lines projected from the elevation.

Step 5 • Draw the lines vertically from the plotted points.

Step 6 • Construct the top and bottom surface on the surface development.

Step 7 • Outline the solid.

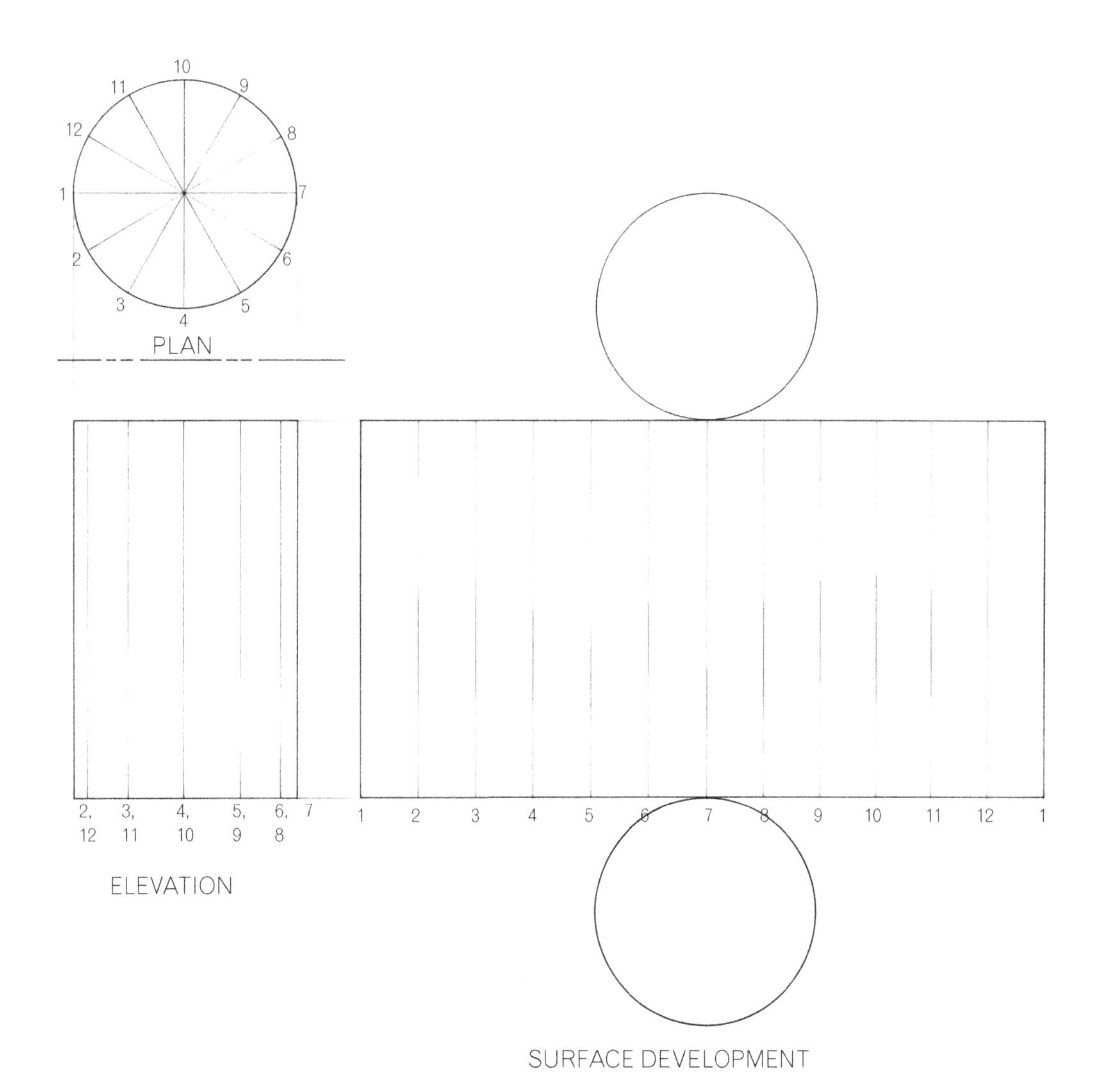

ISBN: 9780170233279

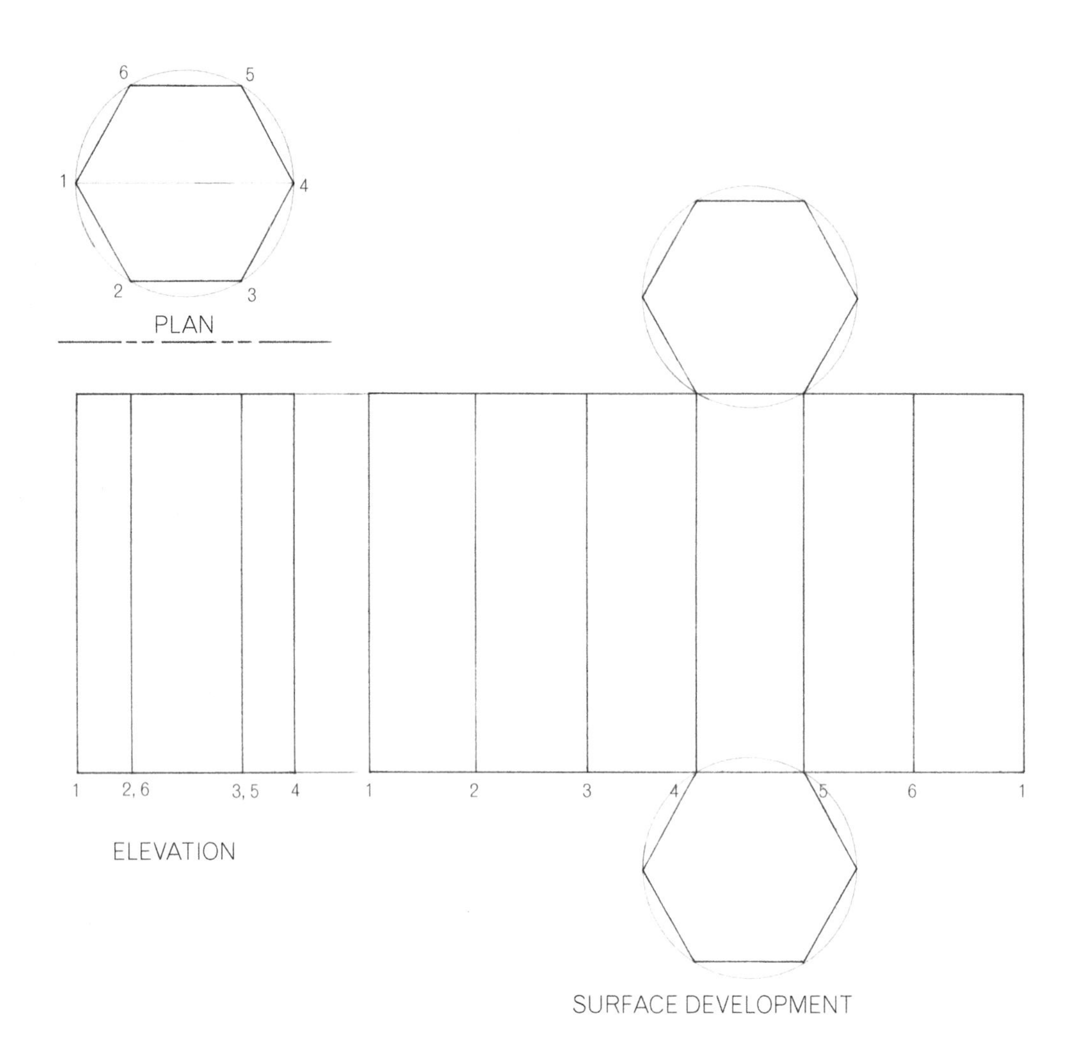

Radial line surface developments

Radial line developments have sides that project to a point. The same process is followed for the construction of a cone or any other pyramid.

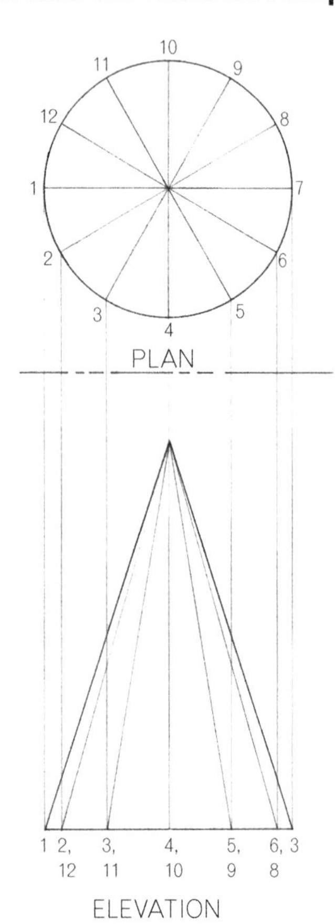

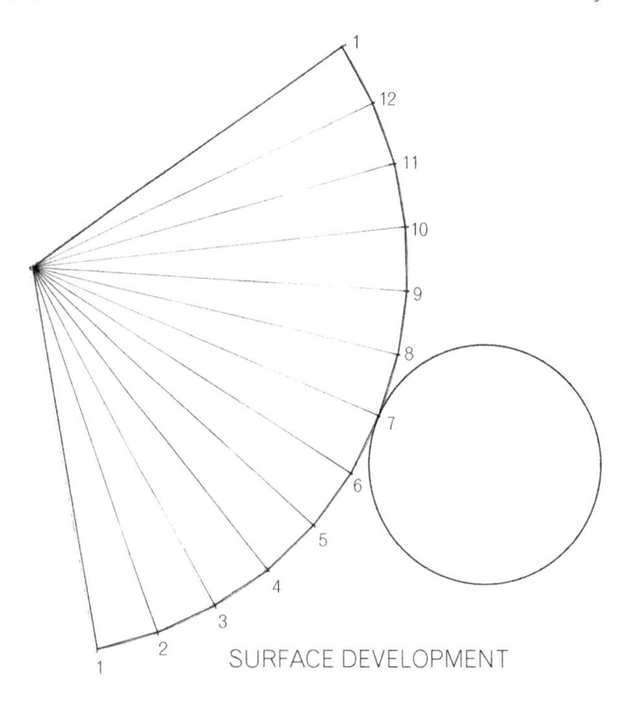

ISBN: 9780170233279

Step 1 • Construct the plan and elevation for the pyramid.

Step 2 • Find the true length line by projecting it as shown in the examples. Join the point where this line crosses the base line to the apex of the elevation.

Step 3 • Arc a line the length of the true length line you have found.

Step 4 • With your compass take the side measurements of the object from the plan.

Step 5 • Plot these side measurements around the arc.

Step 6 • Join the points along the radial line to the centre point.

Step 7 • Construct the bottom of the shape on any side of the base.

Step 8 • Outline the solid.

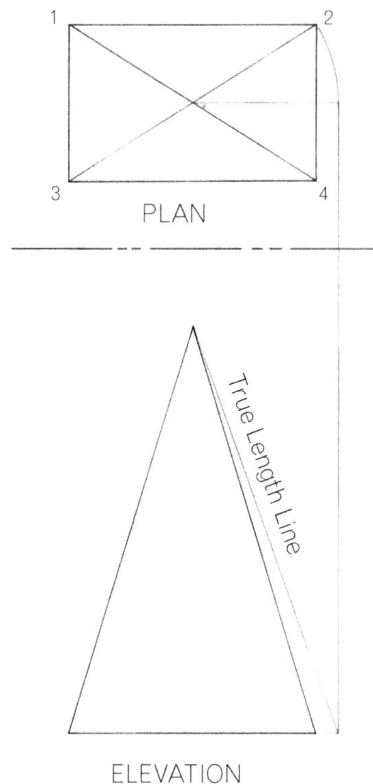

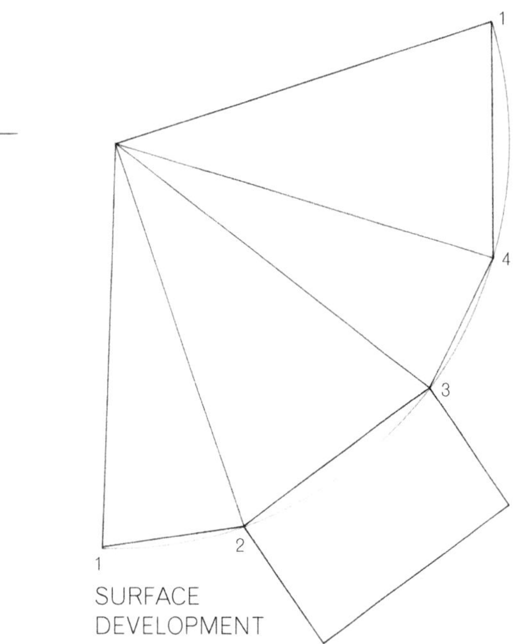

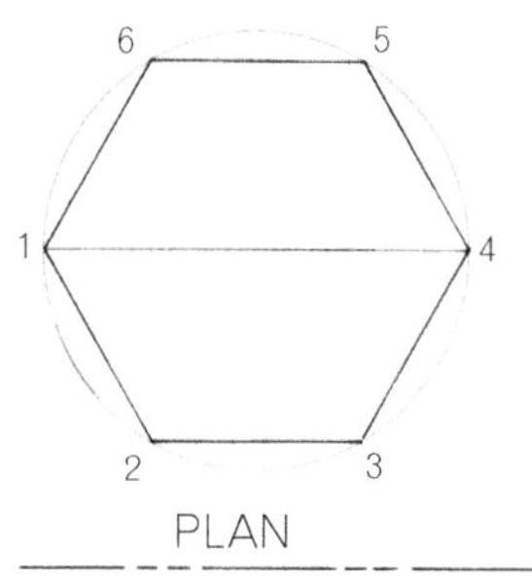

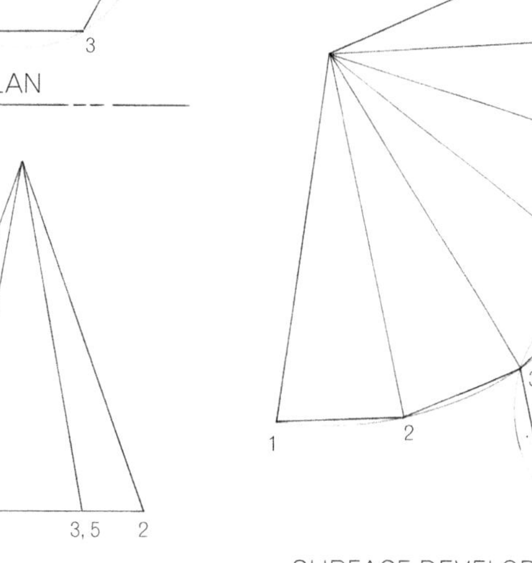

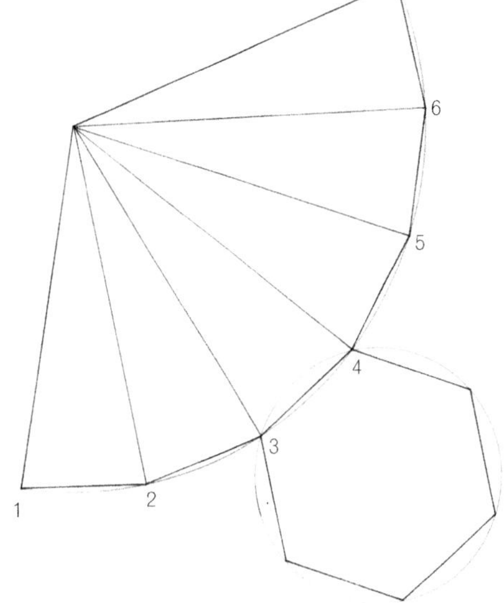

ISBN: 9780170233279

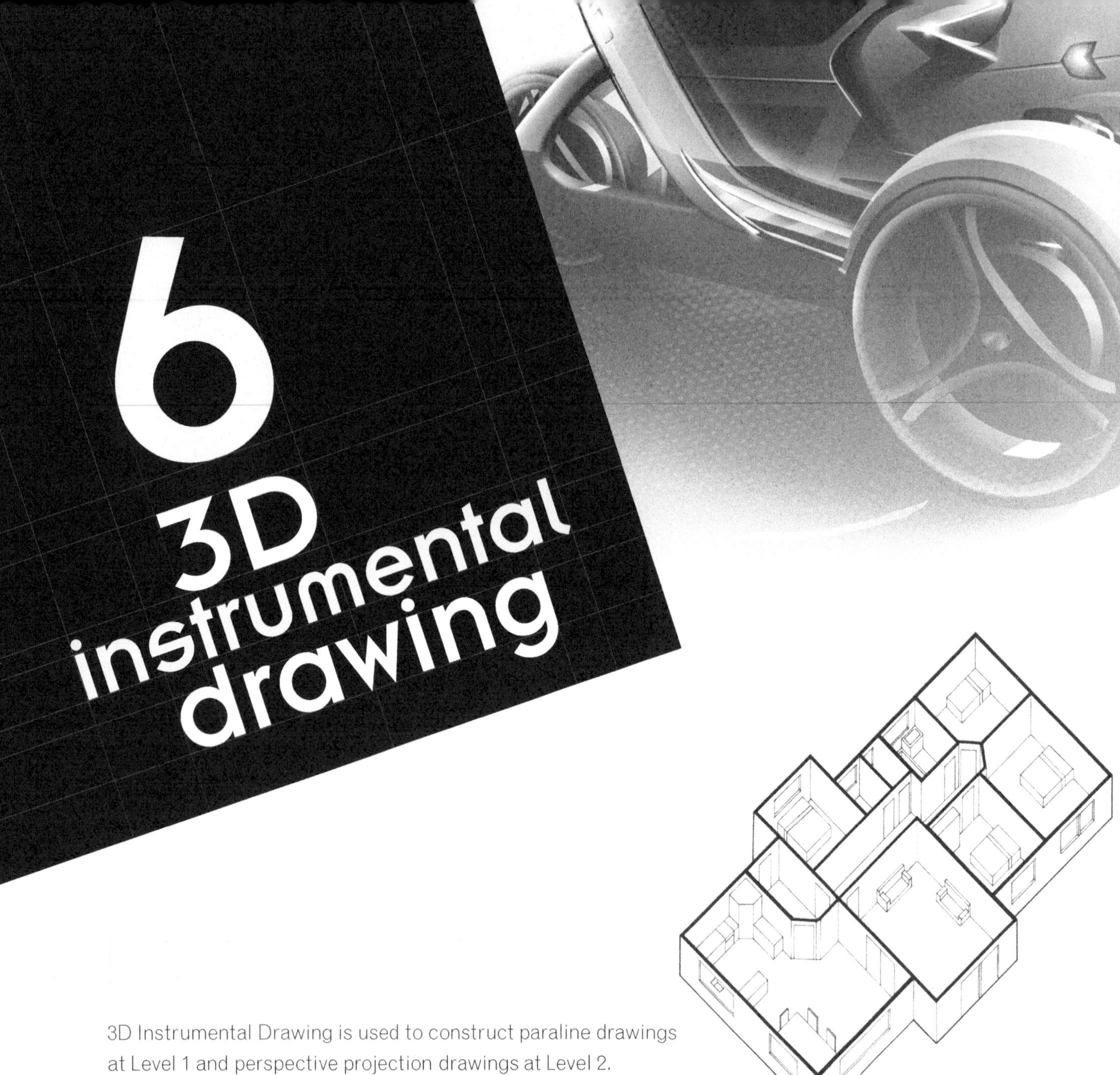

6 3D instrumental drawing

3D Instrumental Drawing is used to construct paraline drawings at Level 1 and perspective projection drawings at Level 2. Paraline drawings and perspective projection drawings are used to communicate design ideas by describing the design features.

Paraline drawings (Level 1) are 3D drawings produced using paraline techniques such as isometric, oblique, planometric, trimetric and diametric. To effectively communicate your design ideas you will need to produce accurate and precise paraline drawings that show indepth information about technical features of a design. In-depth drawings could include exploded or sectional views that explain the design features.

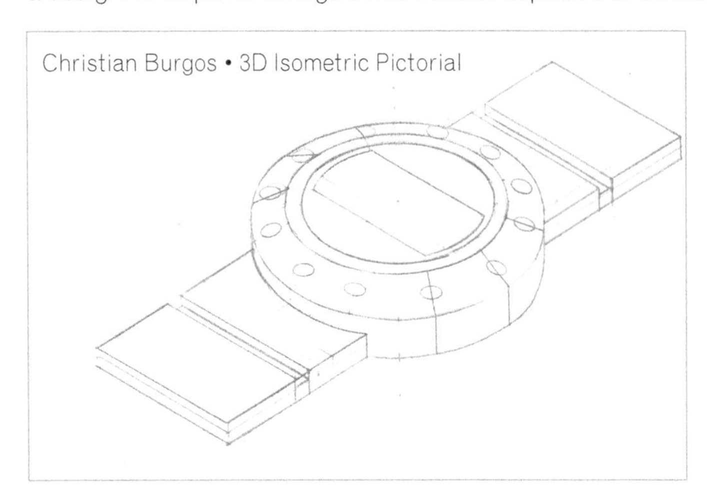

Christian Burgos • 3D Isometric Pictorial

Perspective projection drawings (Level 2) are pictorial drawing methods that involve parallel and/or angular perspective projection. The conventions of these drawings include the use of a picture plane, station point, eye level line, ground line, vanishing point and a height line. You will need to accurately use perspective drawing techniques to show the detail of the design features. To effectively communicate design ideas you will need to select a view point that enables the detail of the design features to be shown.

ISBN: 9780170233279

To construct any objects you need to be able to construct a box and circle in a pictorial view. The same will apply for any type of 3D instrumental drawing method. A box is used as a crate and can be used to construct other geometric solids. The construction of a circle using a pictorial method means that cylinders and cones can be constructed.

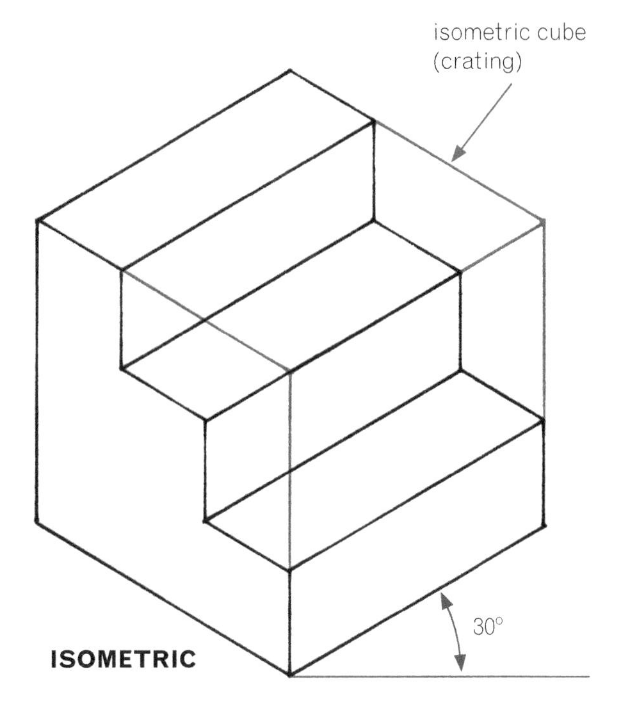

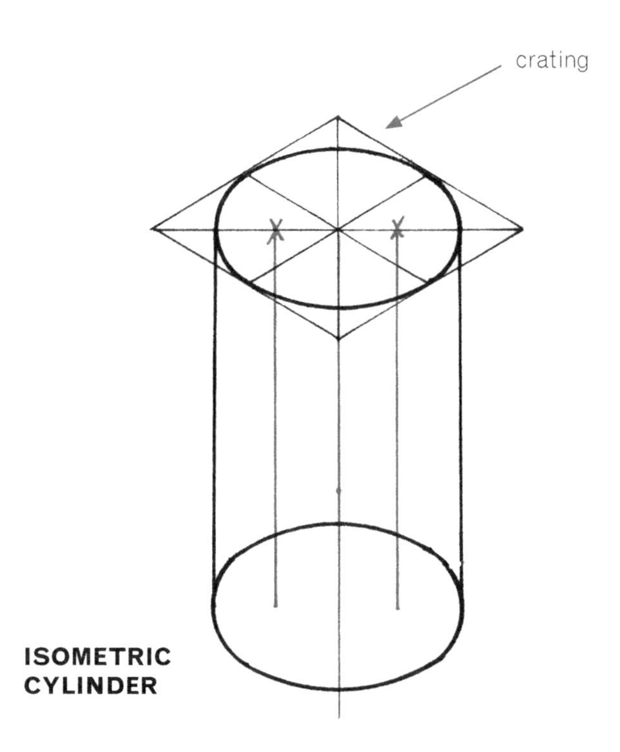

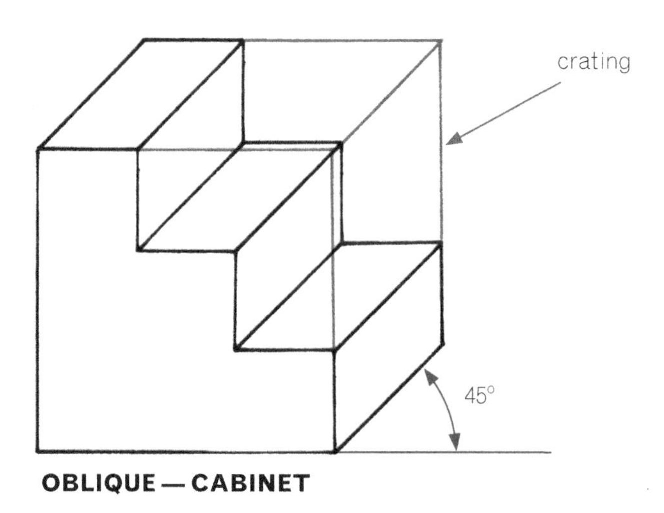

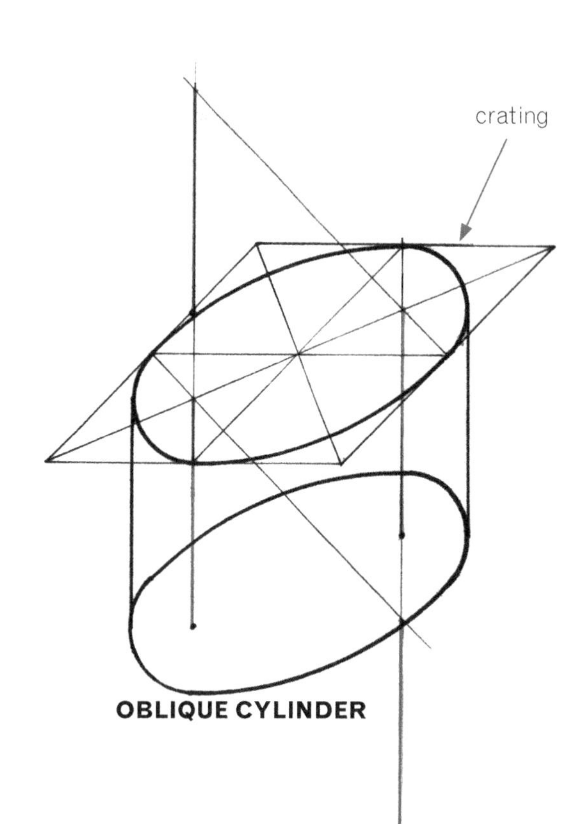

ISBN: 9780170233279

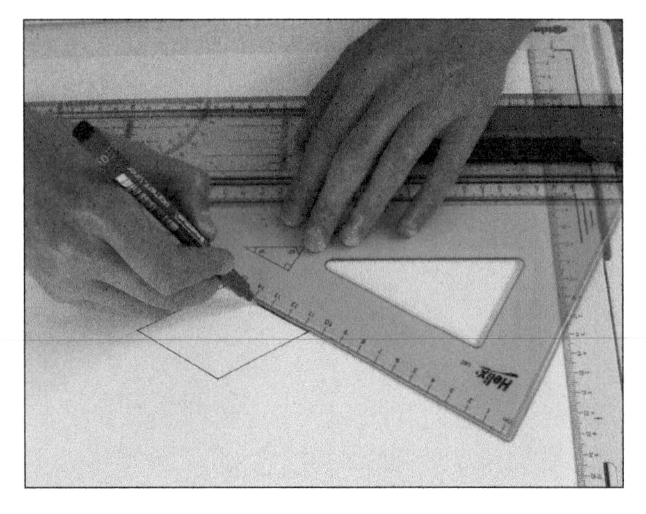

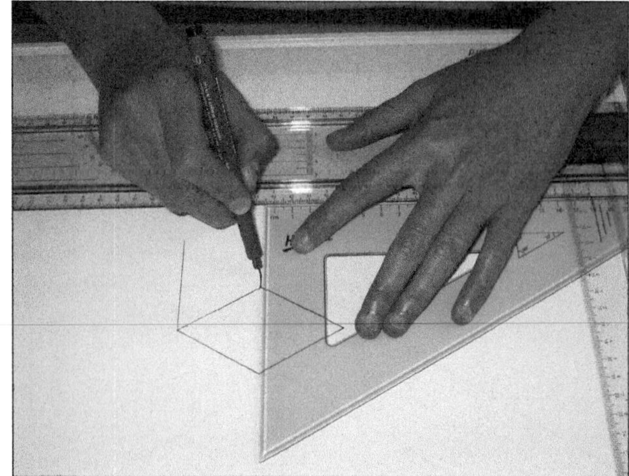

Isometric circle

Step 1 • Draw the outside edges of the square for the circle. The square should be drawn the diameter of the circle. Join the points of each corner together to find the centre, this has been shown with the green lines.

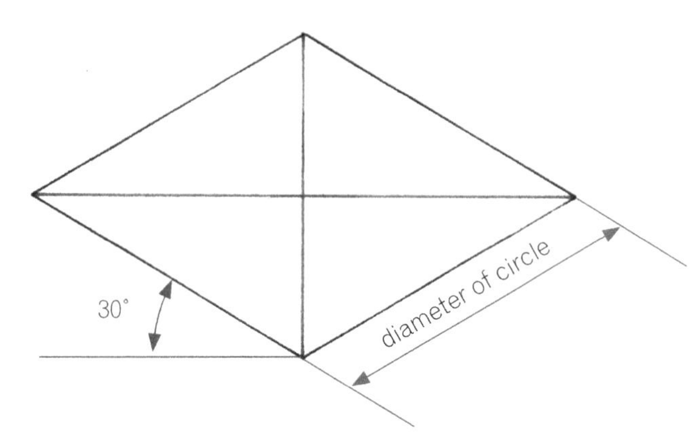

Step 2 • Draw 30° angles through the centre, this has been shown with the blue lines.

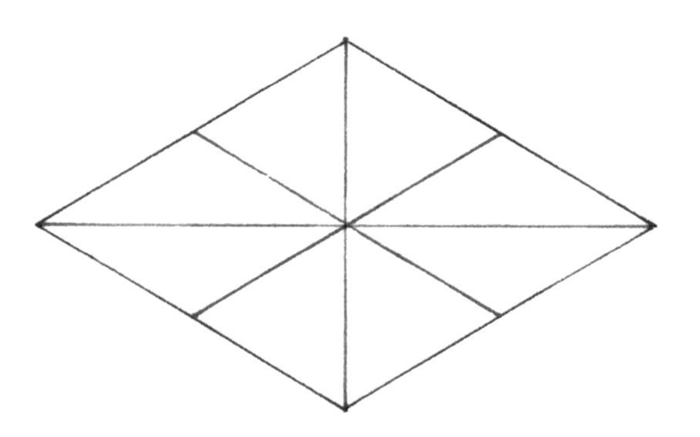

Step 3 • From corner **A** take your compass and scribe across to the other side of the isometric square and scribe an arc. The arc and the point to scribe from have been shown in black.

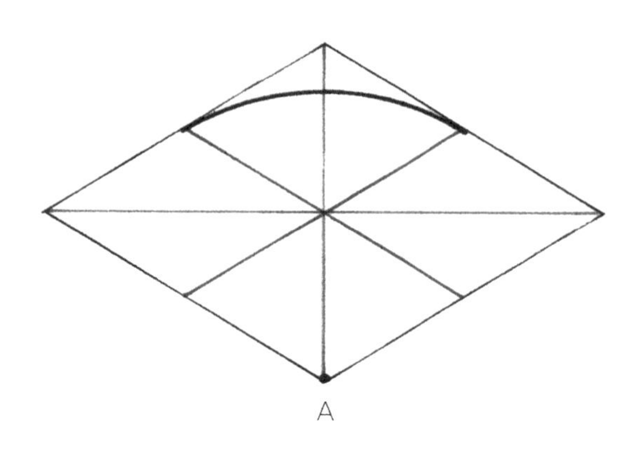

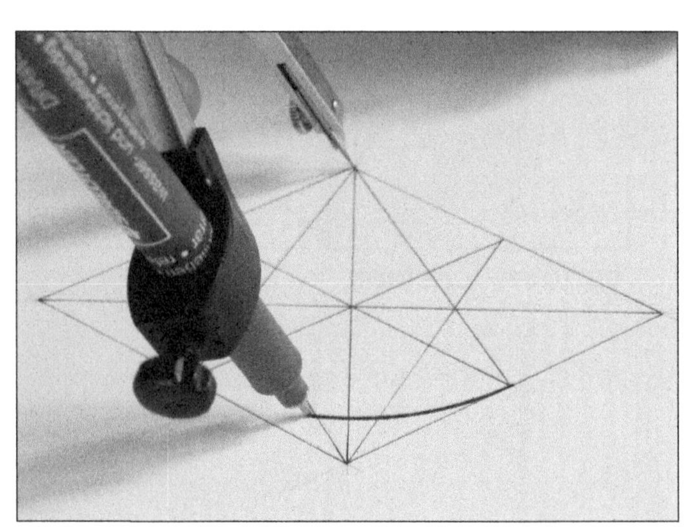

ISBN: 9780170233279

Step 4 • Repeat the process from the other side of the isometric square.

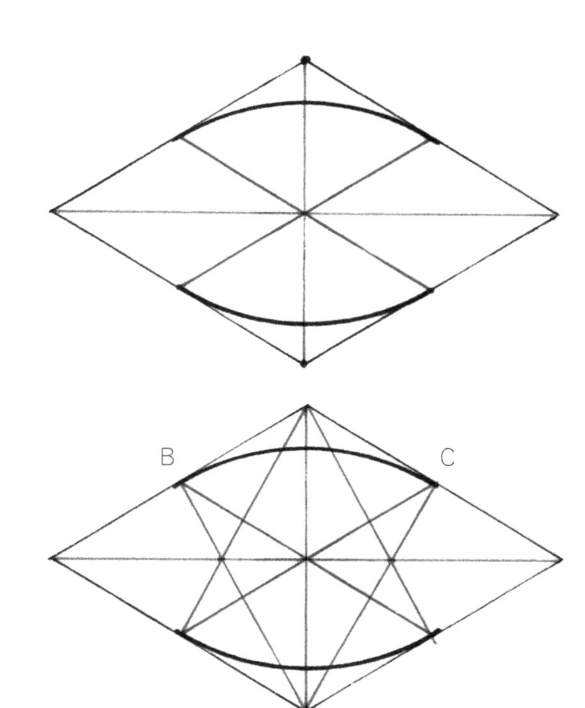

Step 5 • To get the points for the end curve on the isometric circle, draw construction lines from point **A** to **B** and from **A** to **C**. Repeat the same on the opposite side of the isometric square. These construction lines have been shown with red lines.

Step 6 • The points where the red lines cross are used for the end arcs. The end arc is scribed from the closest point. A mistake a lot of students make is to use the point furthest from the end arc. Outline the circle.

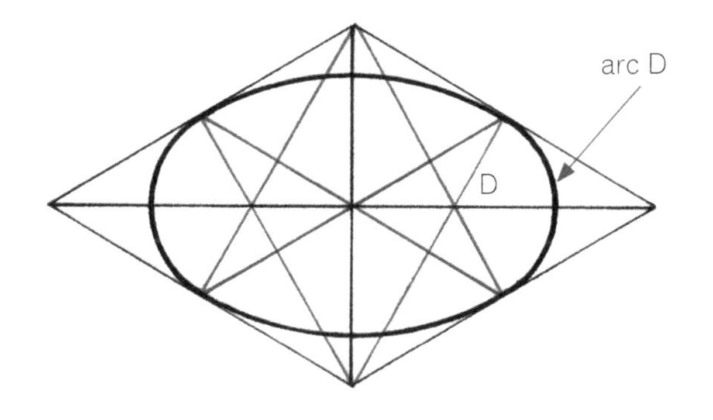

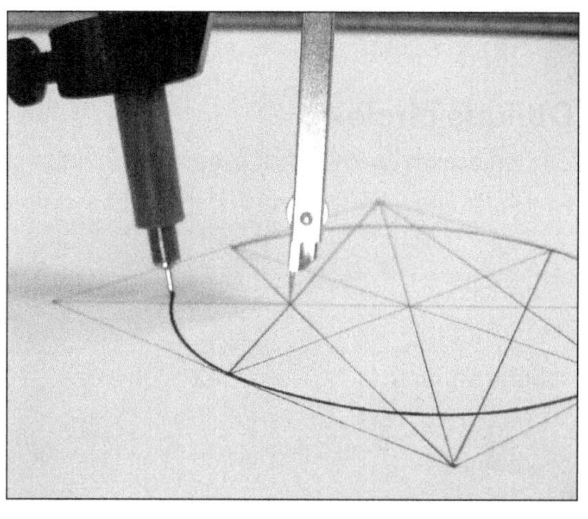

Once you are able to construct an isometric circle you will then be able to construct a cylinder by projecting the points vertically the length the cylinder is intended to be. The projection has been shown with red projection lines. Draw the second isometric circle using the lengths used for the initial circle. To complete the cylinder join the outside edges of both circles.

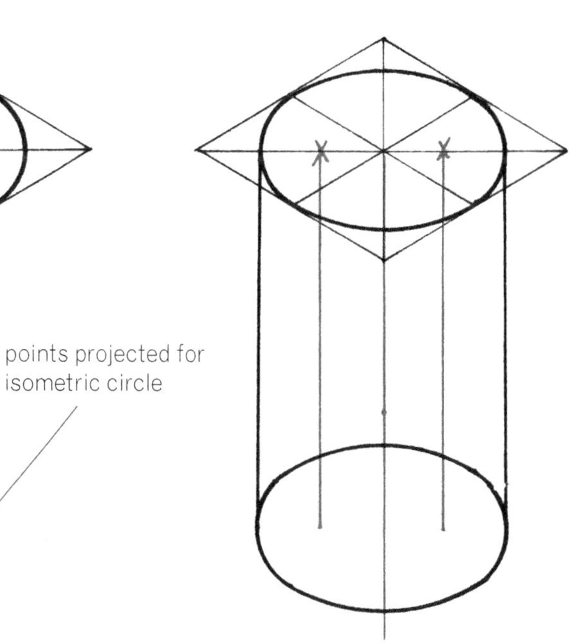

ISBN: 9780170233279

Oblique crating

Oblique is drawn with the viewer able to directly look at the front face and the depth of the object is taken back on a 45° angle. There are two types of oblique drawing: cavalier and cabinet. Cavalier uses the true length of the object for the depth. Cabinet uses half the true length for the depth of an object.

The crating technique is also used when constructing an object in oblique.

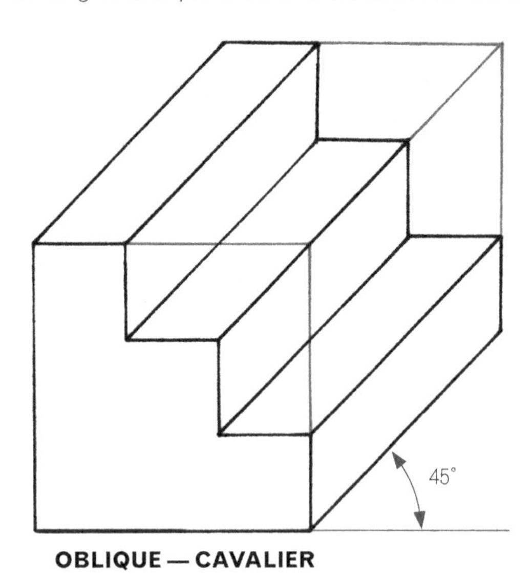

OBLIQUE — CAVALIER

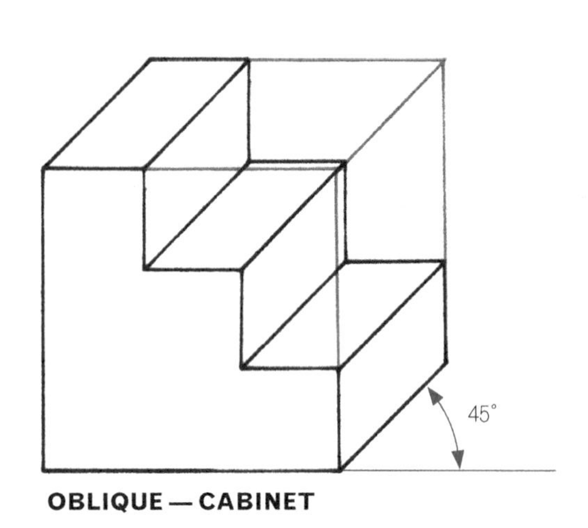

OBLIQUE — CABINET

Oblique circles

Circles drawn on the front face of an oblique view are drawn as a true circle. Circles on the other two surfaces which represent the depth of an object have to be constructed.

Step 1 • Construct the oblique square for the circle. The square needs to be drawn the diameter of the circle. Join the corners of the square together.

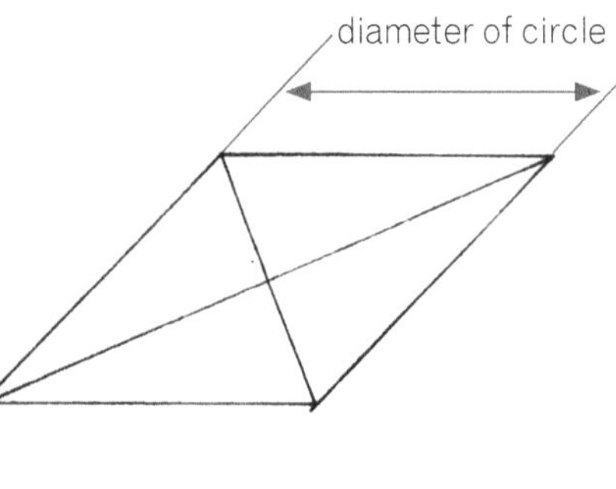

Step 2 • Draw horizontally and on a 45° line through the centre of the square. This has been shown with the green lines.

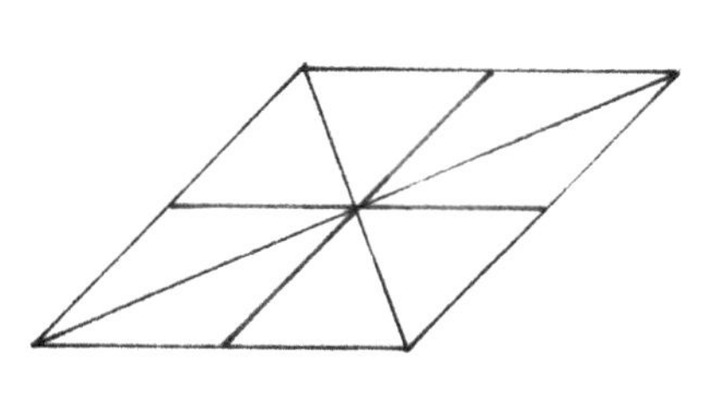

Step 3 • From point **A** draw a line vertically down and from point **C** draw a line vertically up. This has been shown with the red lines.

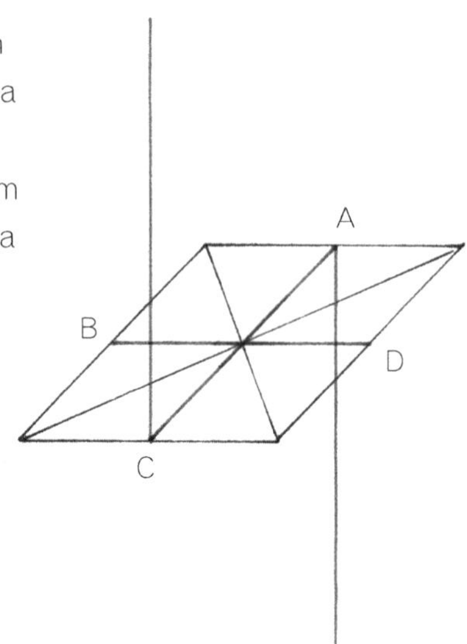

Step 4 • Draw a 45° line down from point **B** and up from point **D**. This has also been shown with the red construction lines. The four points where the red lines cross are to be used for the arcs of the oblique circle. The construction up to this point needs to be accurate otherwise when you scribe the arcs for the oblique circle they will not meet.

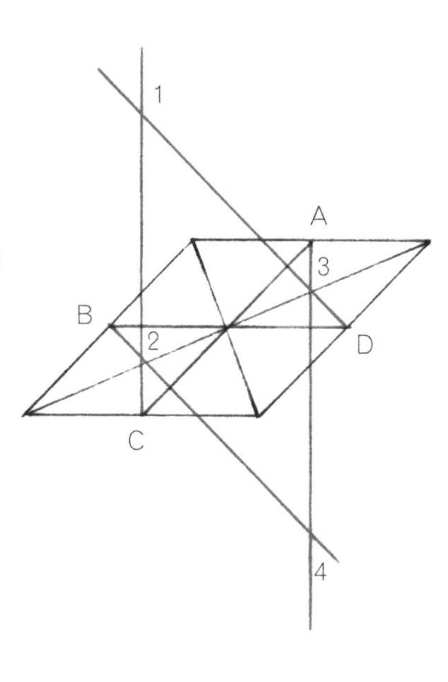

ISBN: 9780170233279

Step 5 • Position the length on the compass from point **4** to **A** and scribe an arc from **A** to **B**. Do the same on the opposite side. Position the compass on point **1** and scribe from **C** to **D**.

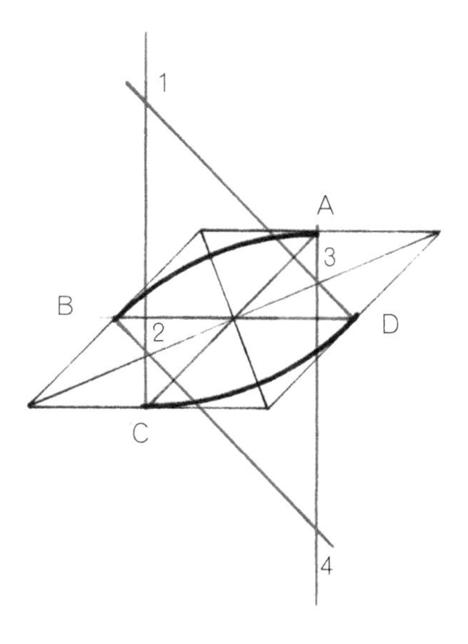

Step 6 • Position the length of the compass from point **2** to **B** and scribe an arc around to point **C**. Do the same on the opposite side. Position the compass on point **3** and scribe from **A** to **D**.

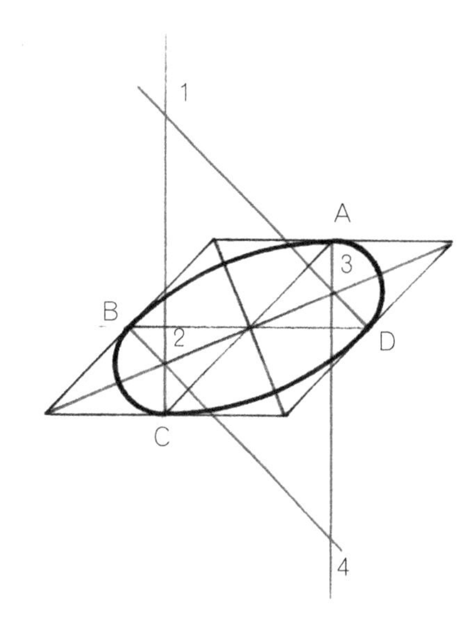

Once you are able to construct an oblique circle you will then be able to construct a cylinder by projecting the points vertically from the circle. All points need to be projected the length of the cylinder. This has been shown with red projection lines. To complete the cylinder draw the second oblique circle and join the outside edge of both the circles.

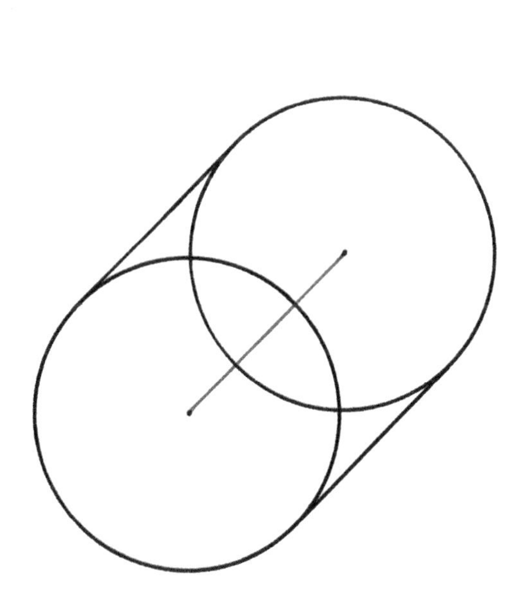

FRONT FACE OBLIQUE CYLINDER

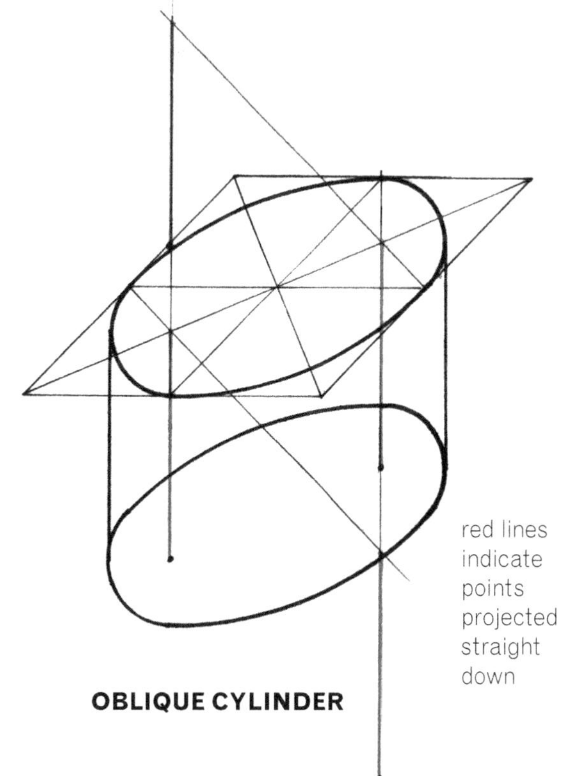

OBLIQUE CYLINDER

ISBN: 9780170233279

Geometric solids

Geometric solids are cylinders, cones, prisms and pyramids.

Isometric solids

Square-based prism

Step 1 • Draw a cube in isometric.

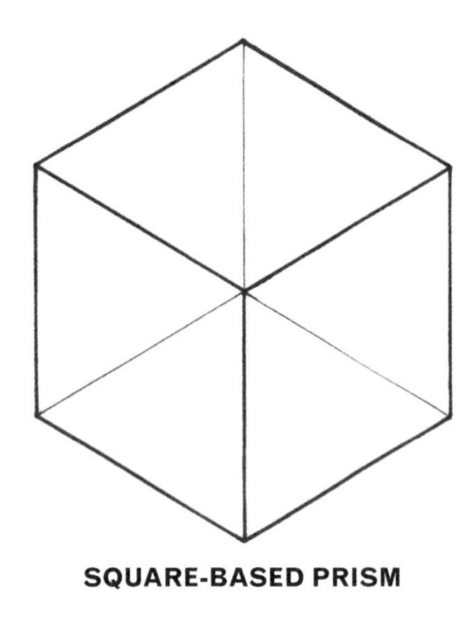

SQUARE-BASED PRISM

Square-based pyramid

Step 1 • Draw a square base.

Step 2 • Draw lines from corner to corner on the base to find the centre.

Step 3 • Draw the height of the pyramid vertically from the centre of the base.

Step 4 • Join the corners of the base to the top of the height line to outline the solid.

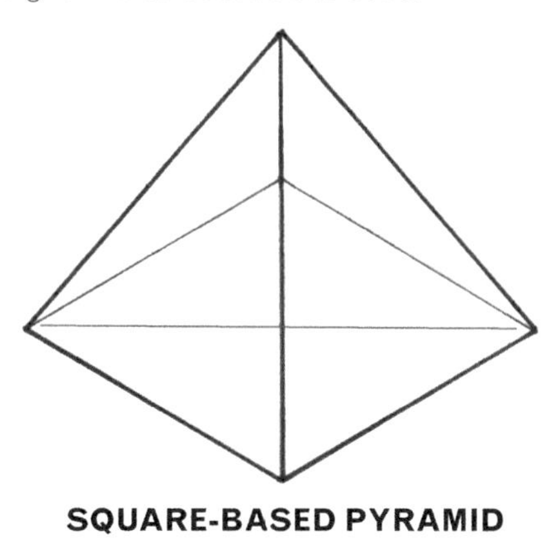

SQUARE-BASED PYRAMID

Cylinder

Step 1 • Draw an isometric circle on the base.

Step 2 • Draw the second isometric circle in line with the base at the distance required for the height.

Step 3 • Complete the cylinder by joining the edges of the cylinder.

Step 4 • Outline the solid.

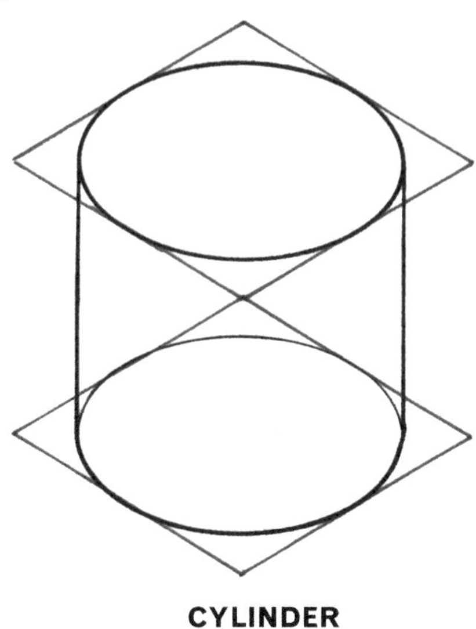

CYLINDER

Cone

Step 1 • Draw an isometric circle on the base.

Step 2 • Draw lines from corner to corner on the base to find the centre.

Step 3 • Draw the height of the cone vertically from the centre of the base.

Step 4 • Join the edge of the circle on the base to the top of the height line.

Step 5 • Outline the solid.

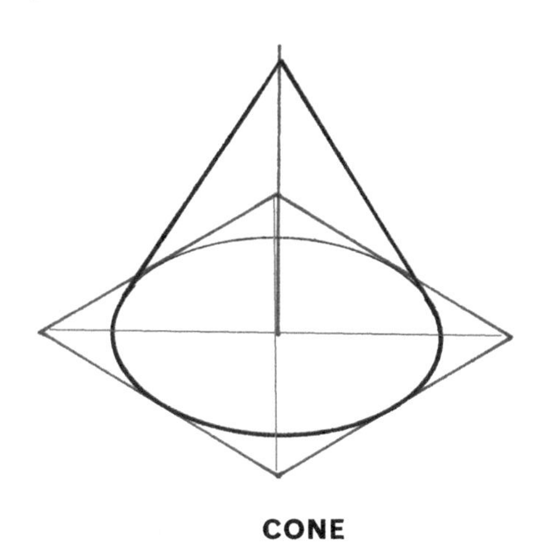

CONE

ISBN: 9780170233279

Hexagonal prism

To construct a hexagonal prism you need to firstly draw the plan view of the hexagon. When a square is drawn around the hexagon it is not the same length on both sides of the square so you firstly need to draw a plan view to find out what those lengths are.

Step 1 • To construct the hexagon in isometric, draw the base square. Take these measurements from the plan view. Measure length **x** and **y**.

Step 2 • To construct the shape of the hexagon on the base again take the measurements from the plan view. Measure length **a** in from the corners on both sides to find these points and plot onto the isometric box.

Step 3 • Measure length **b** in from the corner and plot the length onto the isometric box.

Step 4 • Join the points of the hexagon together on the base.

Step 5 • From each point of the hexagon draw each line vertically the same distance, the required height.

Step 6 • Join the points of the second hexagon which will be the height of each line.

Step 7 • Outline the solid.

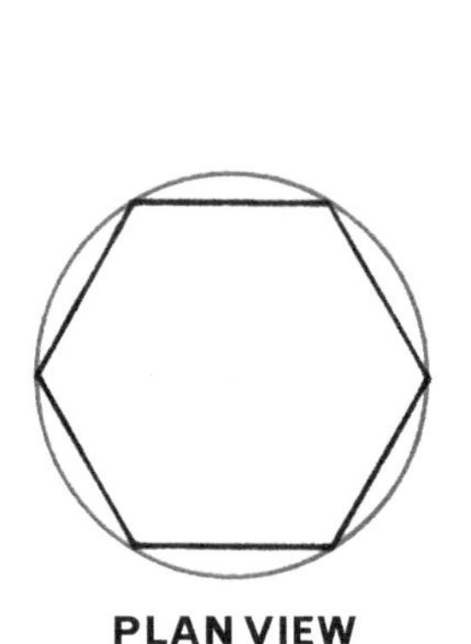

PLAN VIEW

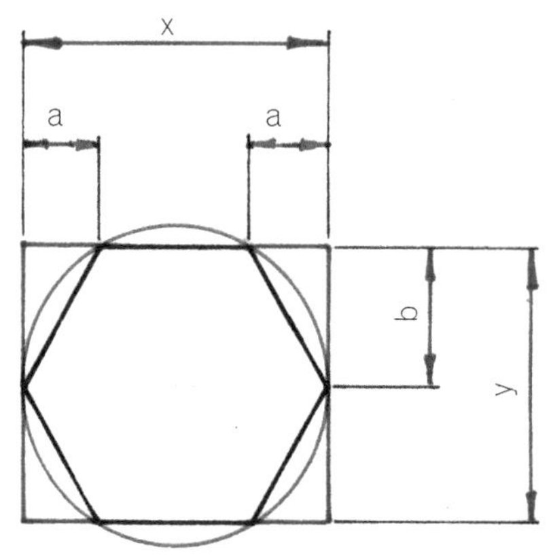

PLOTTING MEASUREMENTS

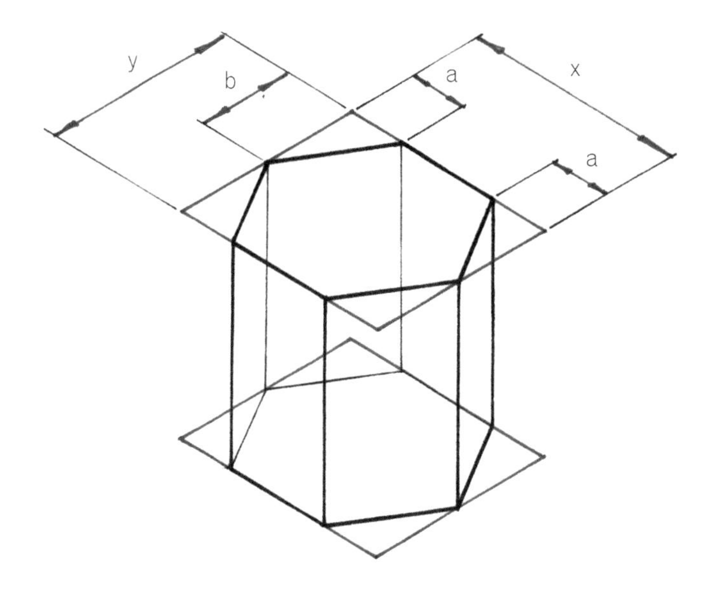

HEXAGONAL PRISM

Hexagonal pyramid

Step 1 • Draw a hexagon on the base.

Step 2 • Draw the height of the pyramid vertically from the centre of the base.

Step 3 • Join the points on the hexagon of the base to the top of the height line.

Step 4 • Outline the solid.

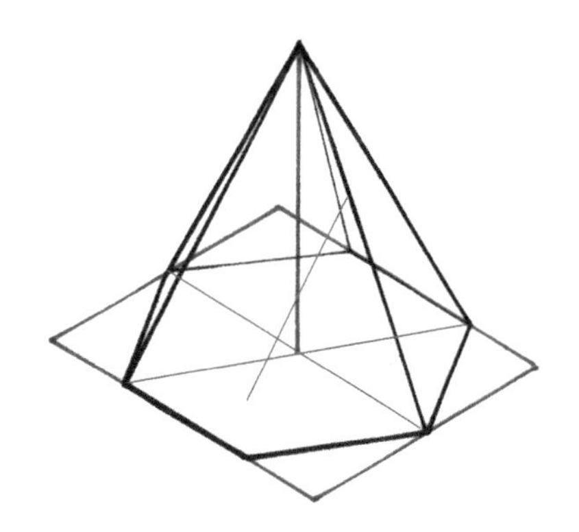

HEXAGONAL PYRAMID

ISBN: 9780170233279

Triangular prism

To draw a triangular prism in isometric you also need to draw the plan view of the triangle to find the lengths that you will plot in isometric.

Step 1 • Draw a box that is the length **x** and **y** on the base.

Step 2 • Measure length **a** from the plan and plot it onto the base.

Step 3 • Join the points of the base together to get the base of the triangle.

Step 4 • From each point of the triangle draw each line vertically the required height.

Step 5 • Outline the solid.

PLAN VIEW

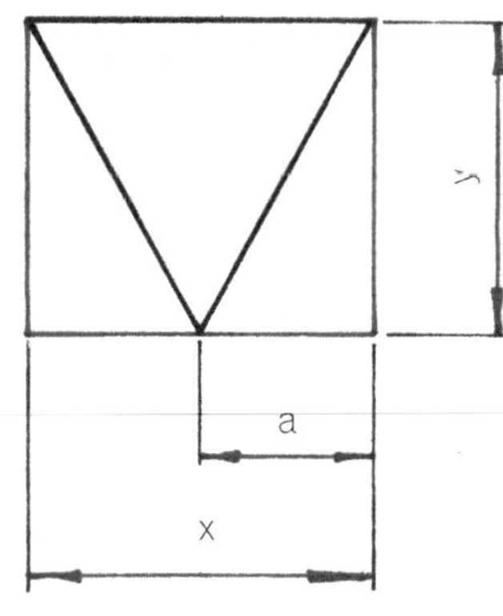

PLOTTING MEASUREMENTS

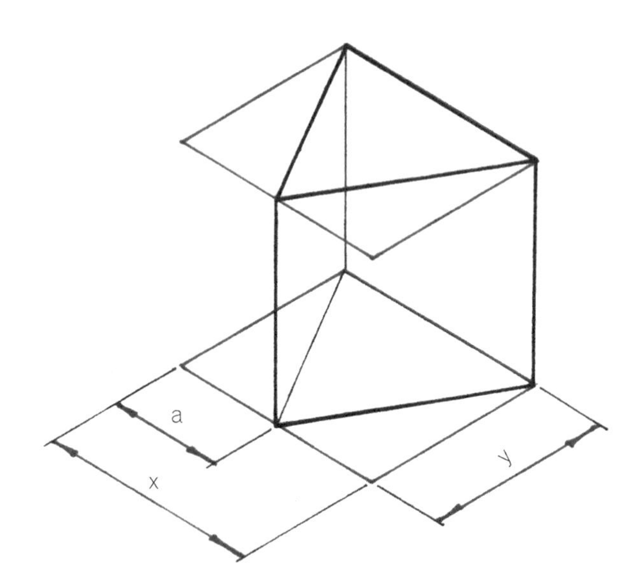

TRIANGULAR PRISM

Triangular pyramid

Step 1 • Draw a triangle on the base.

Step 2 • Draw the height of the pyramid vertically from the centre of the base.

Step 3 • Join the points on the triangle of the base to the top of the height line.

Step 4 • Outline the solid.

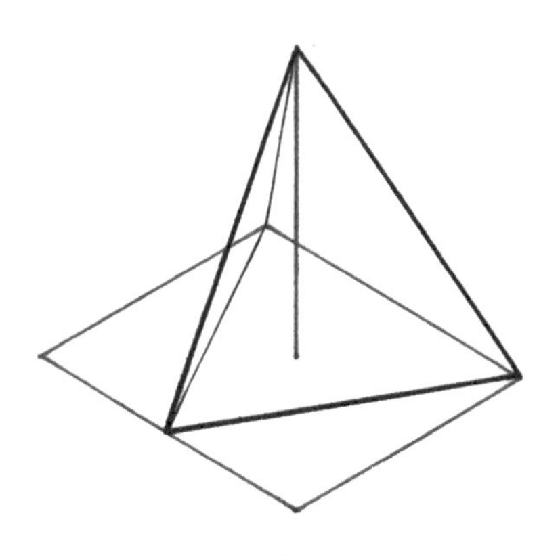

TRIANGULAR PYRAMID

ISBN: 9780170233279

Oblique solids

Square-based prism

Step 1 • Draw a cube in oblique.

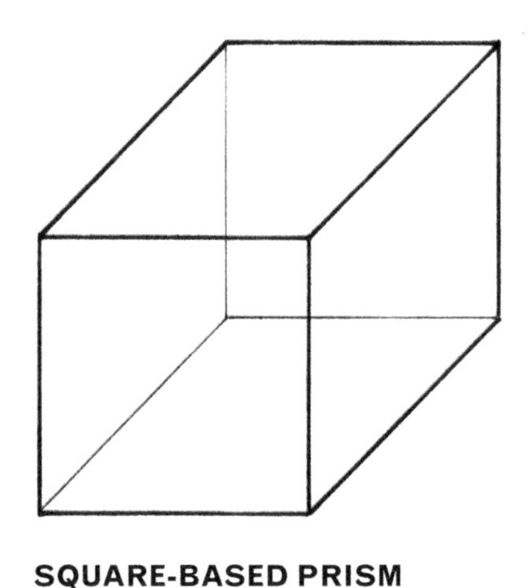

SQUARE-BASED PRISM

Square-based pyramid

Step 1 • Draw a square base.

Step 2 • Draw the height of the pyramid vertically from the centre of the base.

Step 3 • Join the corners of the base to the top of the height line.

Step 4 • Outline the solid.

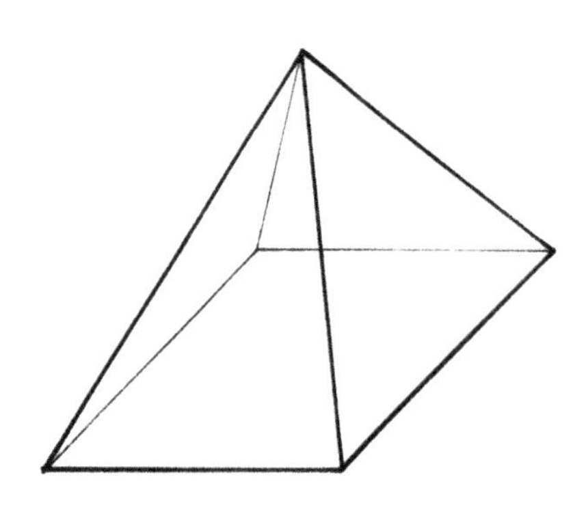

SQUARE-BASED PYRAMID

Cylinder

Step 1 • Draw an oblique circle on the base.

Step 2 • Draw the second oblique circle in line with the base at the distance required for the height away from the base.

Step 3 • Complete the cylinder by joining the edges of the cylinder.

Step 4 • Outline the solid.

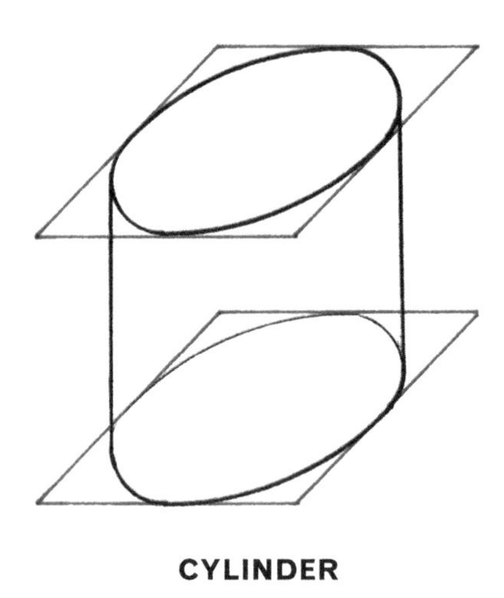

CYLINDER

Cone

Step 1 • Draw an oblique circle on the base.

Step 2 • Draw the height of the cone vertically from the centre of the base.

Step 3 • Join the edge of the circle on the base to the top of the height line.

Step 4 • Outline the solid.

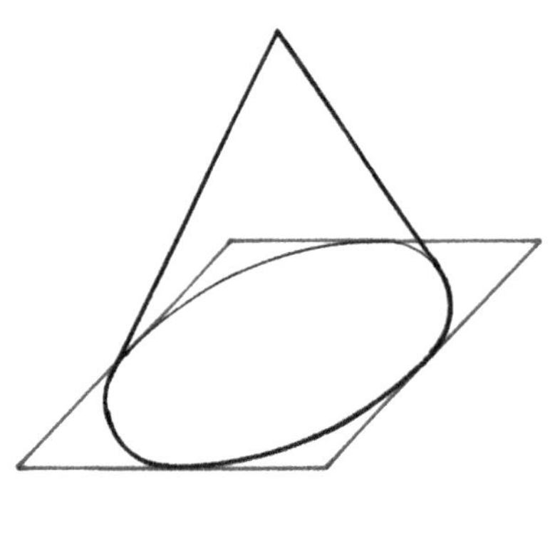

CONE

ISBN: 9780170233279

Hexagonal prism

To construct a hexagonal prism you need to firstly draw the plan view of the hexagon. When a square is drawn around the hexagon it is not the same length on both sides of the square so you firstly need to draw a plan view to find out what those lengths are.

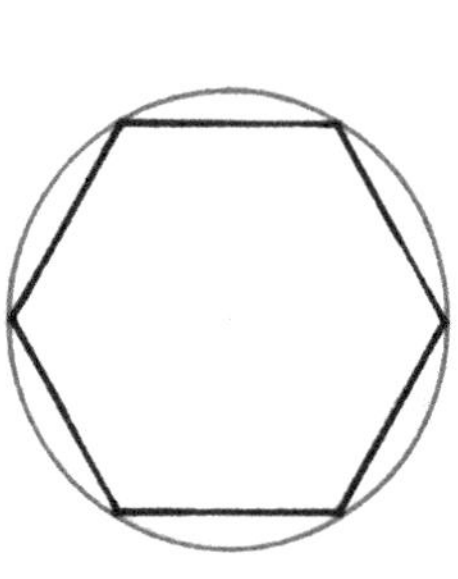

PLAN VIEW

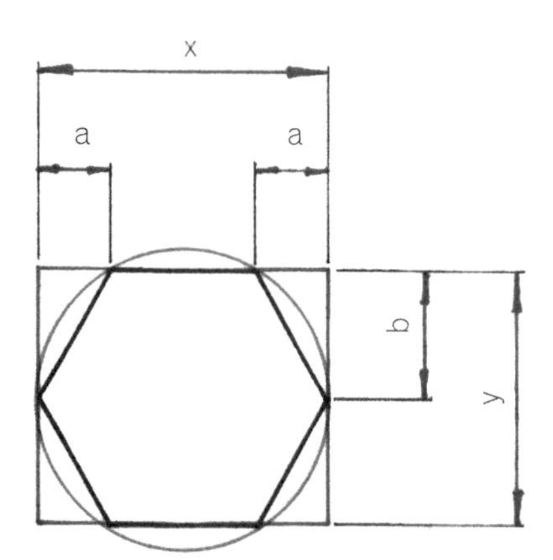

PLOTTING MEASUREMENTS

Step 1 • To construct the hexagon in oblique, draw the base square. These measurements should be taken from the plan view. Measure length **x** and **y**.

Step 2 • To construct the shape of the hexagon on the base again take the measurements from the plan view. Measure length **a** in from the corners on both sides to find these points and plot onto the oblique box.

Step 3 • Measure length **b** in from the corner and plot the length onto the oblique box.

Step 4 • Join the points of the hexagon together on the base.

Step 5 • From each point of the hexagon draw each line vertically the same distance, to the required height.

Step 6 • Join the points of the second hexagon which will be the height of each line.

Step 7 • Outline the solid.

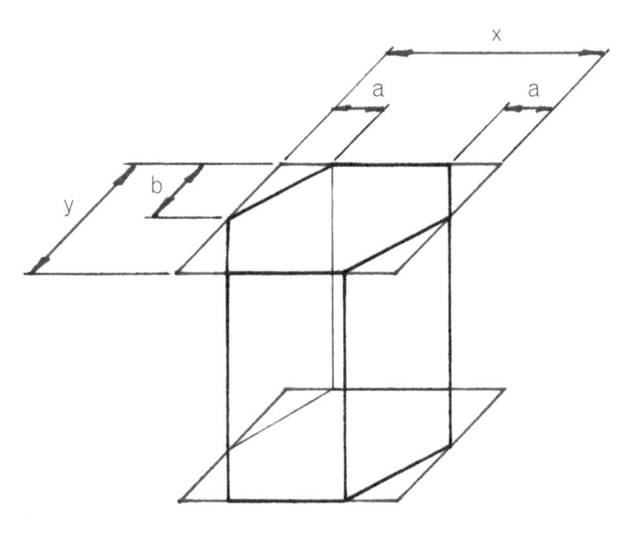

HEXAGONAL PRISM

Hexagonal pyramid

Step 1 • Draw a hexagon on the base.

Step 2 • Draw the height of the pyramid vertically from the centre of the base.

Step 3 • Join the points on the hexagon of the base to the top of the height line.

Step 4 • Outline the solid.

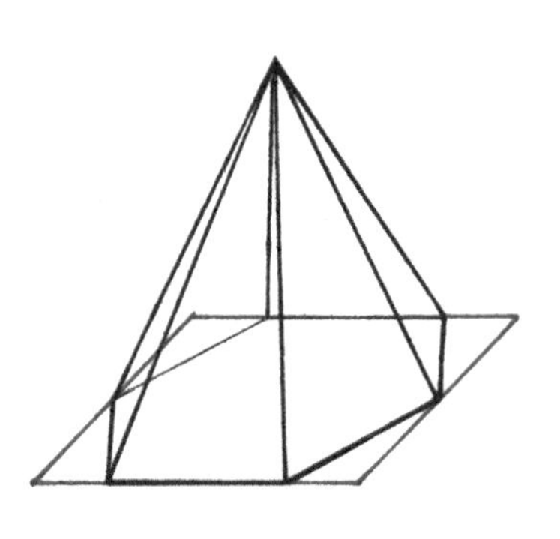

HEXAGONAL PYRAMID

ISBN: 9780170233279

Triangular prism

To draw a triangular prism in oblique you also need to draw the plan view of the triangle to find the lengths that you will plot in oblique.

PLAN VIEW

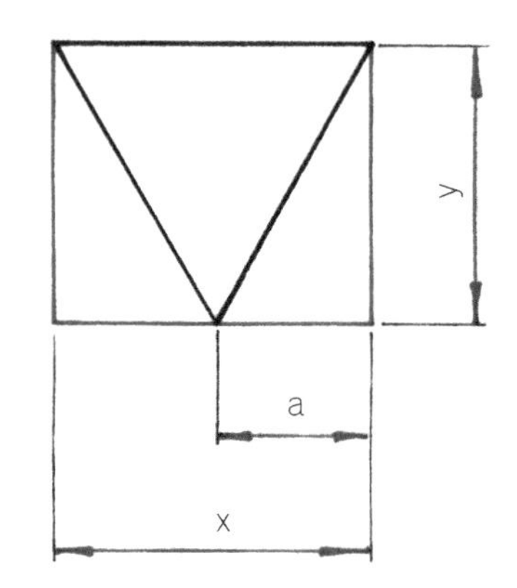

PLOTTING MEASUREMENTS

Step 1 • Draw a box that is the length **x** and **y** on the base.

Step 2 • Measure length **a** from the plan and plot it onto the base.

Step 3 • Join the points of the base together to get the base of the triangle.

Step 4 • From each point of the triangle draw each line vertically the required height.

Step 5 • Outline the solid.

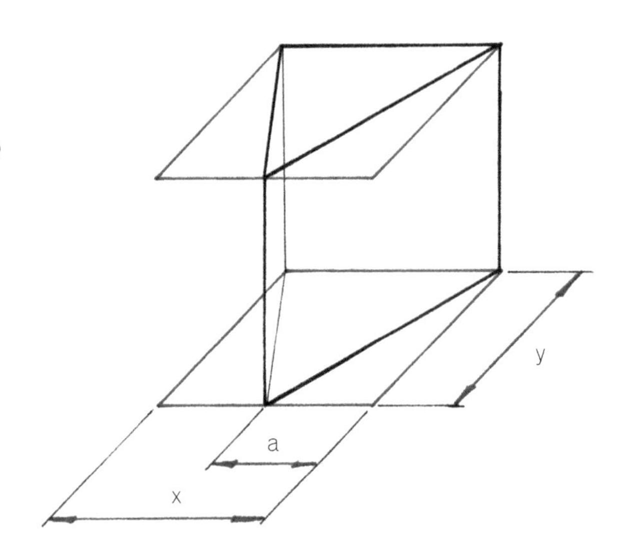

TRIANGULAR PRISM

Triangular pyramid

Step 1 • Draw a triangle on the base.

Step 2 • Draw the height of the pyramid vertically from the centre of the base.

Step 3 • Join the points on the triangle of the base to the top of the height line.

Step 4 • Outline the solid.

When using these geometric solids in your design work remember to identify the solid you have constructed. Identifying the solid means naming the solid you have used.

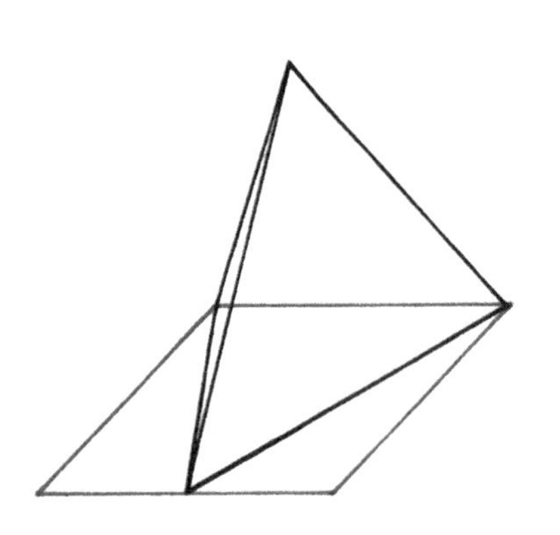

TRIANGULAR PYRAMID

ISBN: 9780170233279

Truncated solids

If a solid has been truncated this information is to be plotted onto the pictorial view from a plan and elevation using the crating method.

Truncated square-based prism

The overall box is drawn and the heights are taken directly from the elevation and plotted. The prism is then outlined.

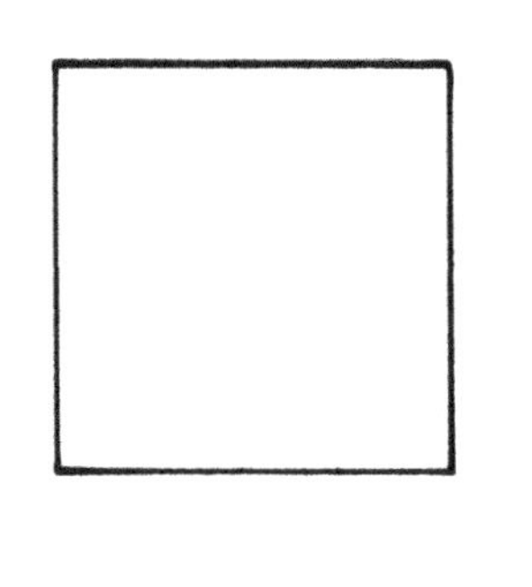

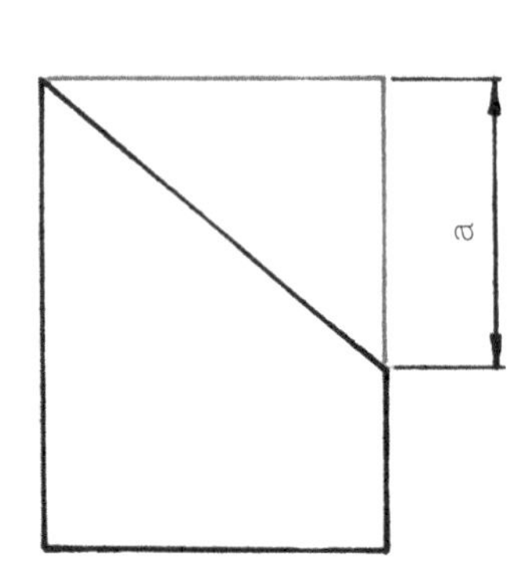

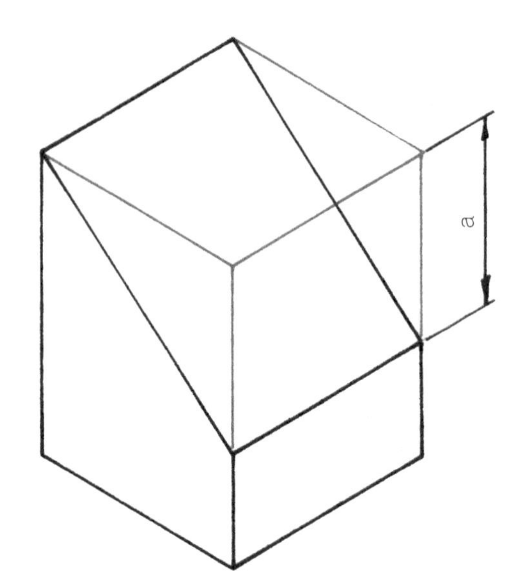

Hexagonal truncated prism

The heights are plotted from the elevation onto the pictorial. The prism is then outlined. Be careful to plot the points accurately.

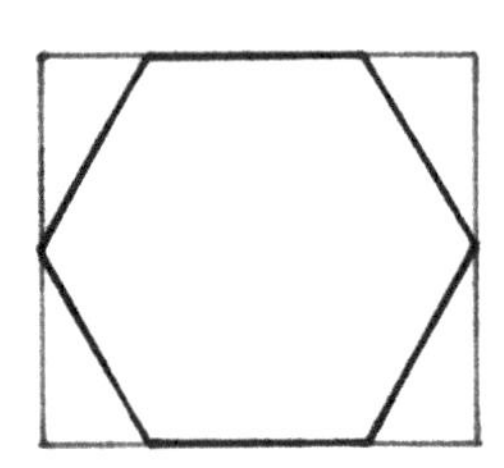

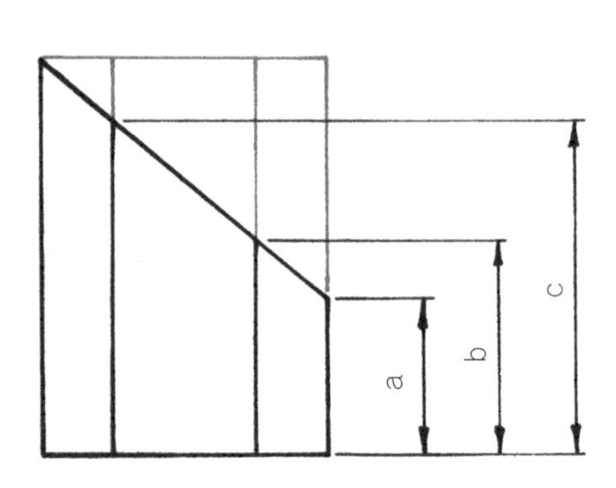

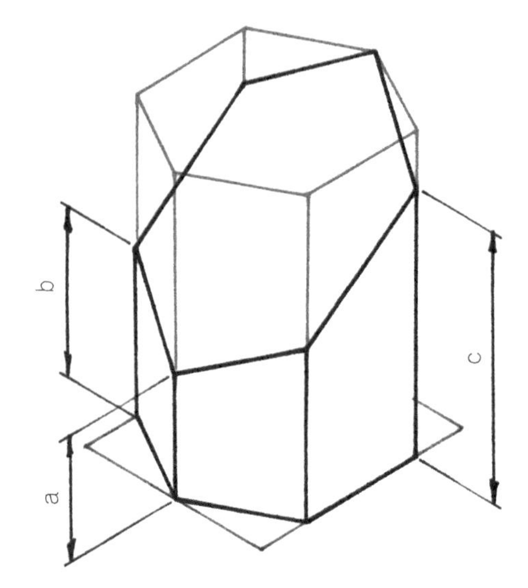

ISBN: 9780170233279

Square-based pyramid

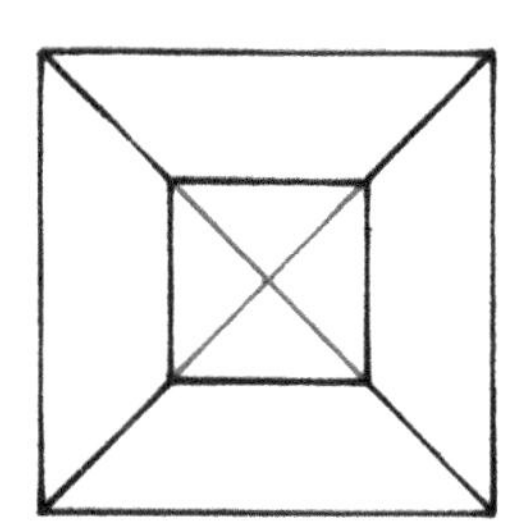

The heights from the elevation cannot be directly transferred to the pictorial as the sides are on an angle joined to the apex and are not true length. The prism must be constructed in a box and then all lines and points transferred to the crating around the pyramid. This has been shown with the green construction lines

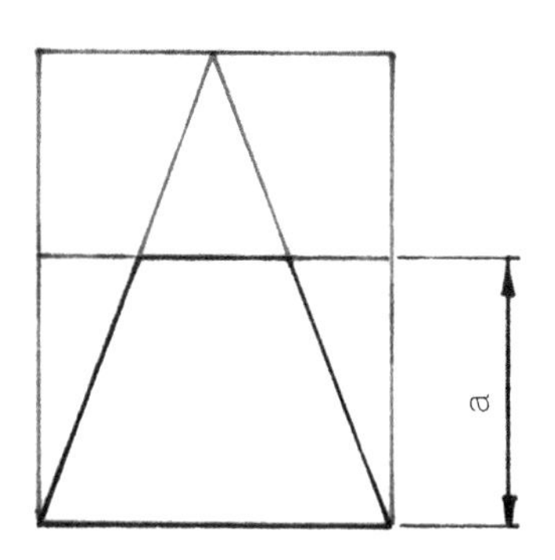

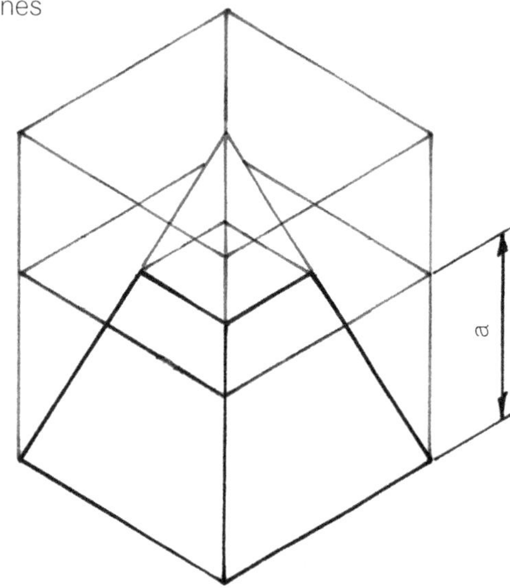

Cylinder

A truncated cylinder is a little more complex to construct.

Step 1 • Divide the plan view into twelve sections.

Step 2 • Draw lines across the circle where the twelve points cross the circle. These lines have been shown in green.

Step 3 • Plot these distances onto the base of the cylinder.

Step 4 • Where the lines cross the circle on the base of the cylinder project lines straight up. The height of each of these lines should be taken from the corresponding lines on the elevation. This has been shown with the vertical green lines on the cylinder.

Step 5 • Outline the shape of the slanted surface.

Step 6 • Outline the rest of the cylinder.

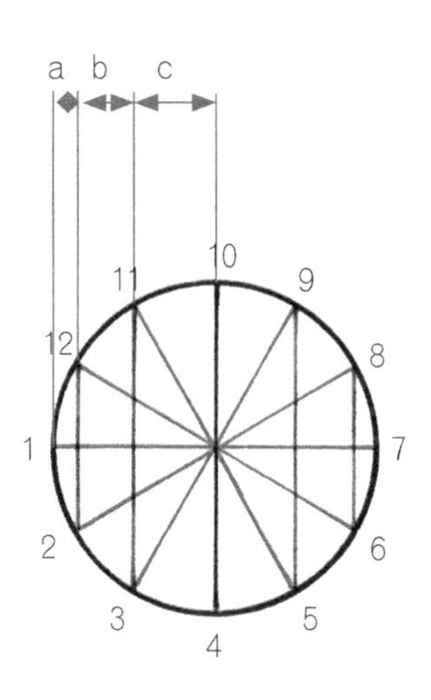

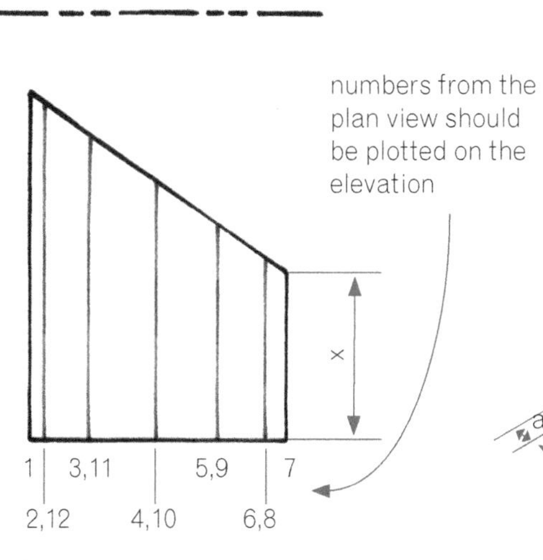

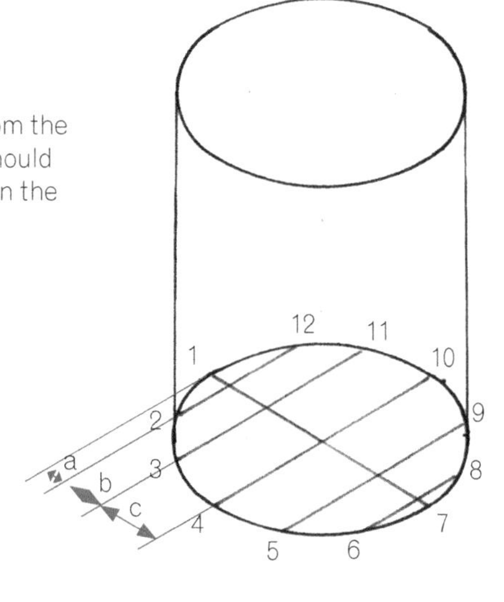

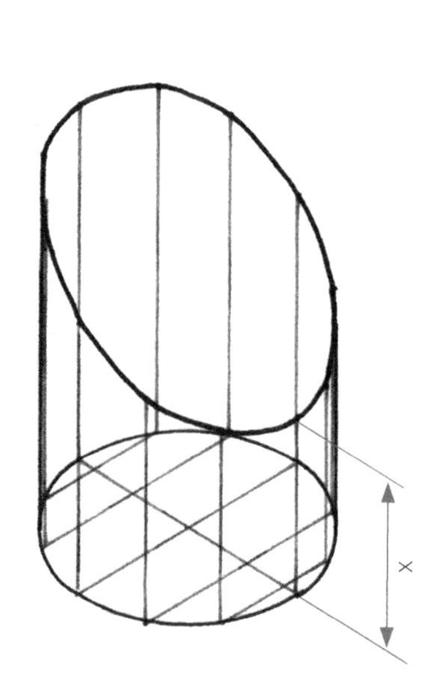

ISBN: 9780170233279

Paraline drawings are 3D drawings where real dimensions are used along the axes. The drawing techniques are isometric, oblique, planometric, trimetric and diametric.

Isometric

The isometric method of pictorial drawing is constructed with a corner of the object facing you and by using two 30° angles, as demonstrated earlier in this chapter. Below is an example of a truck and camera drawn using the isometric technique.

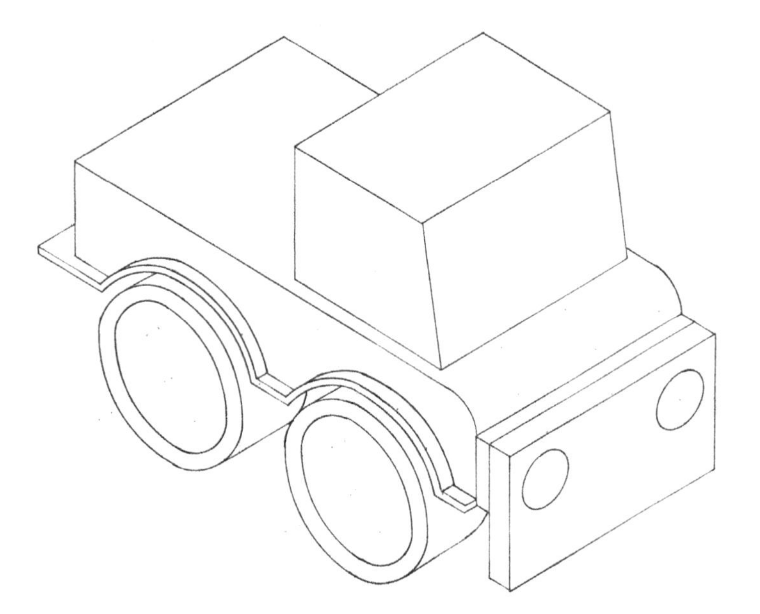

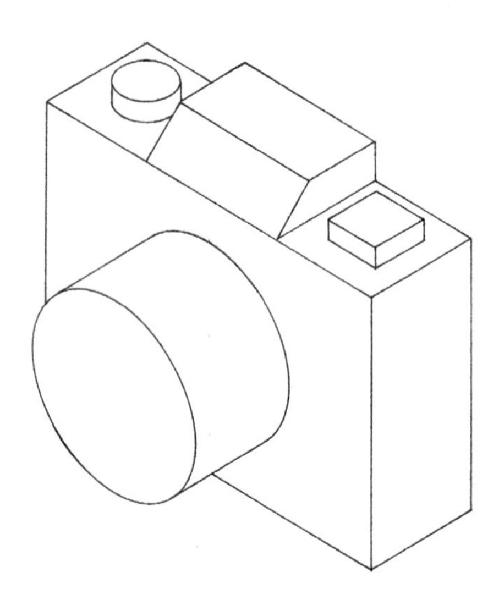

Oblique

The oblique method of pictorial drawing is constructed with a front view of the object facing you and by using an angle of 45°, as demonstrated earlier in this chapter. Below is an example of a truck and camera drawn using the oblique technique.

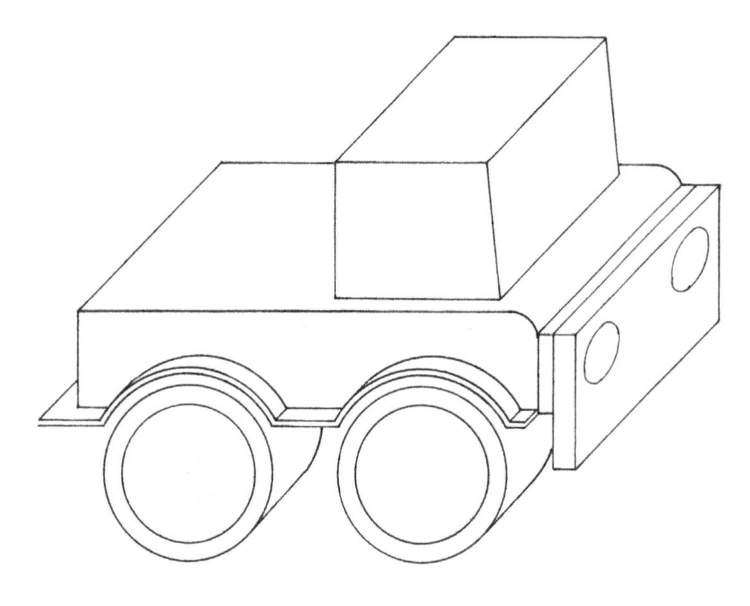

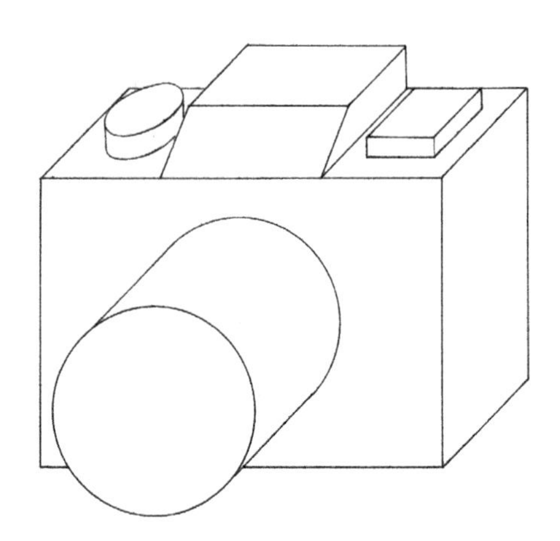

ISBN: 9780170233279

Trimetric

Unlike the isometric method of pictorial drawing shown previously, which is based on one scale, trimetric uses three different scales for:

the height (z axis), the width (X axis), the depth (y axis)

This drawing style is very similar to isometric but does allow the angles to be different rather than both at 30°. A trimetric projection is where no two axes form equal angles with the plane of projection. Each of the three axes and the lines parallel to them have different ratios for foreshortening.

The advantage of trimetric drawing is that it is the closest drawing to perspective of all the paraline views and is the most acceptable to the eye. Any angles may be used in trimetric.

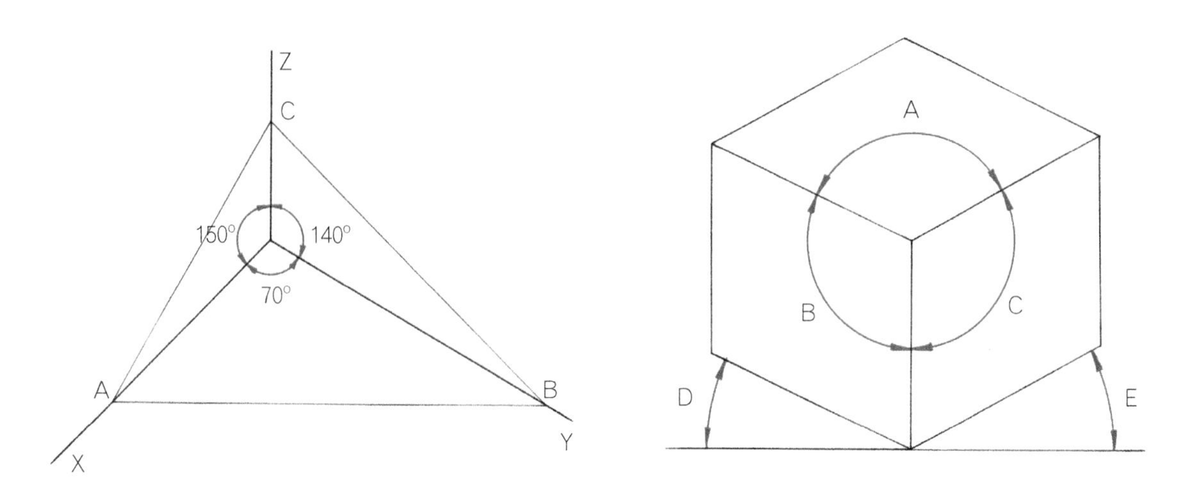

The following demonstrates how the three scales are constructed for the trimetric method of pictorial drawing:

Step 1 • Draw lines at 90° to the angles (Z-90°, X-21°, Y-33°) to form a triangle. The pictorial will be constructed within this space.

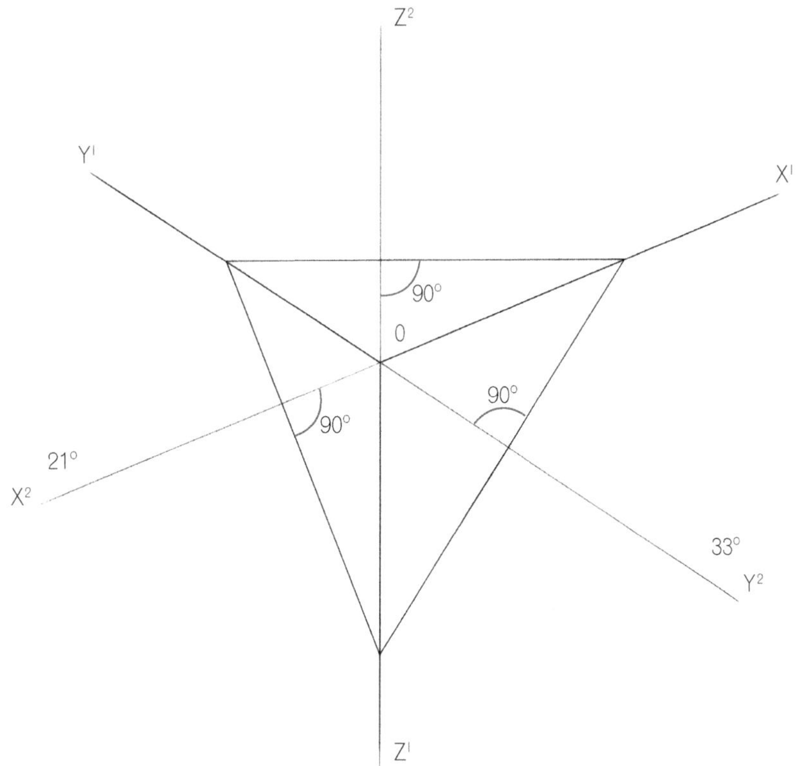

ISBN: 9780170233279

Step 2 • Draw lines parallel to the lines you have drawn in step 2, e.g. A to B, leaving adequate space (the plan, main view and side view will be drawn in this area).

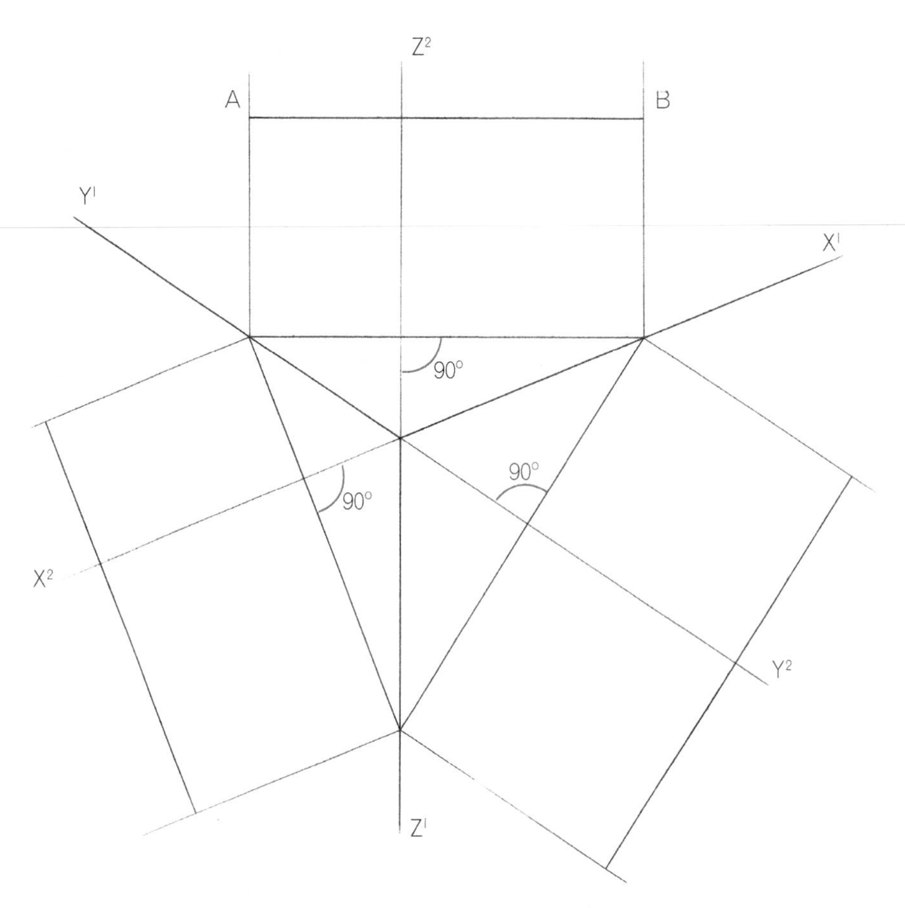

Step 3 • Bisect each of the lines drawn in step 2 and scribe an arc between these points, e.g. A to B. Where the Z, X and Y axis intersect the arc join to the ends of the line drawn at step 3, e.g. Z^2 to A and B. These lines will create a 90° angle.

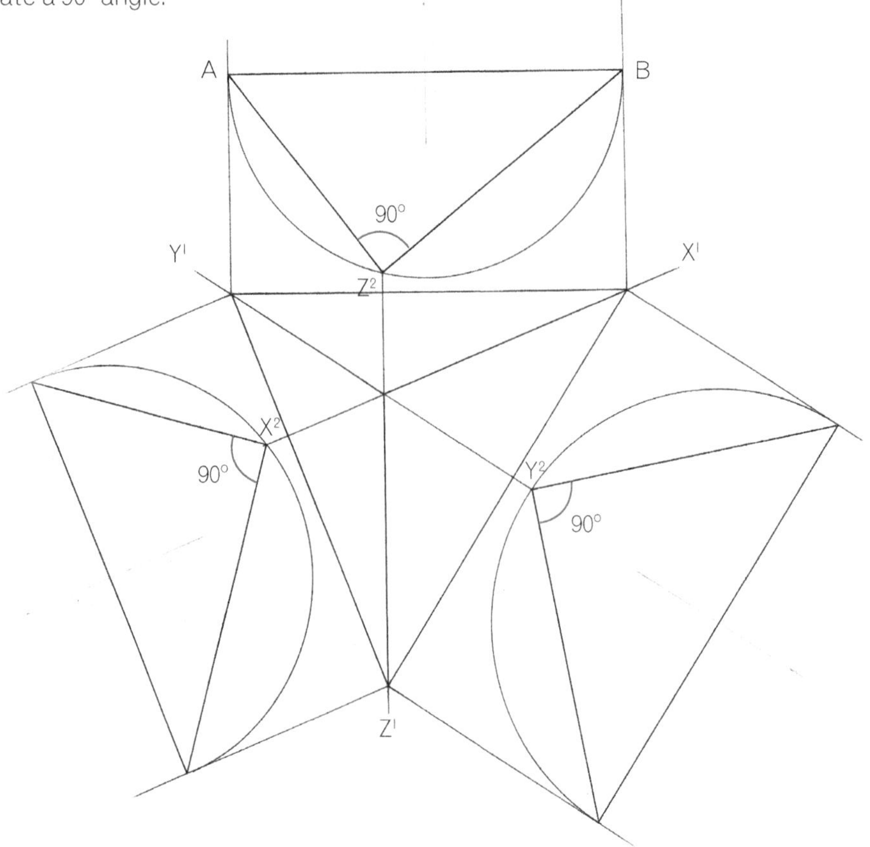

ISBN: 9780170233279

Step 4 • As shown within each of the 90° angles constructed at step 5, draw the plan view, main view and side view.

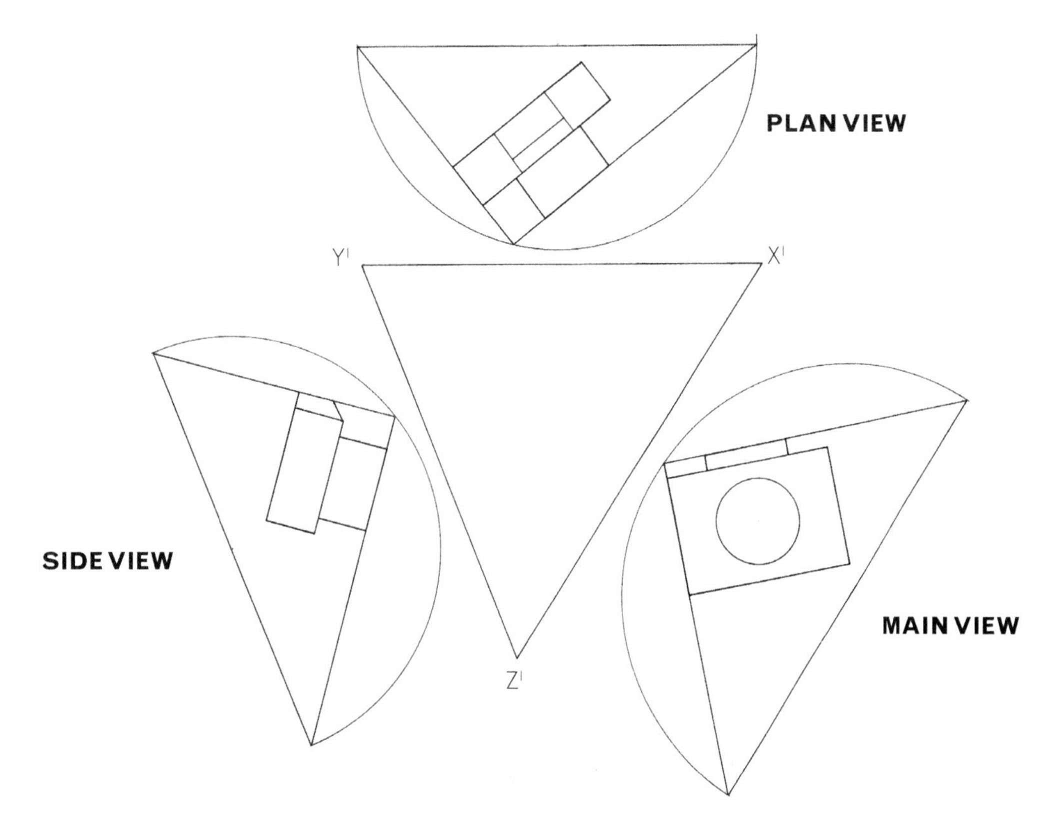

Step 5 • To construct the camera in trimetric project lines from the plan view perpendicular to line $Y^1 - X^1$, from the main view perpendicular to line $X^1 - Z^1$ and from the main view perpendicular to line $Z^1 - Y^1$. You will need to crate the object as you do when constructing any other pictorial such as isometric or oblique.

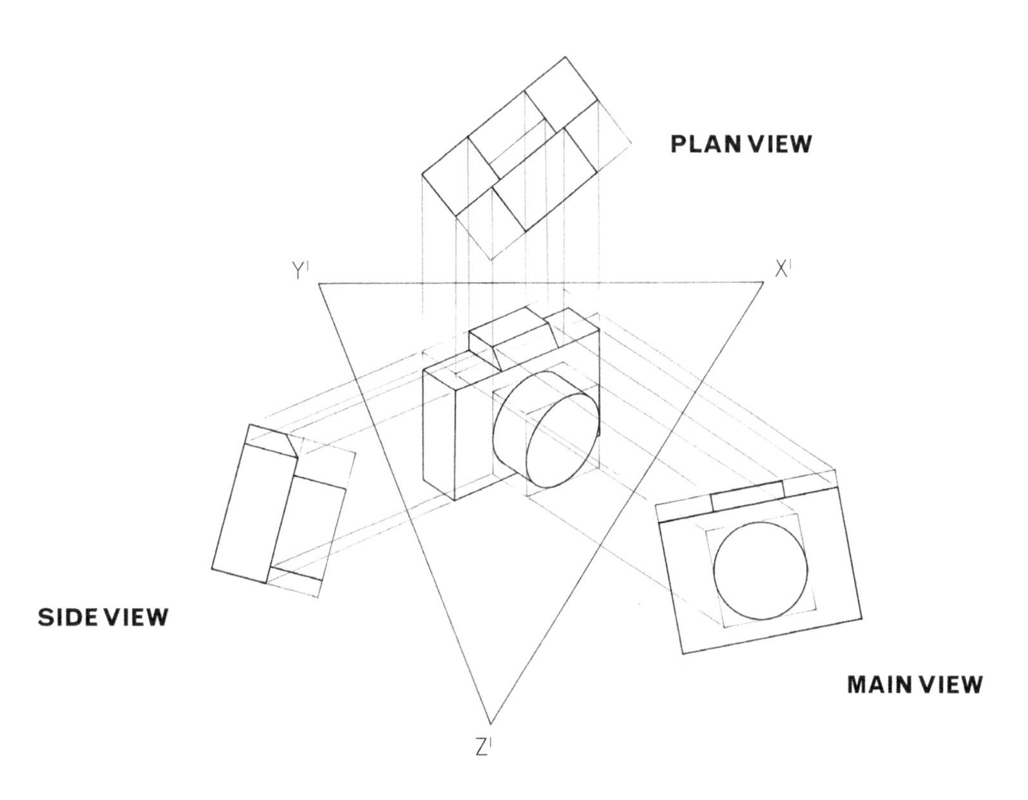

ISBN: 9780170233279

Diametric

A diametric drawing is a projection where two of an object's axes make equal angles with the plane of projection and the third angle is larger or smaller than the other two.

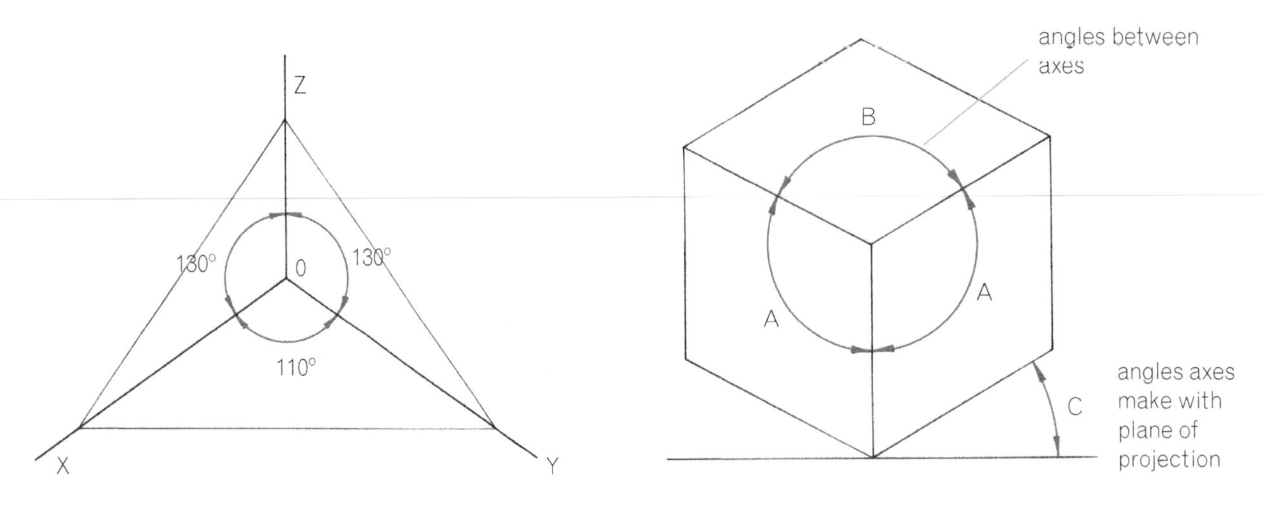

Planometric

Planometric is useful to explain the interior of a room or house. Planometric is a useful drawing to present your final idea for a spatial design brief as you can use your final floor plan, move it to the correct angle, trace the floor plan and then project the heights vertically.

There are two types of planometric view. The floor plan can be drawn on a 45° angle or on a 30°/60° angle. Planometric drawn on a 45° angle reduces the real life height to three-quarters of the original height otherwise the walls would appear too high and you would not effectively be able to see the interior.

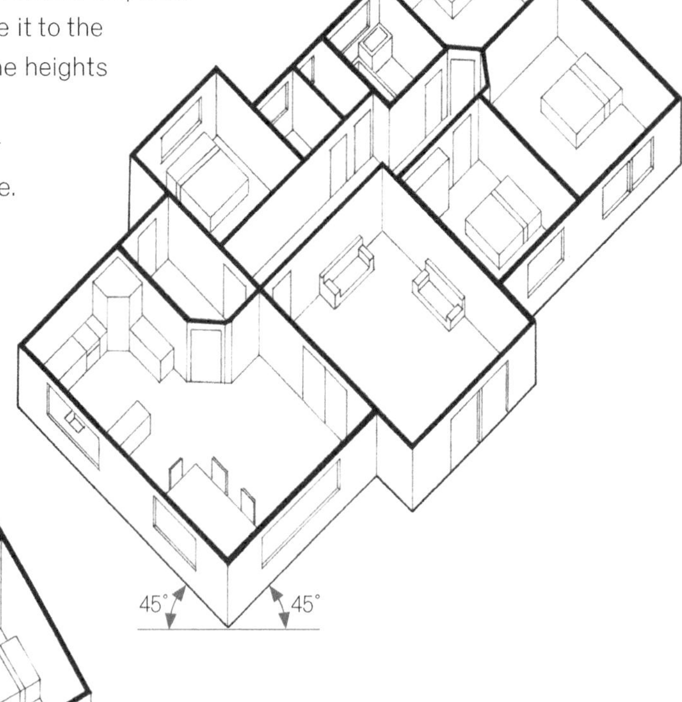

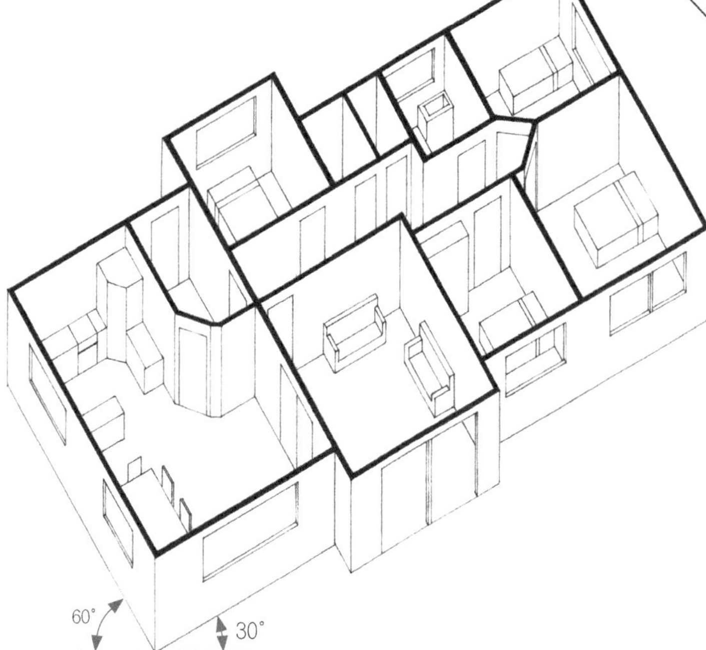

Planometric drawn on 30° and 60° angles uses the actual height. The advantage of 30°/60° planometric is that you can see more of the main exterior side of the house as well as the interior which could be important if you were presenting detailing on the exterior of the building.

ISBN: 9780170233279

Paihere Tims — St Peter's College

The following work produced by Paihere Tims demonstrates the use of a floor plan to construct a planometric view of the interior of a kitchen, customer space and exterior for a cafe design.

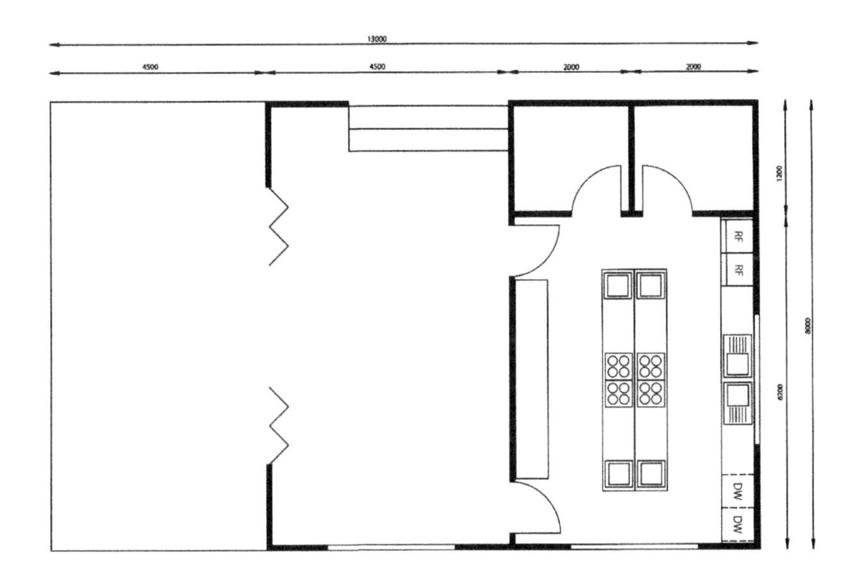

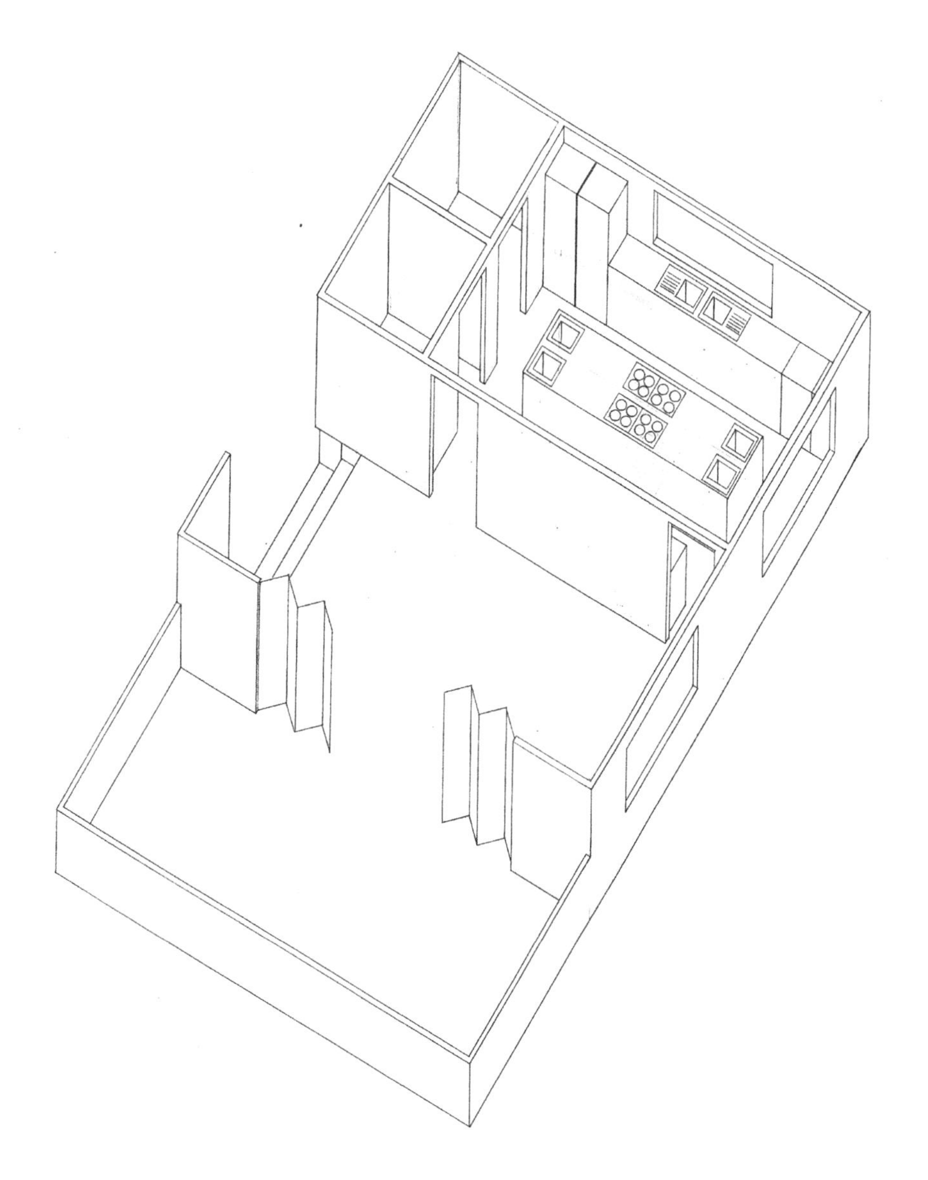

ISBN: 9780170233279

Ordinate method

If a shape is made up of complex curves such as tangent curves and you want to construct it in isometric you will need to use the ordinate method to plot out the points required to make the shape.

Step 1 • Draw an elevational view of the shape, in this case a bottle.

Step 2 • Mark even points along the centre line.

Step 3 • Draw the construction lines horizontally across the elevation.

Step 4 • You can plot these lengths in isometric by drawing a square the same width and heights of the elevation.

Step 5 • Mark out the even points down the centre line and draw them across on a 30° angle.

Step 6 • Plot points along these lines using the measurements from the centre on the elevation.

Step 7 • Join the points together.

Step 8 • From the points created, draw 30° lines from each segment for the depth of the object.

Step 9 • Plot the depth on each of these lines.

Step 10 • Join the points and outline the object.

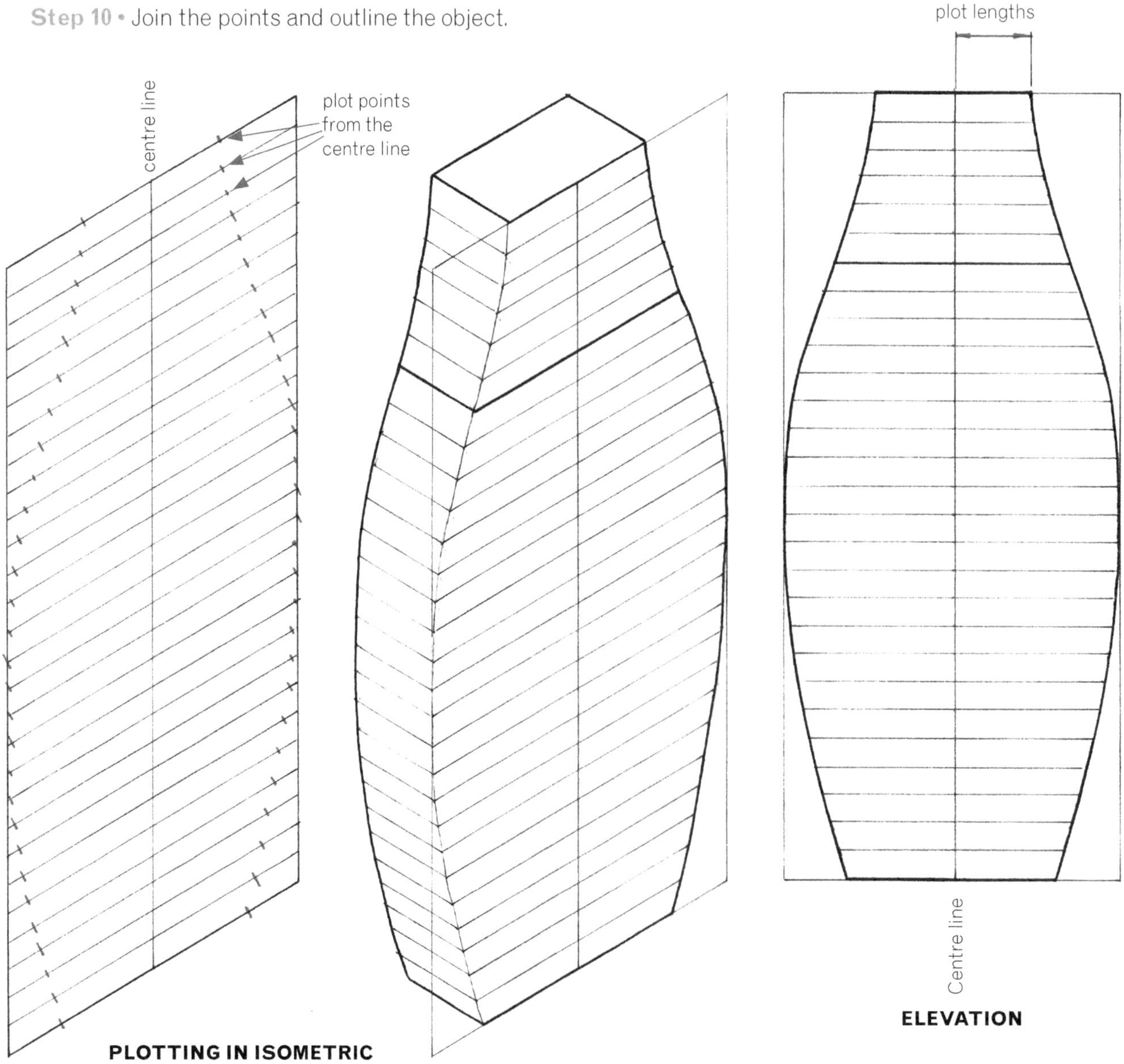

PLOTTING IN ISOMETRIC

ELEVATION

ISBN: 9780170233279

Curved surfaces

You will notice in everyday life that most products have some sort of radius on the edge of the product. The following images explain how these are constructed.

Example **A** explains where the isometric circles are to be positioned if they are to be used for the radius of a corner of an object.

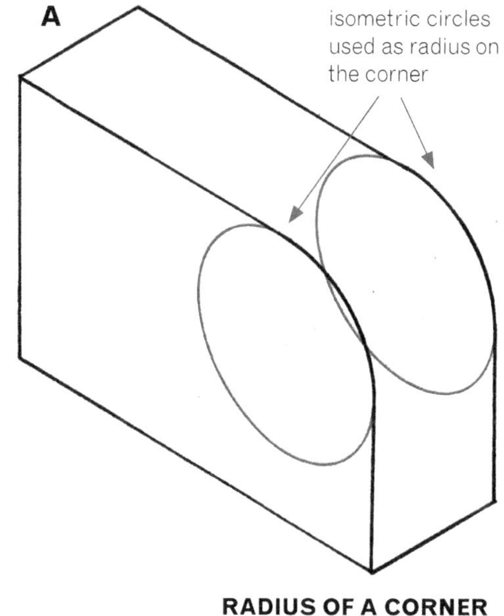

RADIUS OF A CORNER

Example **B** explains where the isometric circles may be positioned if each side of the object is curved.

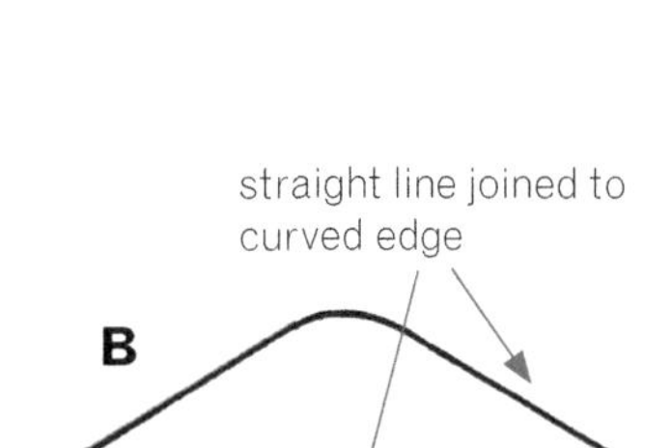

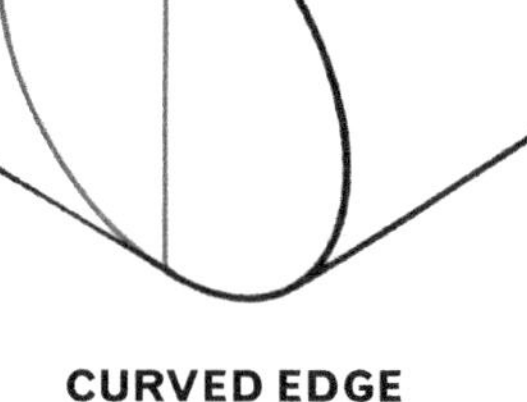

CURVED EDGE

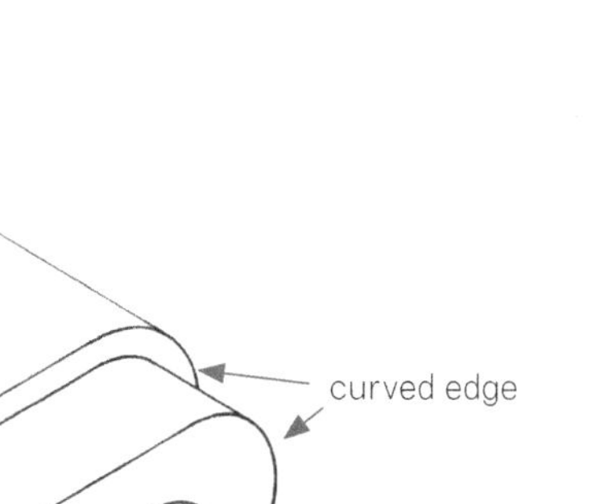

ISOMETRIC WITH CURVED SURFACES

The pictorial of the pen and cap uses curved edges and corners with a radius to construct the image in isometric.

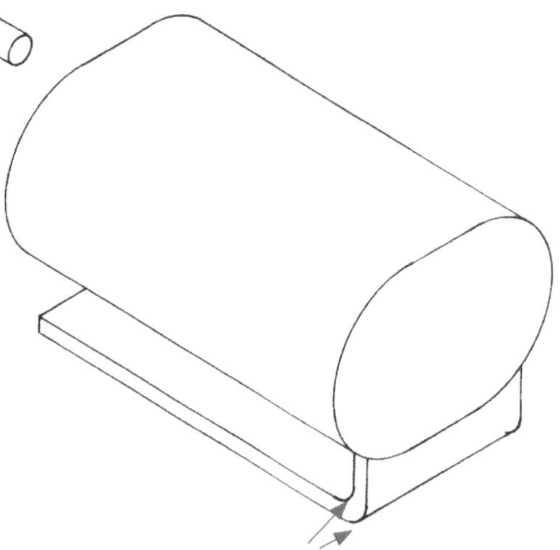

ISBN: 9780170233279

Detailed drawings

Sectioned pictorial

One way to present a pictorial that explains the interior of an object is to draw a sectioned pictorial. The example presents a pictorial view of what the bottle top would look like if it was cut in half. Visually this is an effective way of communicating information.

Exploded pictorial

To explain the interior of an object (how the components fit together) you can use any type of drawing as an exploded pictorial. You need to make sure that you construct the pictorial accurately. All parts need to be projected from where they are meant to be positioned. The image below shows the projection of each part with the green lines. Start by drawing the main component and projecting the other parts from this.

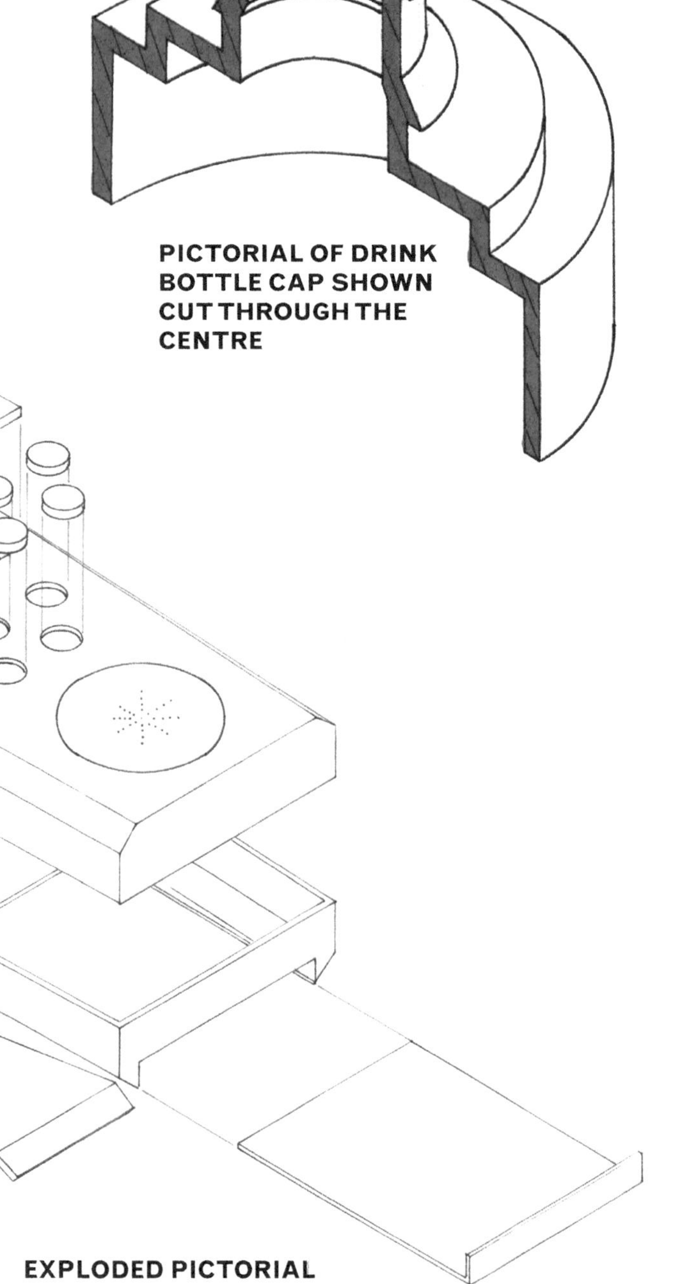

PICTORIAL OF DRINK BOTTLE CAP SHOWN CUT THROUGH THE CENTRE

EXPLODED PICTORIAL

ISBN: 9780170233279

Thanya Chansouk — St Peter's College

The following work produced by Thanya Chansouk demonstrates the use of two different types of paraline drawing techniques to construct the instrumental pictorials of the shelter. He has used isometric and planometric drawing techniques. Thanya has produced an exploded view of the components to demonstrate more detailing and show how the parts fit together.

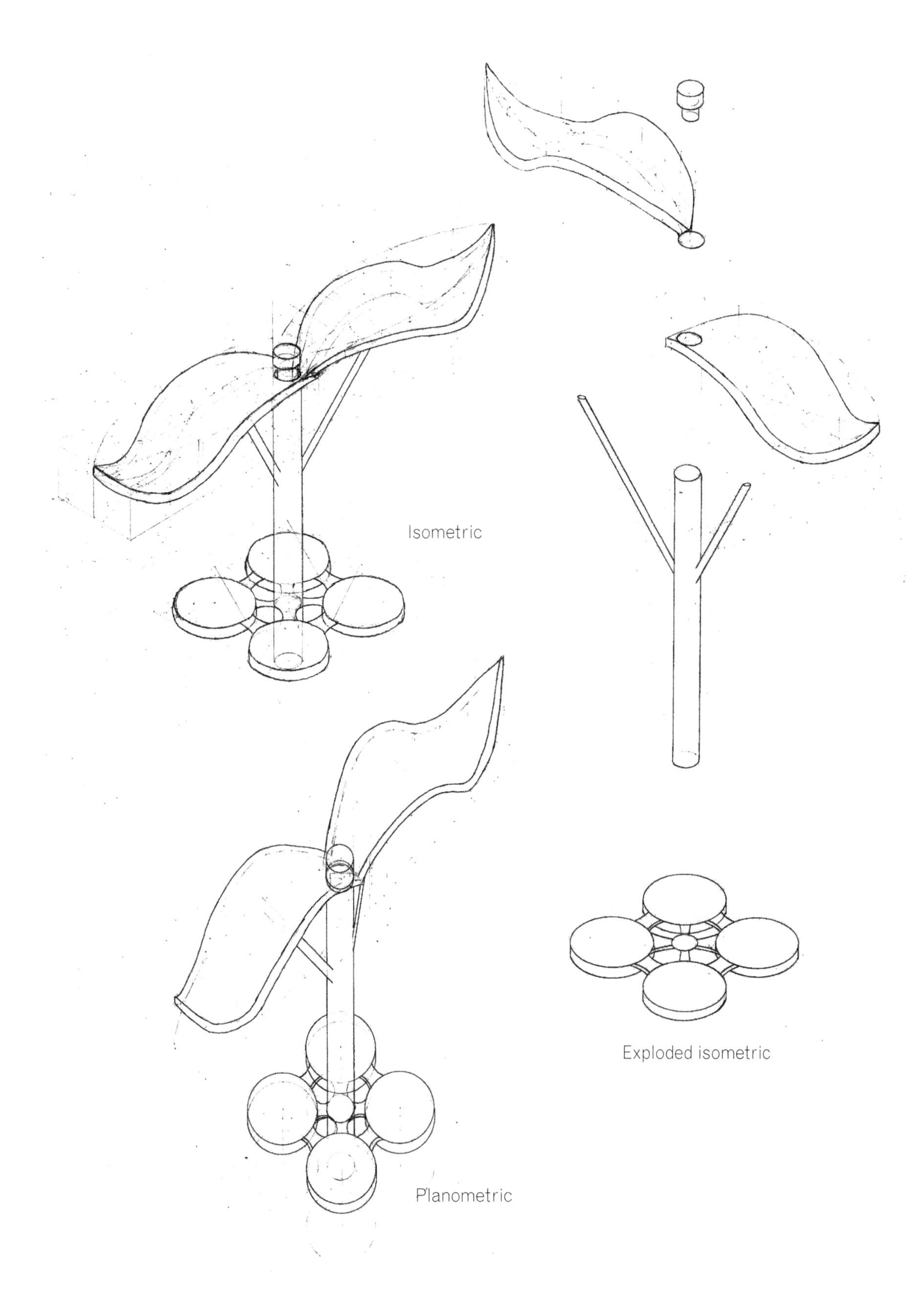

ISBN: 9780170233279

Perspective sketching has been discussed previously in this text. For Level 2 Design and Visual Communication measured perspective is introduced. To present final design ideas students can use instruments to formally draw their ideas in parallel/angular perspective projection.

There is new terminology used to construct a parallel/angular perspective projection. Students will be familiar with some of the terms from completing 2D instrumental working drawings and perspective sketching.

Plan – a view looking at the top of an object. This view is also used for the construction of an orthographic projection.

Elevation – a view looking at the side of an object. This view is also used for the construction of an orthographic projection.

Picture Plane – the line that the lengths of an object are plotted onto. These lengths are then projected onto the pictorial view. The positioning of this line can alter the size of the pictorial constructed. (Abbreviation – **PP**)

Horizon Line – this is the eye level line. The positioning of this line can alter the view of your pictorial. (Abbreviation - **HL**)

Ground Line – the ground line is the base line of the object. (Abbreviation - **GL**)

Height Line – all heights from the elevation are projected onto the height line. The heights are then projected back to the vanishing points. This vertical line is drawn from the ground line.

Station Point – the distance of the viewer from the picture plane line. (Abbreviation - **SP**)

Vanishing Points – points on the horizon line where the lines appear to meet. (Abbreviation - **VP**)

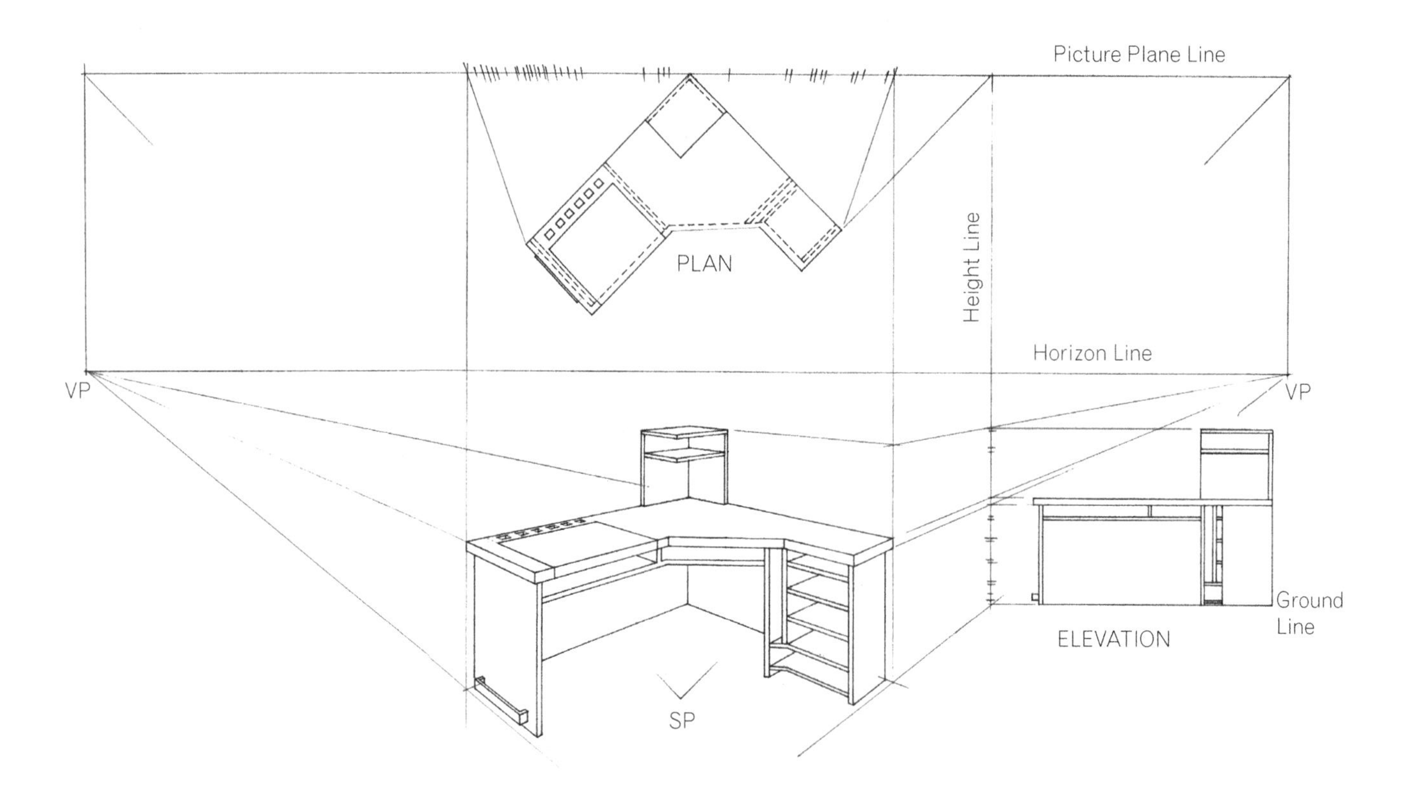

ISBN: 9780170233279

Step-by-step construction

Step 1 • Draw the plan view of the steps on a 45° angle on the picture plane line.

Step 2 • Draw the elevation of the steps on the ground line.

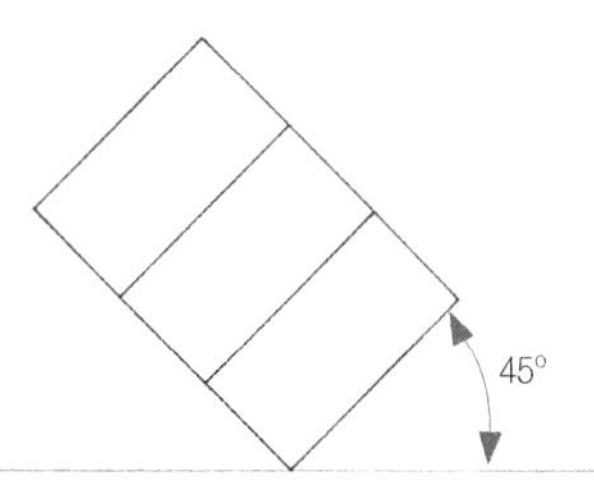

Step 3 • Set the station point up on the page. This point can be positioned anywhere but the positioning will affect the view of the end pictorial.

Step 4 • From the stationary point draw two lines which are perpendicular to one another and are lined up to cross through the picture plane line. It is best if the points that cross the picture plane line from the station point (SP) are as far away from each other as possible. The positioning of where the SP lines cross the picture plane line (PP) line will affect the positioning of the vanishing points (VP) on the horizon line (HL).

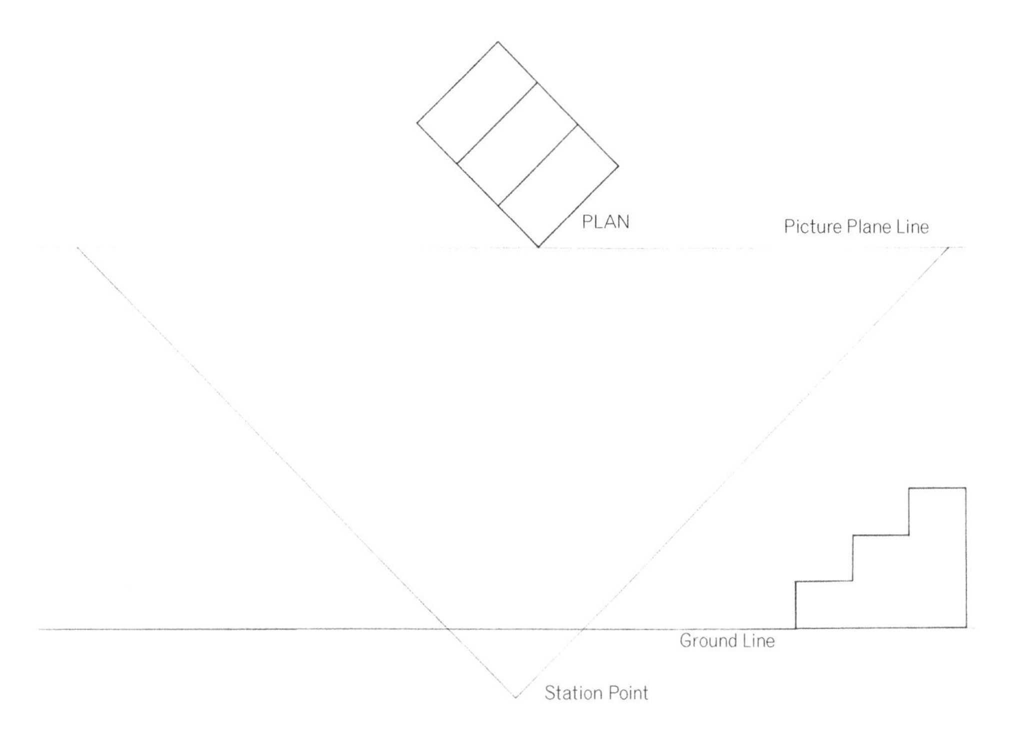

ISBN: 9780170233279

Step 5 • Set the horizon line up. This can be positioned where you choose. Again the positioning of the horizon line will affect the view of the object.

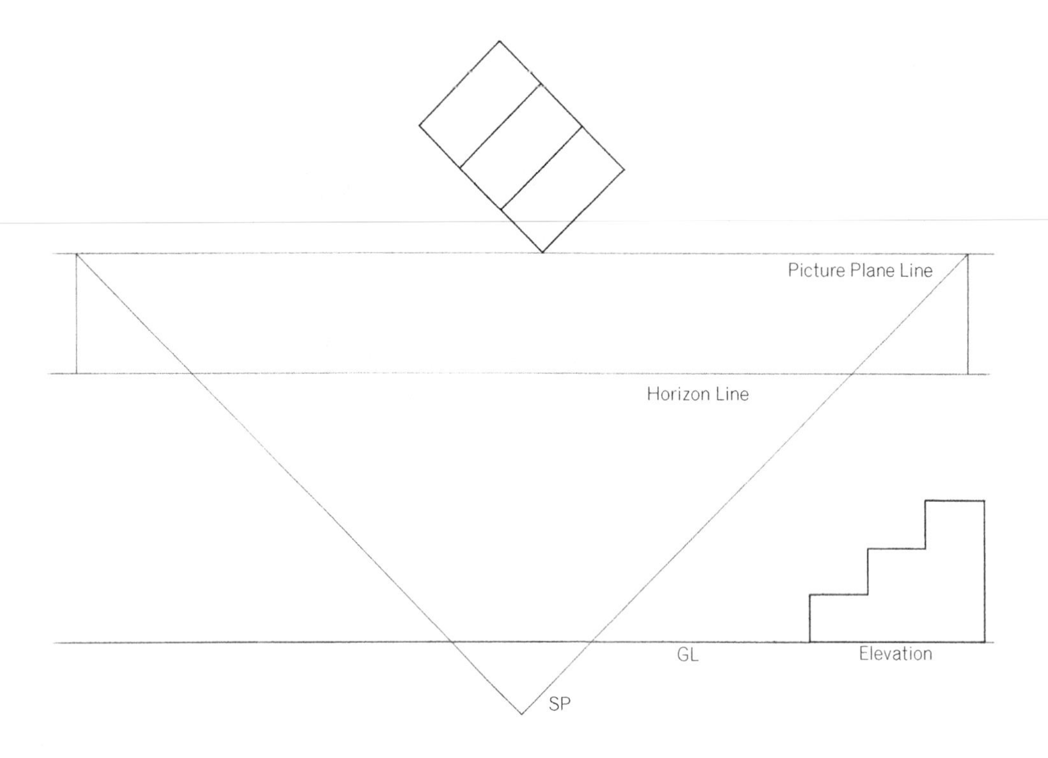

Step 6 • Project the height line straight down from the corner of the plan that is on the picture plan line.

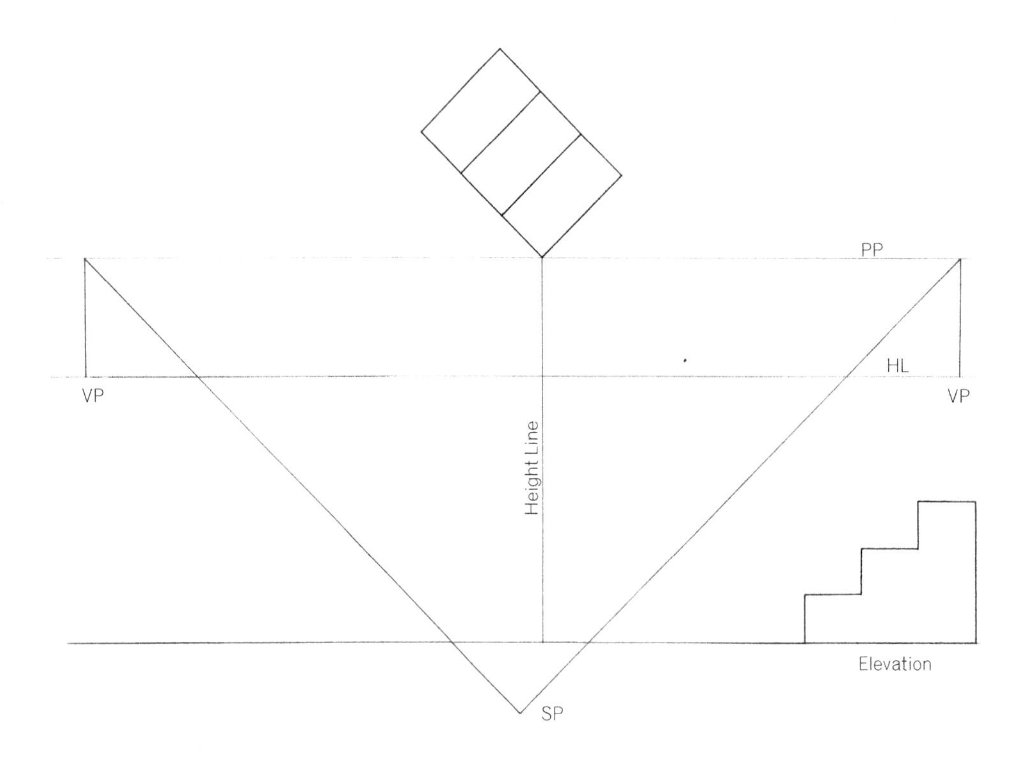

ISBN: 9780170233279

Step 7 • Project the overall height of the object from the elevation horizontally to the height line. These lines have been presented with the red lines.

Step 8 • Take the points where the projected red lines cross the height to the vanishing points. These lines have been presented with the blue lines.

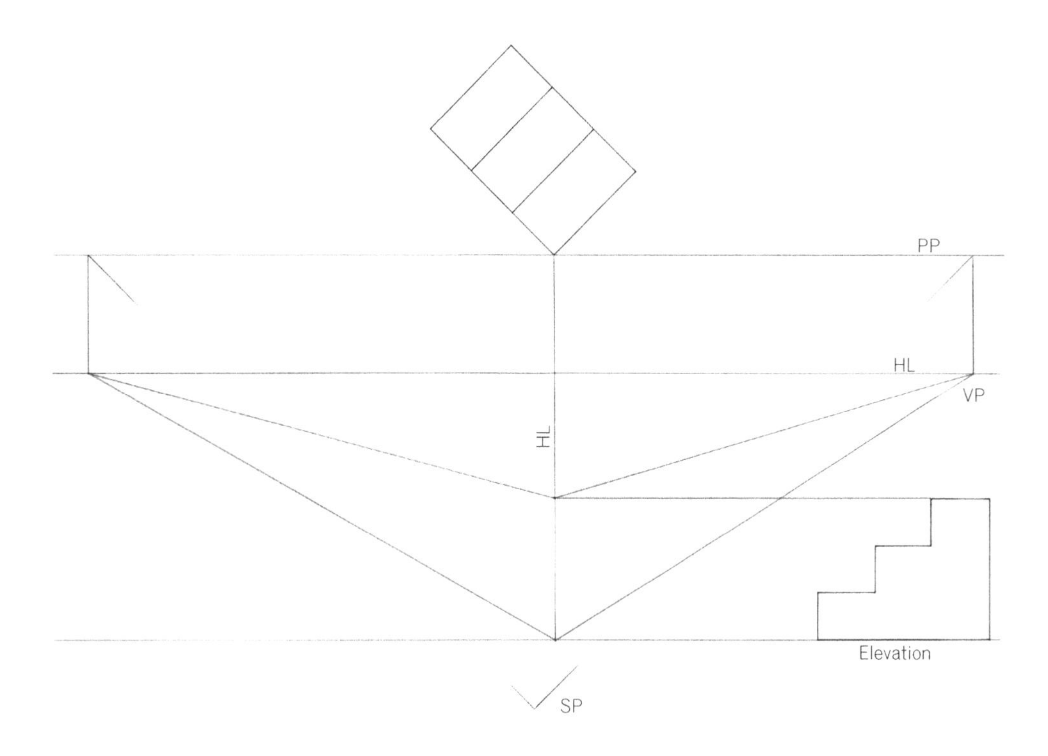

Step 9 • To get the length of the crating for the steps firstly line up the SP with the corners of the plan view. These lines have been presented with the green lines.

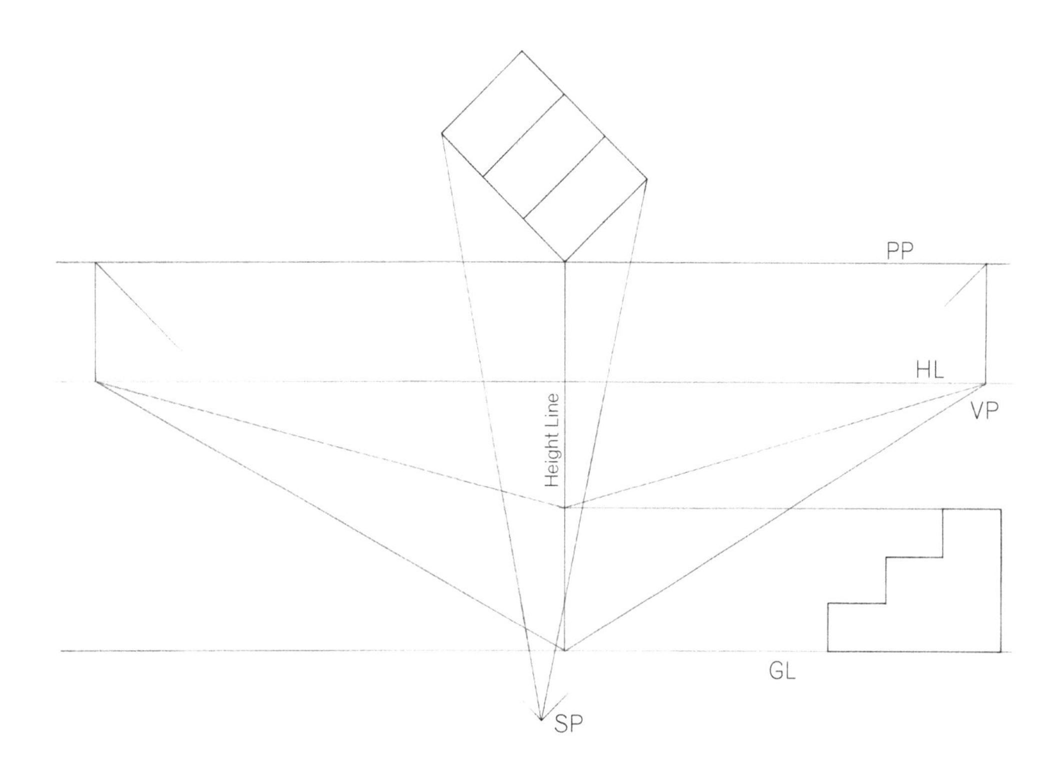

ISBN: 9780170233279

Step 10 • Where the green lines cross the picture plane line, project straight down to get the outside crating of the box for the steps. These lines have also been shown in green.

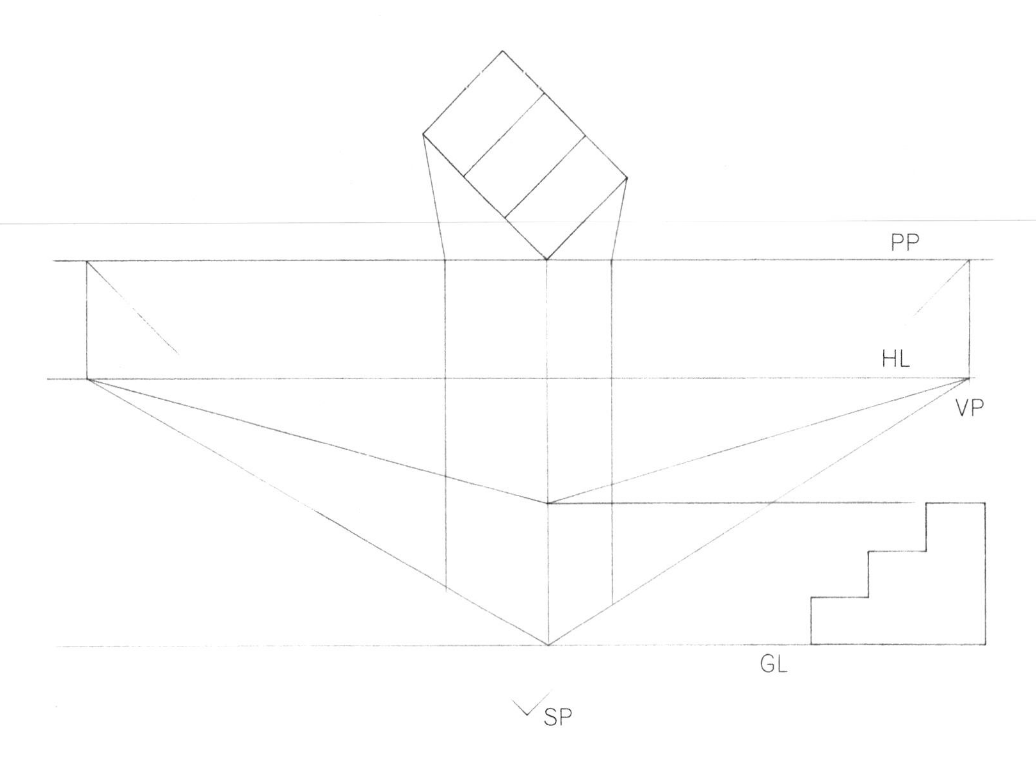

Step 11 • Take the point created by the green line crossing the blue line to the VP. This will give the box for the steps.

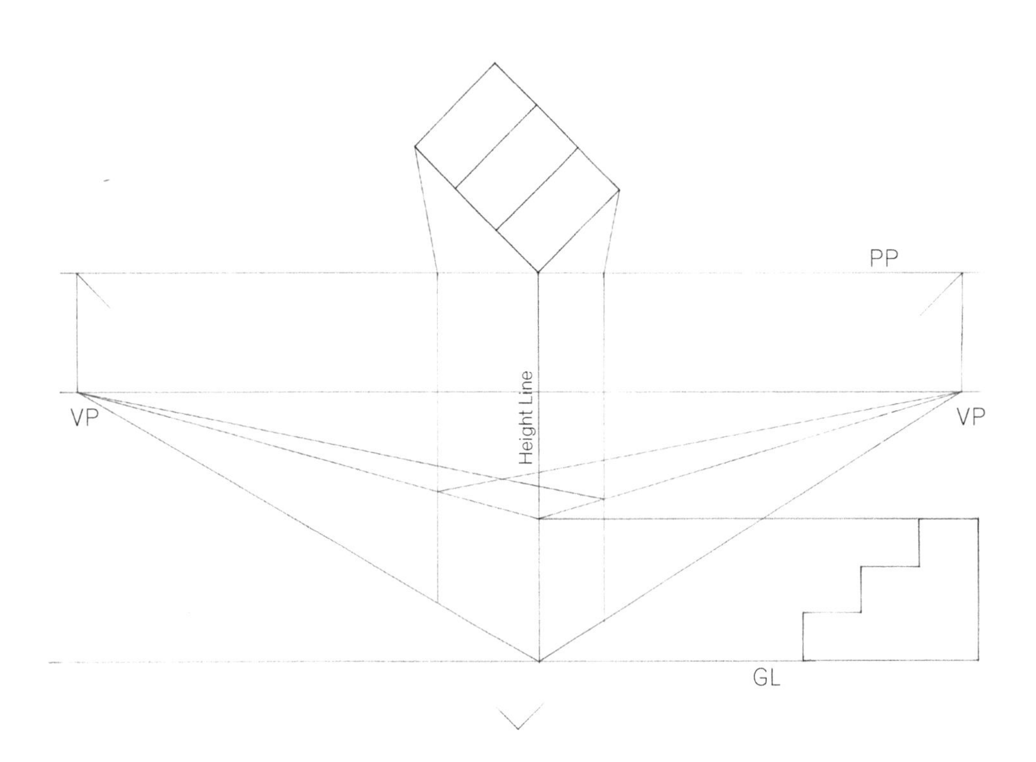

ISBN: 9780170233279

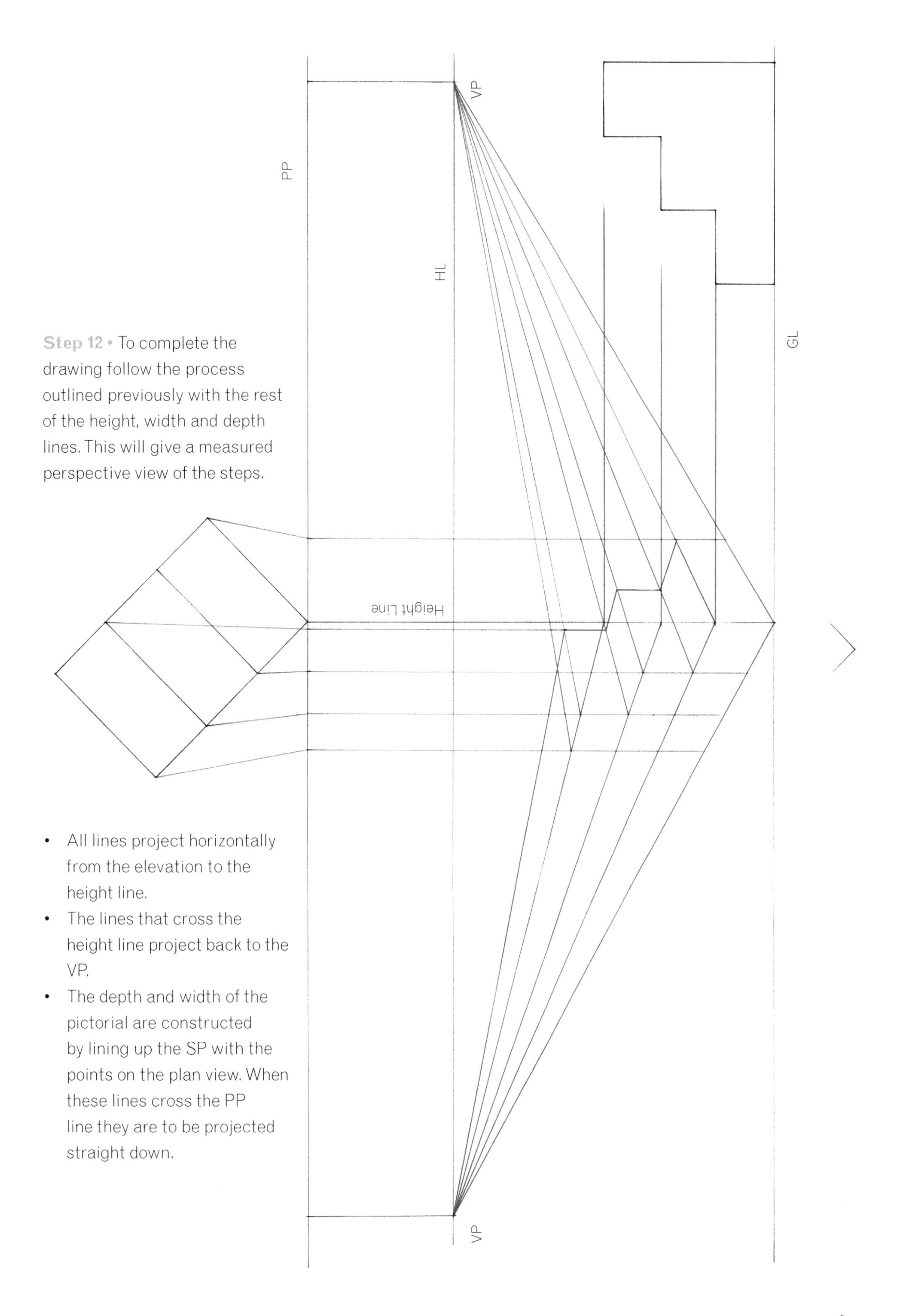

Step 12 • To complete the drawing follow the process outlined previously with the rest of the height, width and depth lines. This will give a measured perspective view of the steps.

- All lines project horizontally from the elevation to the height line.
- The lines that cross the height line project back to the VP.
- The depth and width of the pictorial are constructed by lining up the SP with the points on the plan view. When these lines cross the PP line they are to be projected straight down.

ISBN: 9780170233279

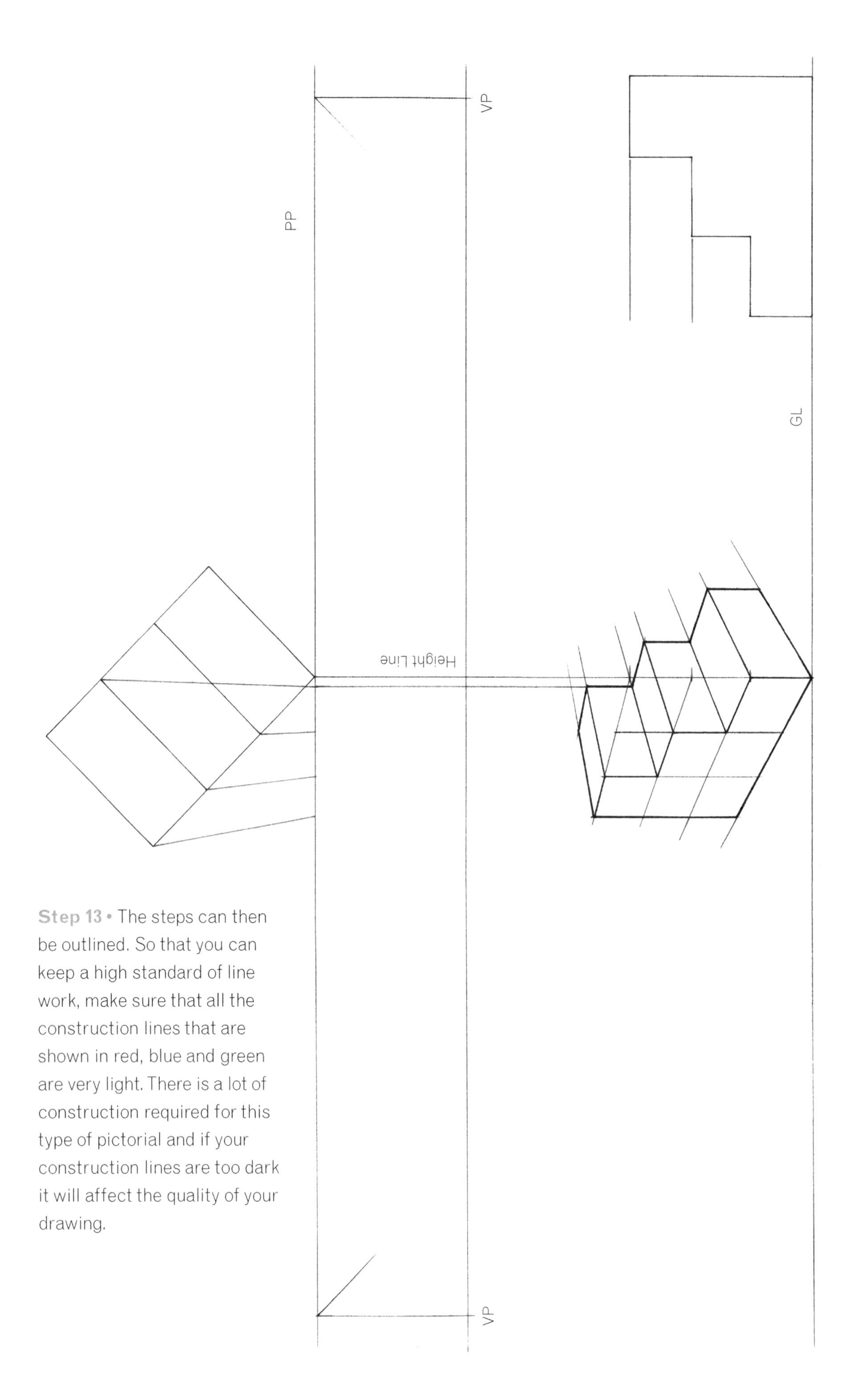

Step 13 • The steps can then be outlined. So that you can keep a high standard of line work, make sure that all the construction lines that are shown in red, blue and green are very light. There is a lot of construction required for this type of pictorial and if your construction lines are too dark it will affect the quality of your drawing.

ISBN: 9780170233279

Cylinder construction

The construction of a cylinder is complex. You will need to plot several points to construct a circle in perspective. Drawing the crated box only for the cylinder and sketching the circles required into this space will not give enough accuracy.

Step 1 • The plan, elevation, picture plane line, stationary point and vanishing points are set up. In this example the picture plane line is drawn through the back corner of the plan.

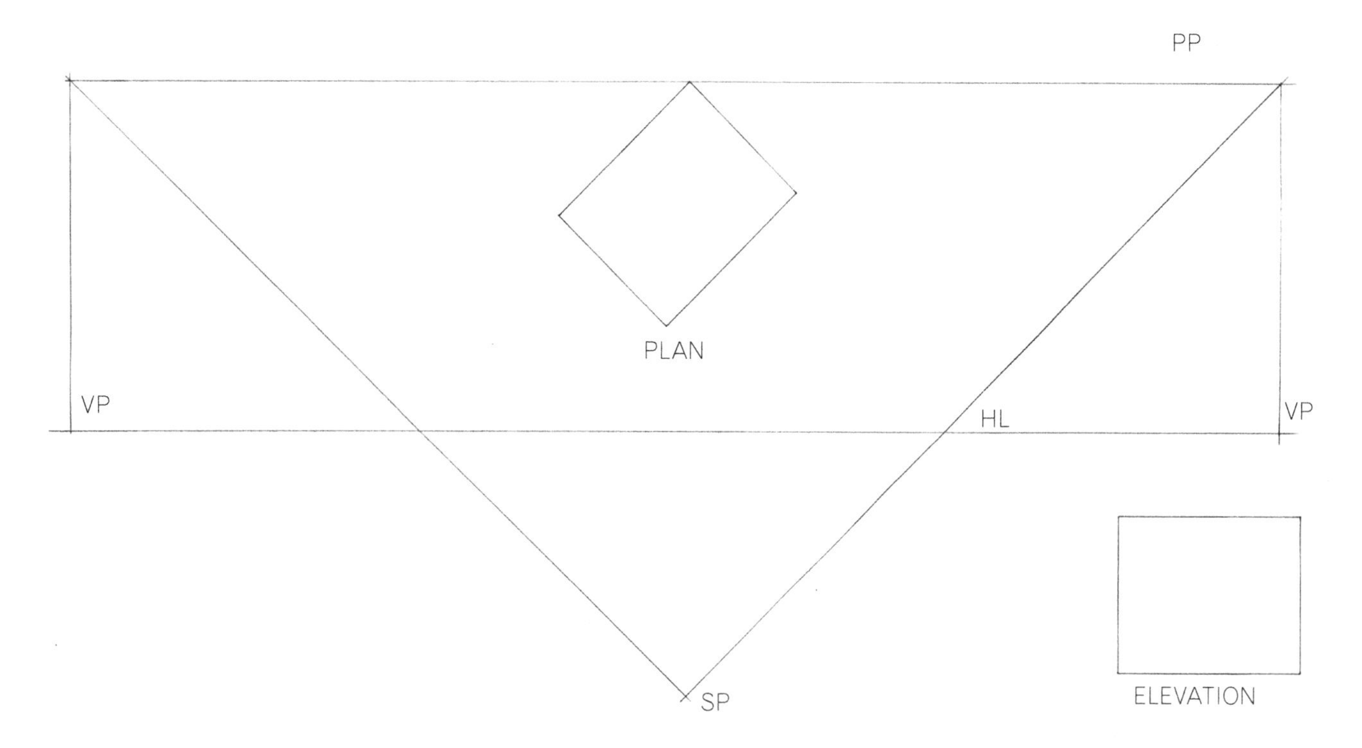

Step 2 • The cylinder needs to be divided into sections. This is achieved by drawing a circle from the front face of the elevation and the plan. The circles are then divided using each side of the 30°/60° set square.

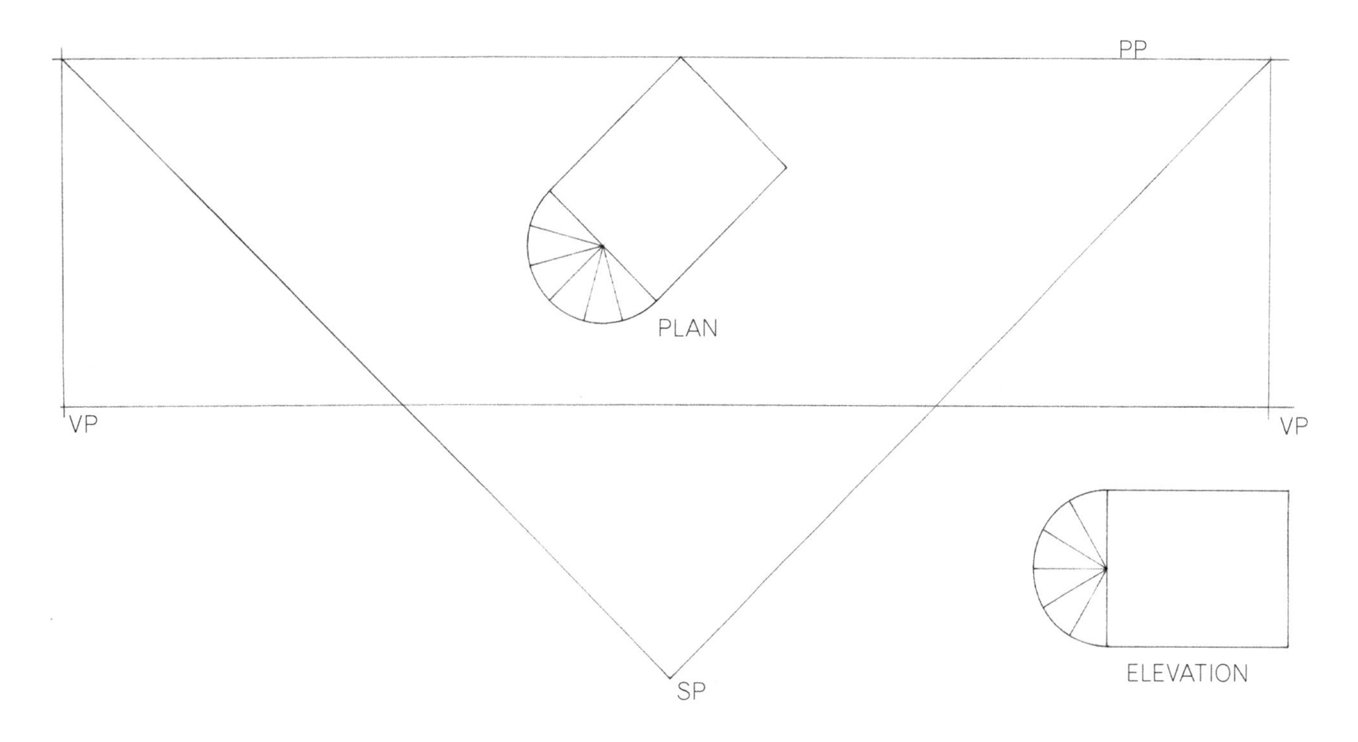

ISBN: 9780170233279

Step 3 • The points that are created by these lines crossing the circle are projected horizontally onto the elevation and at a 45° angle onto the plan. The red lines indicate this on the elevation and the green lines indicate this on the plan view.

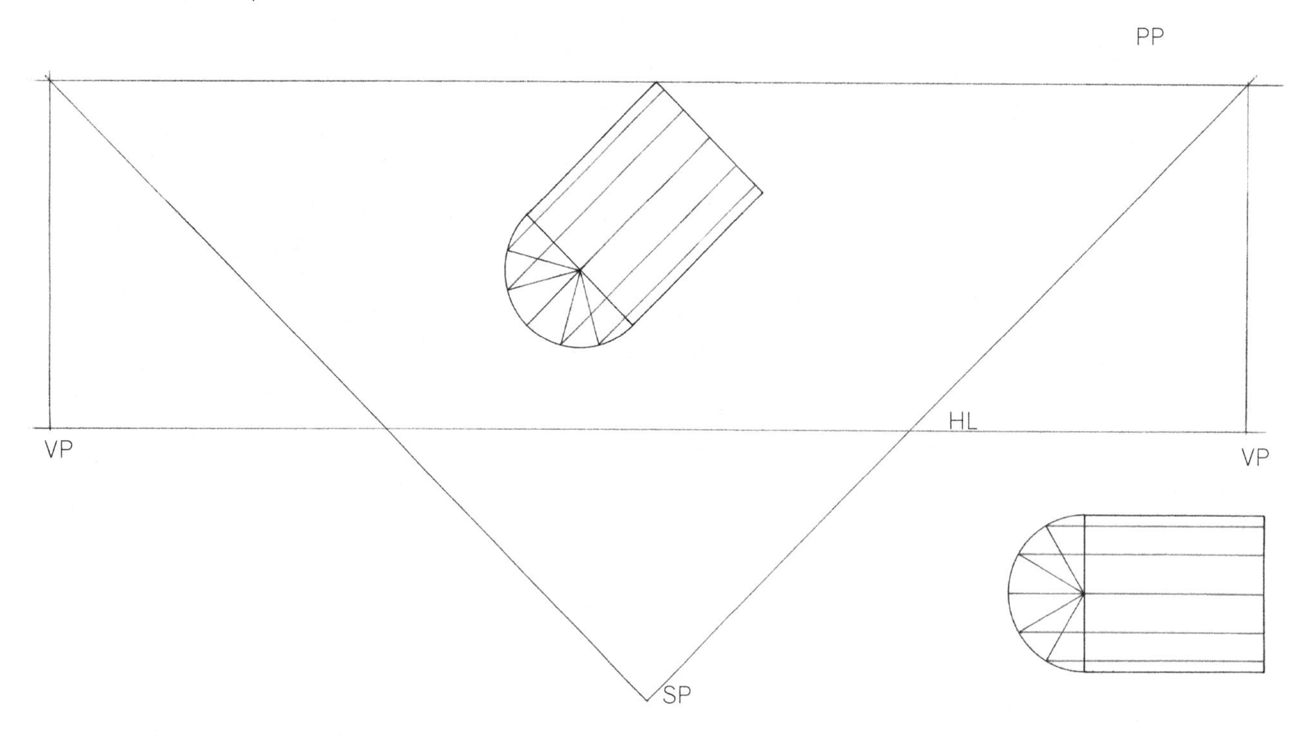

Step 4 • Since the picture plane line is set up on the back point of the plan, the side of the plan needs to be extended to the picture plane line. The point where this line crosses the picture plan is projected for the height line. These lines have been shown in red.

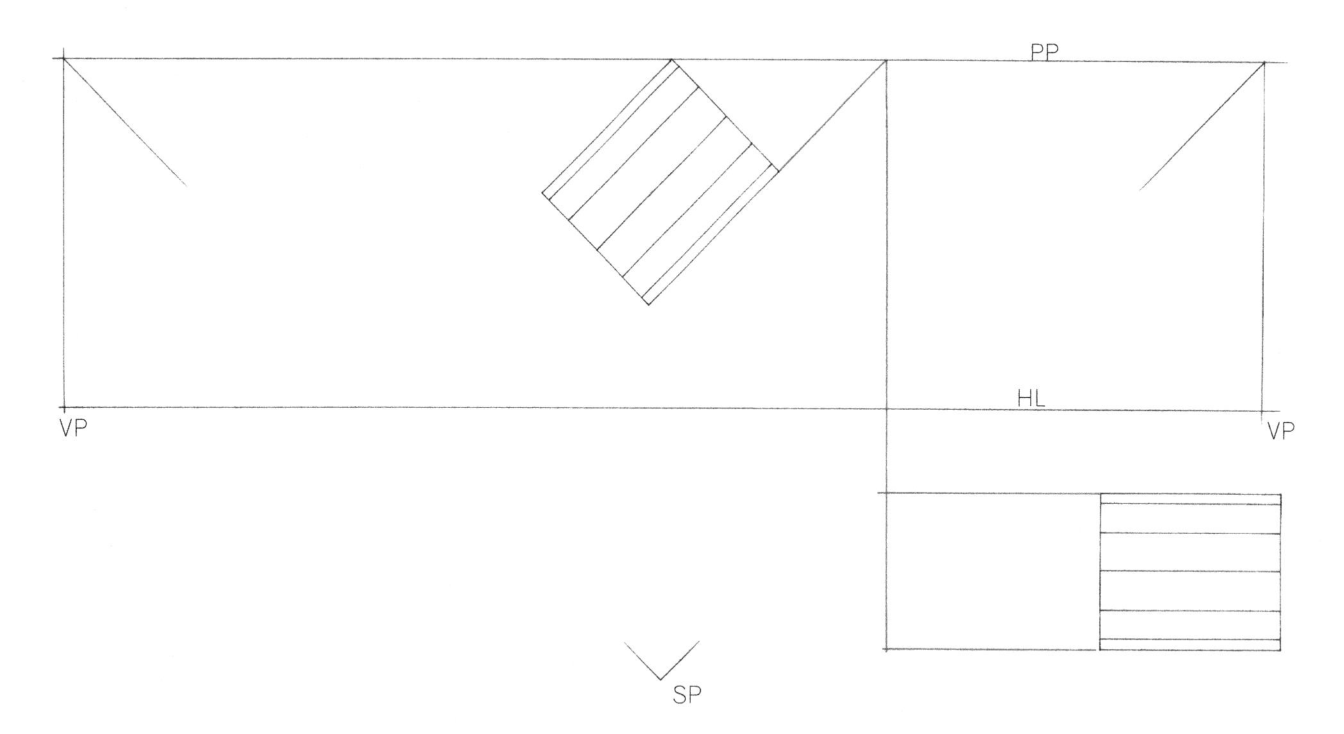

ISBN: 9780170233279

Step 5 • The main points on the plan are lined up with the station point and projected to the picture plane line. These lines are shown in green.

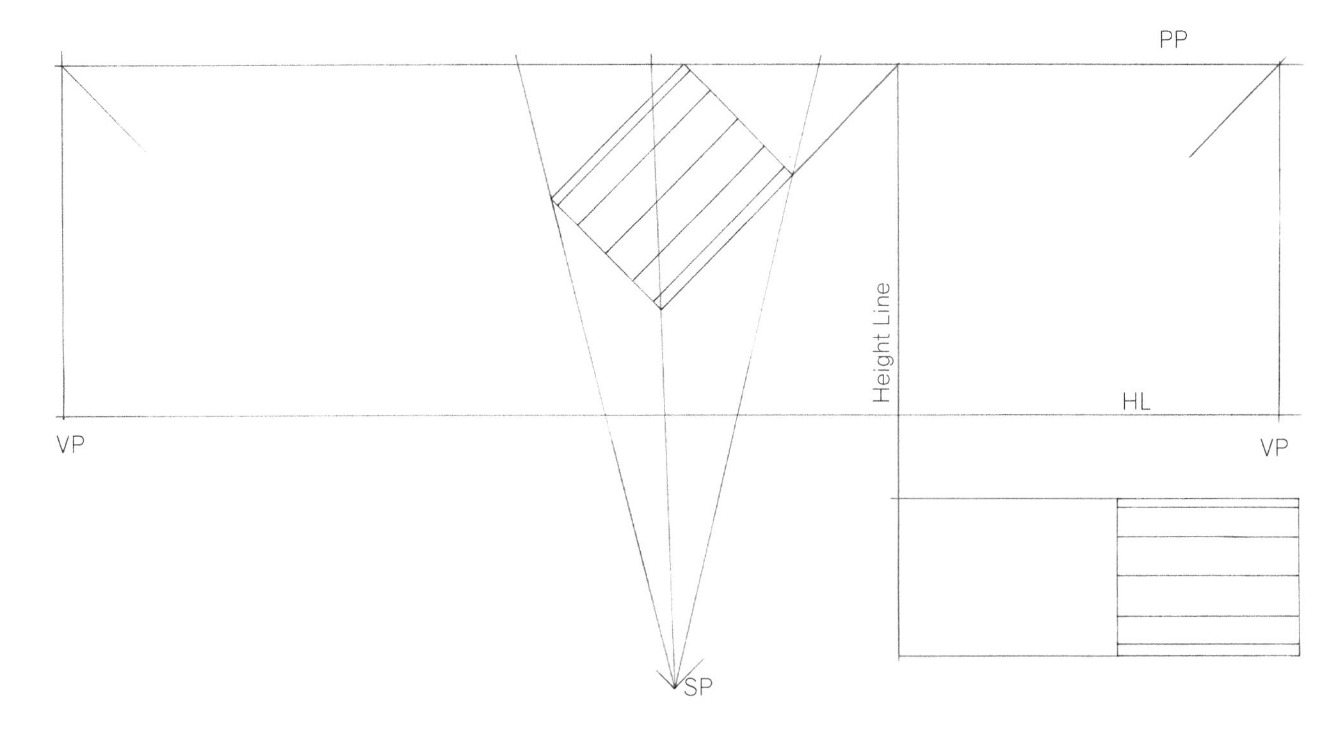

Step 6 • The box for the cylinder is crated by projecting the heights of the elevation onto the height line. These lines are shown in red.

Step 7 • The points that are created are lined up with the vanishing points and projected.

Step 8 • The depth and width of the crated box are created by projecting vertically the lines that have been drawn onto the picture plane line. These lines are shown in green.

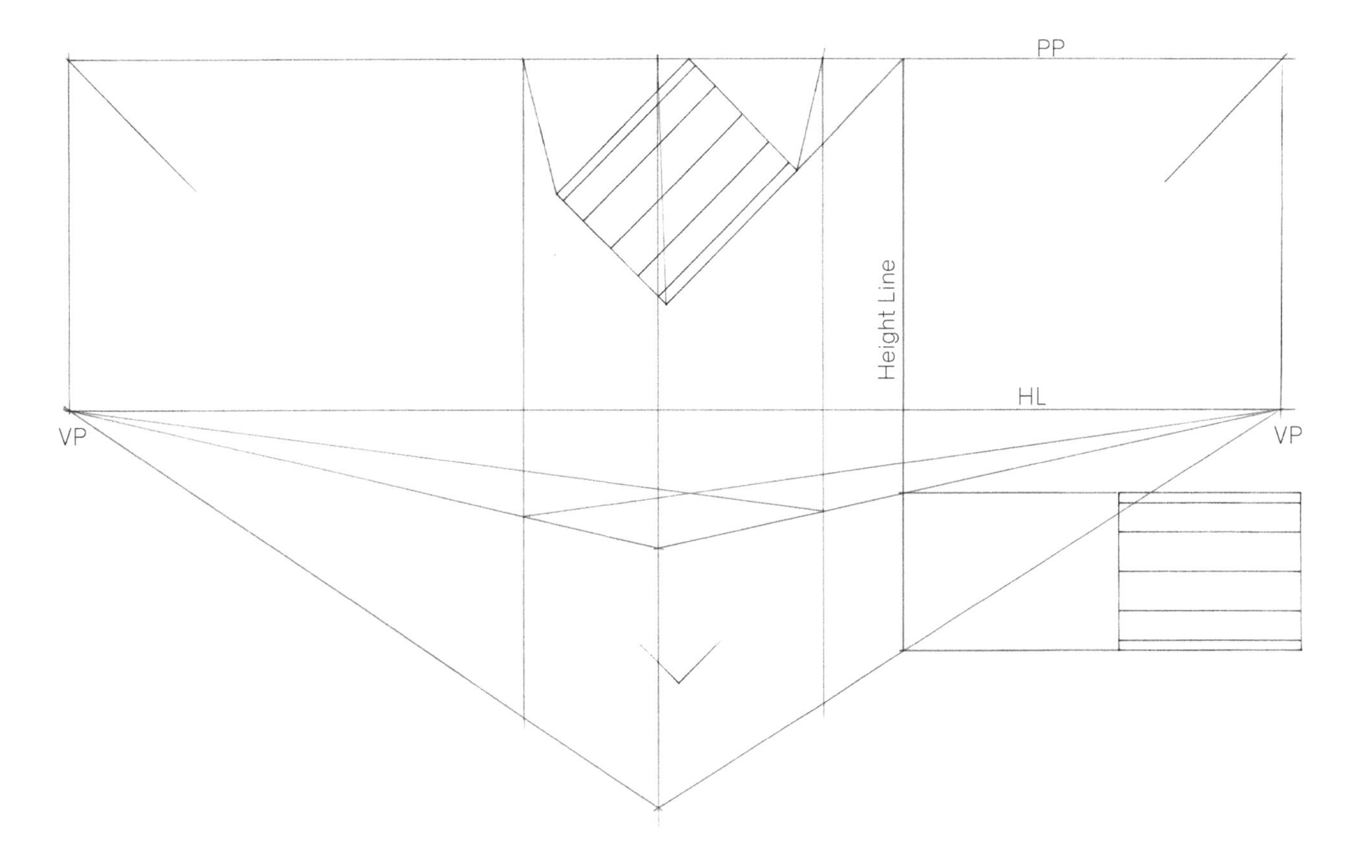

ISBN: 9780170233279

Step 9 • The divisions of the front surface of the cylinder on the elevation view are projected to the height line. These lines are shown in red.
Step 10 • The points created are lined up with the vanishing point and projected. These lines are also shown in red.
Step 11 • The points created by the projected lines from the vanishing point that cross the front edge of the box are then projected to the other vanishing point. These lines are shown in blue.

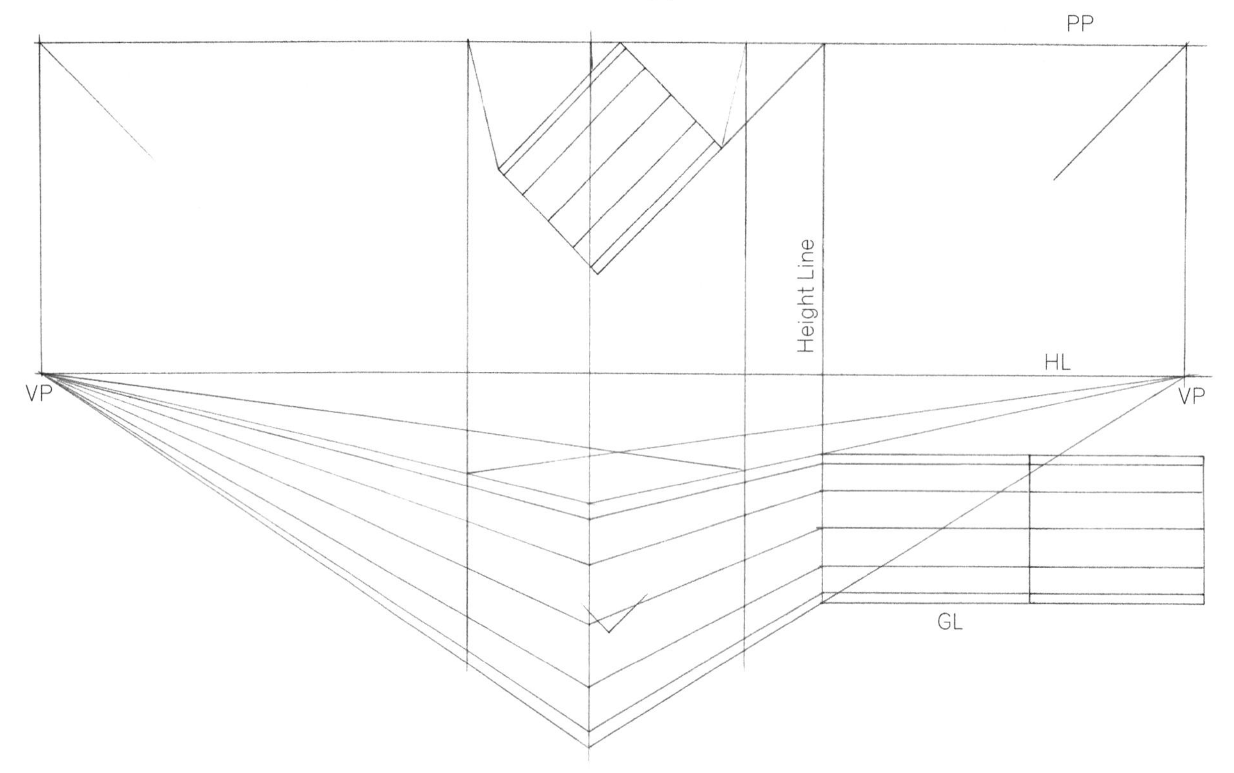

Step 12 • The divisions of the front surface of the cylinder on the plan are lined up with the SP point and projected to the picture plane line. These lines are shown in green.
Step 13 • The points that cross the picture plane line are projected onto the pictorial. These lines are shown in green.

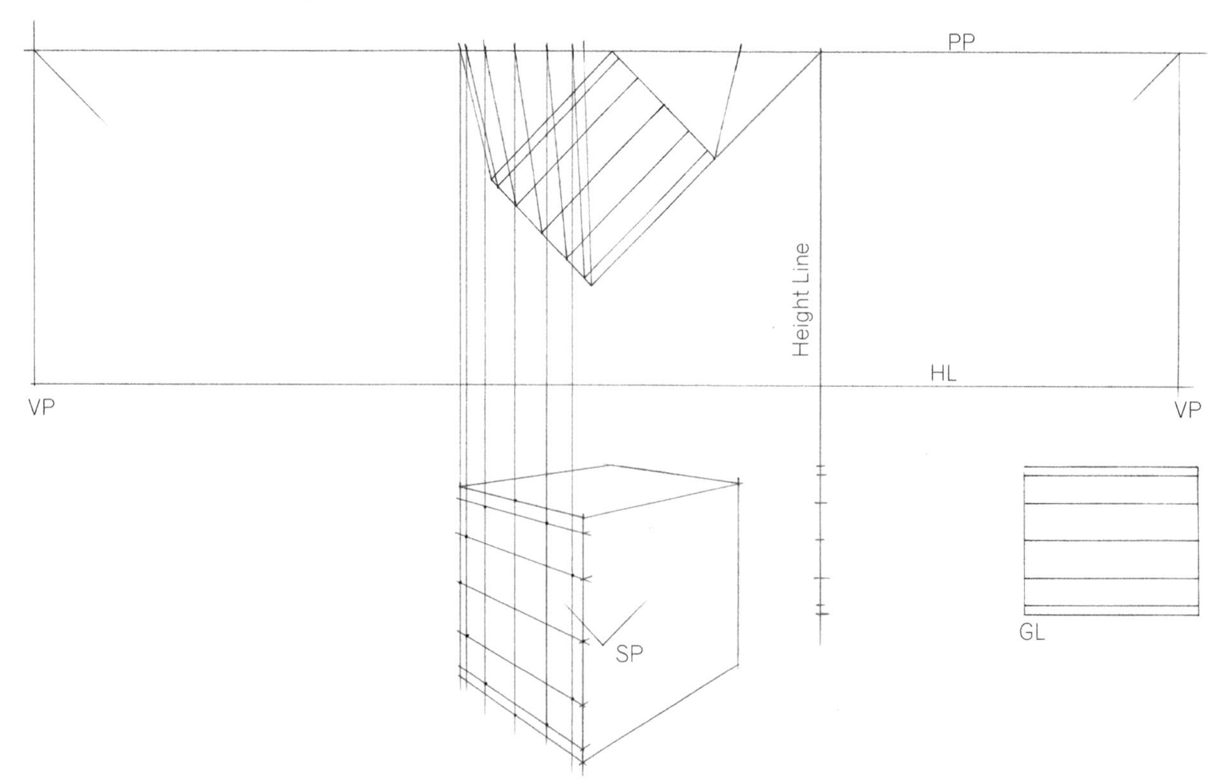

ISBN: 9780170233279

Step 14 • To create the points for the circle on the back surface of the box the same process is followed.

Step 15 • The projected heights are shown in red.

Step 16 • The projected points from the vanishing point through the height line are shown in blue.

Step 17 • The projected lines from the plan view are shown in green.

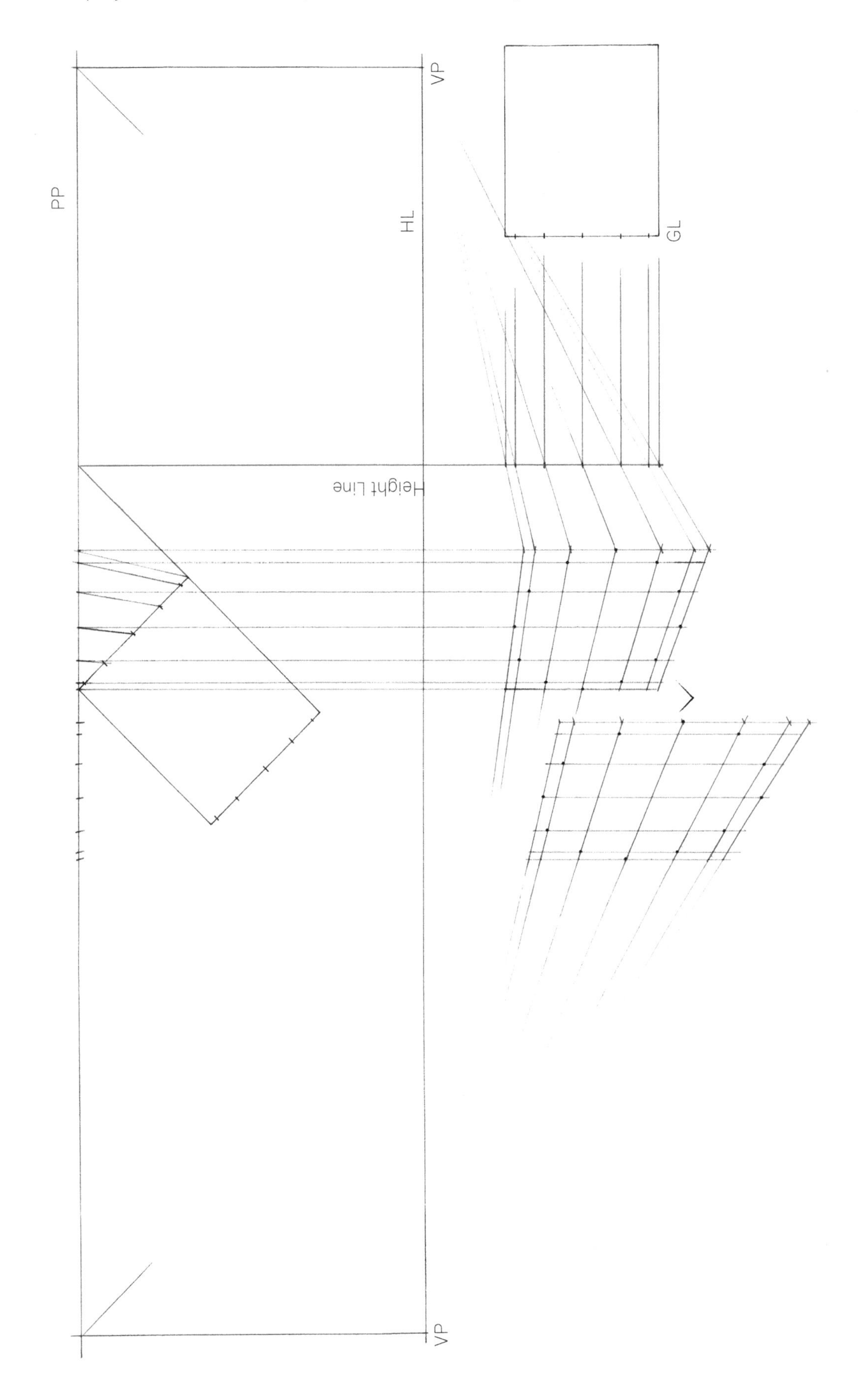

ISBN: 9780170233279

Step 18 • The points created for the two circles can now be joined together as tidily as possible.
Step 19 • The edges of both circles are joined together. These lines should also line up with the vanishing point.

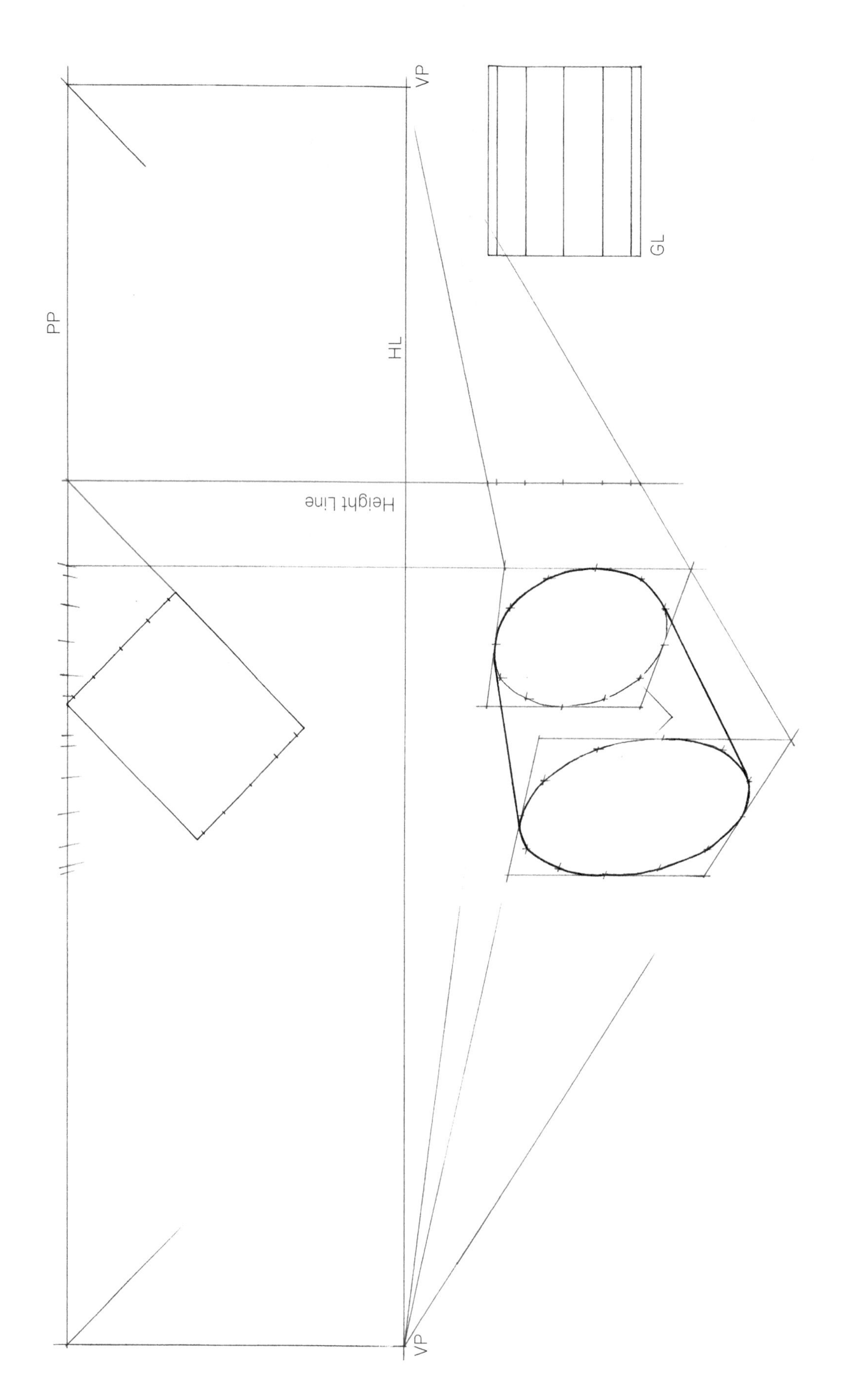

ISBN: 9780170233279

Paihere Tims — St Peter's College

The following work produced by Paihere Tims demonstrates the use of a perspective projection to construct his design for a chair. It is a difficult object to construct since each surface is curved or circular.

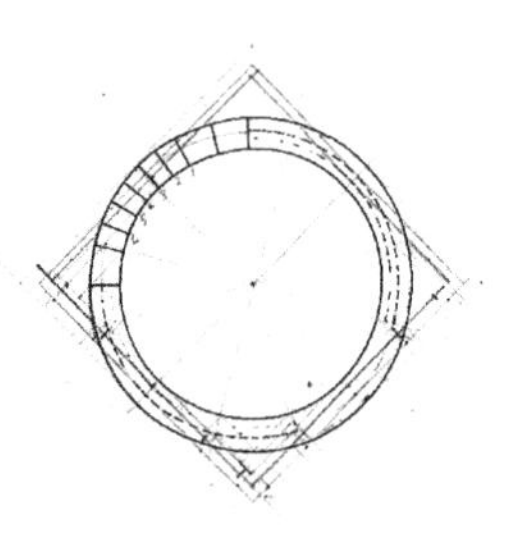

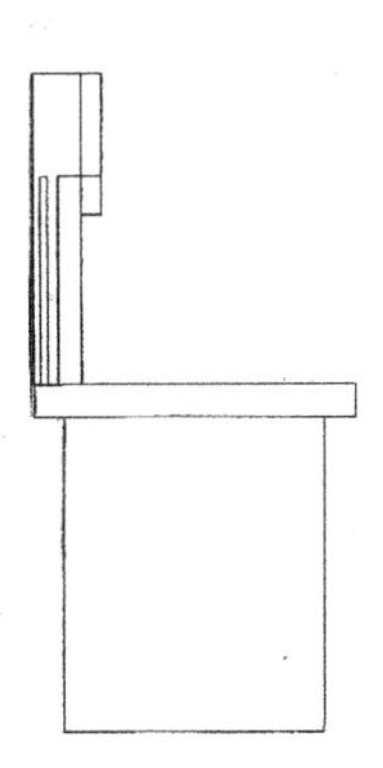

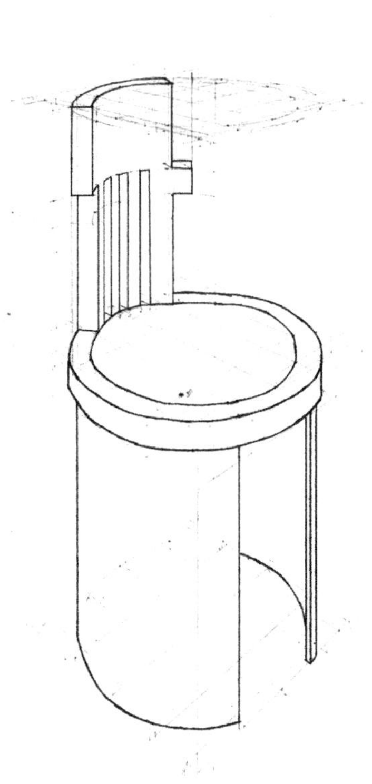

ISBN: 9780170233279

Positioning of the picture plane

The positioning of the picture plane will alter the size of the perspective pictorial. You will need to consider this when you are setting up your page.

The pictorial of the steps below has the picture plane positioned in front of the plan view. This means that the pictorial will be reduced in size.

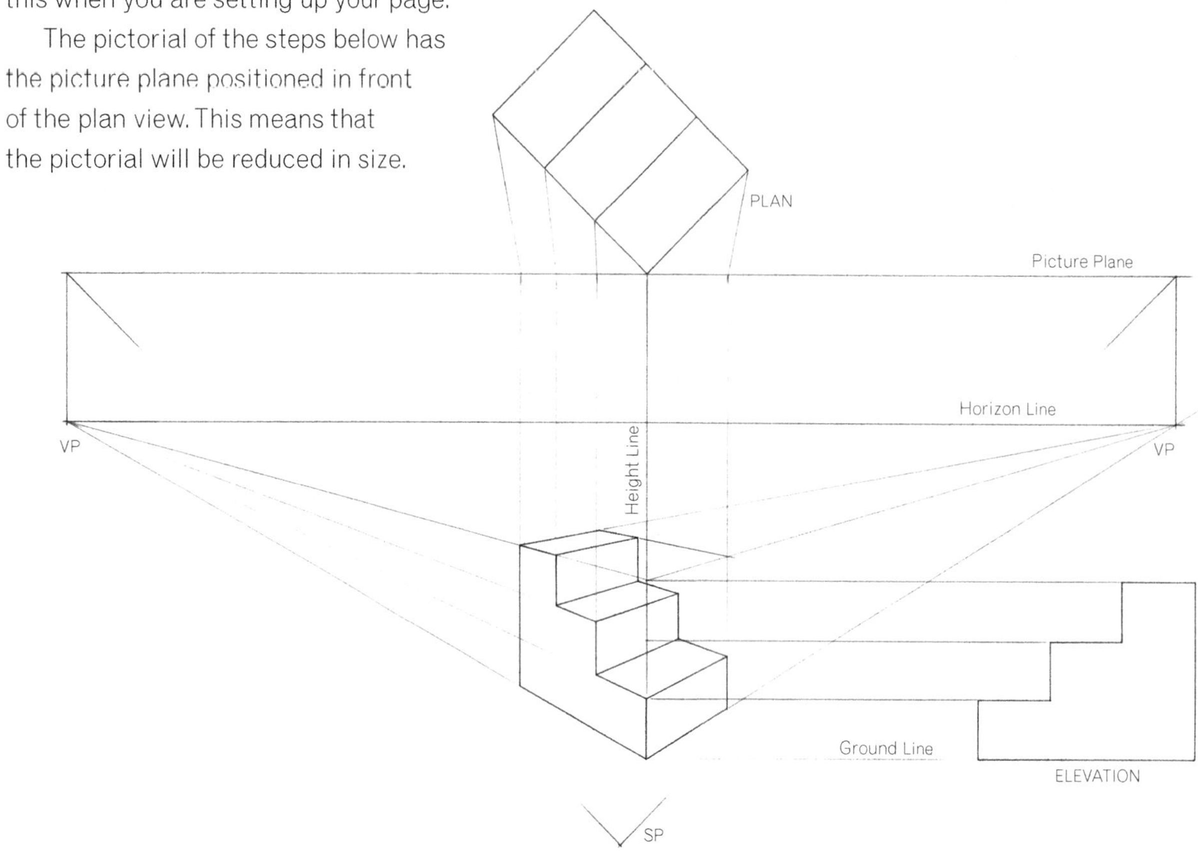

The pictorial of the steps below has the picture plane positioned passing through the plan view. This means that the pictorial will be reduced in size behind the ground level and enlarged in front of the ground level.

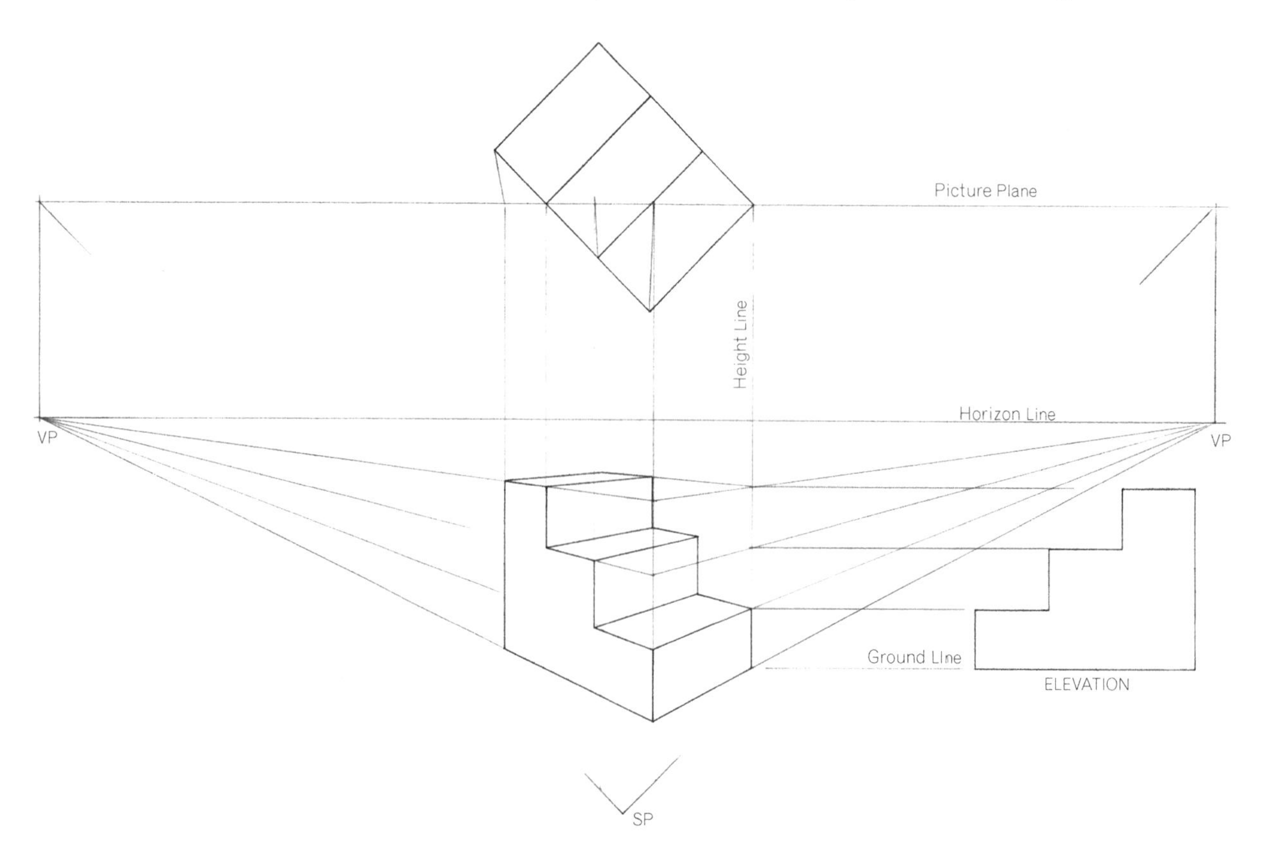

ISBN: 9780170233279

The pictorial of the steps below has the picture plane positioned behind the elevation. This means that the pictorial will be increased in size.

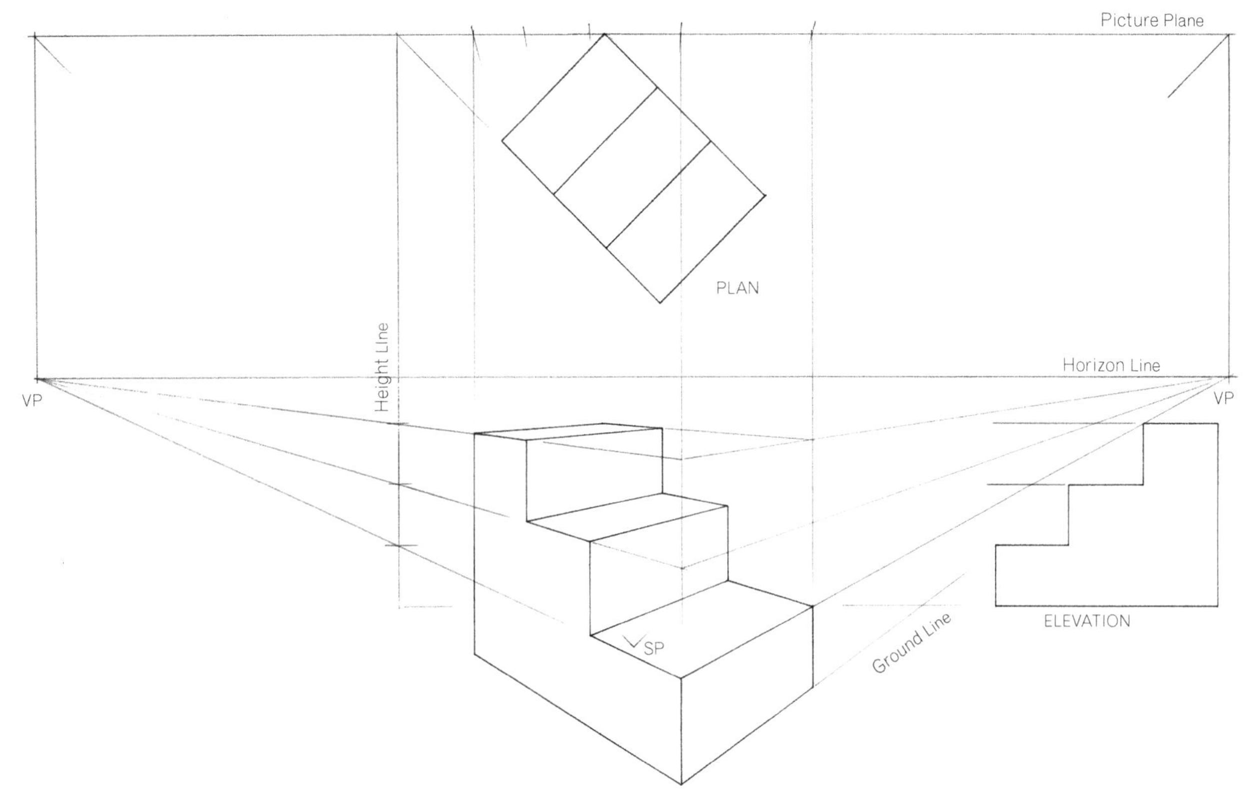

Positioning of the horizon line

The positioning of the horizon line will make a difference to how the object is viewed. The horizon line acts as your eye level line. The positioning of this line can alter the part of the object you can see and can dictate the perceived scale of the object.

The pictorial of the camera below shows the viewer looking down on the object. This is because the horizon line is set above the object.

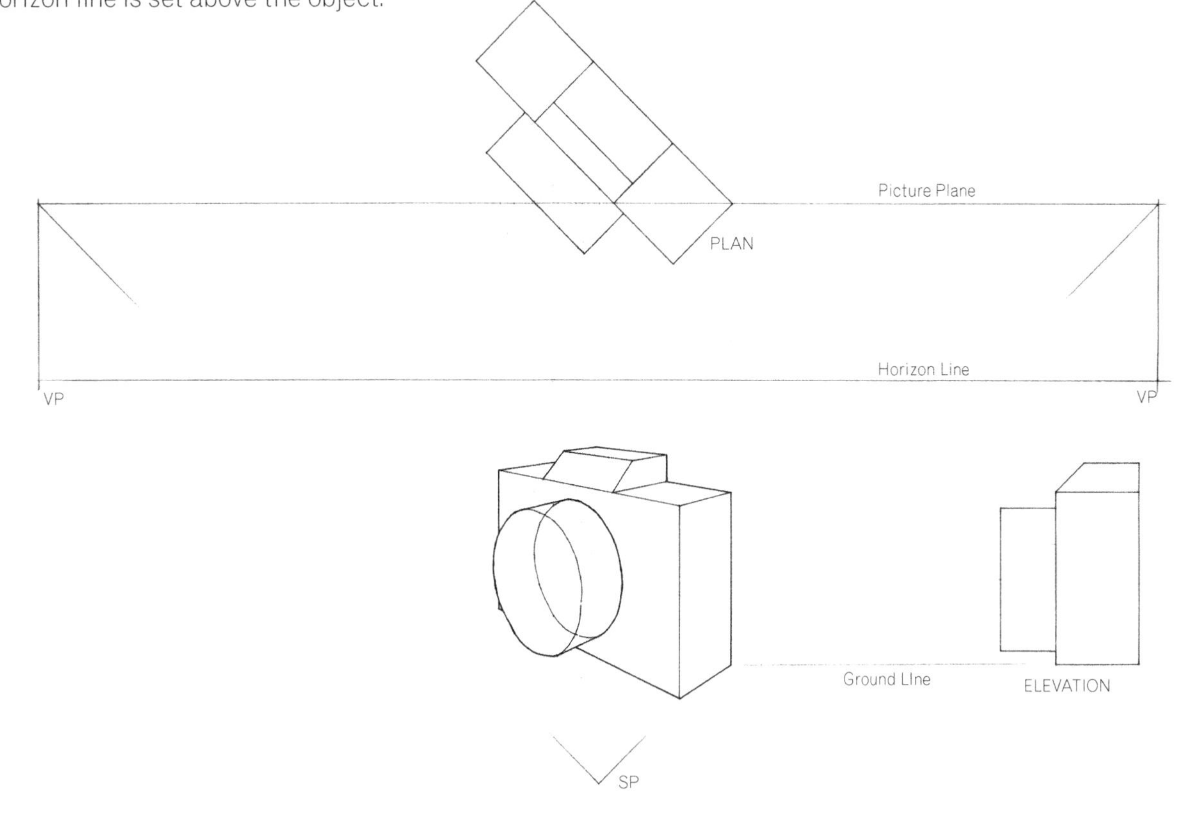

ISBN: 9780170233279

The pictorial of the camera below shows the viewer looking up and down at the object. This is because the horizon line is set in the middle of the object.

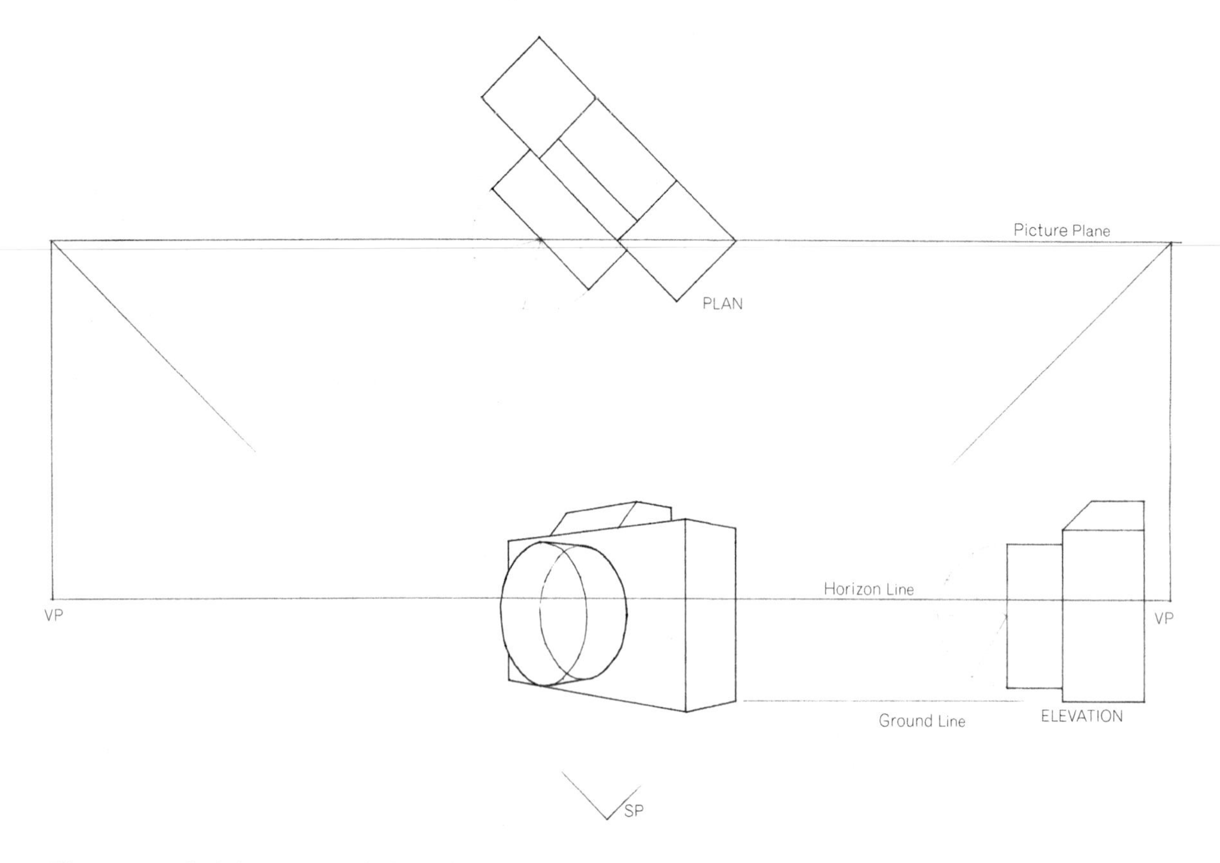

The pictorial of the camera below shows the viewer looking up at the object. This is because the horizon line is set below the object.

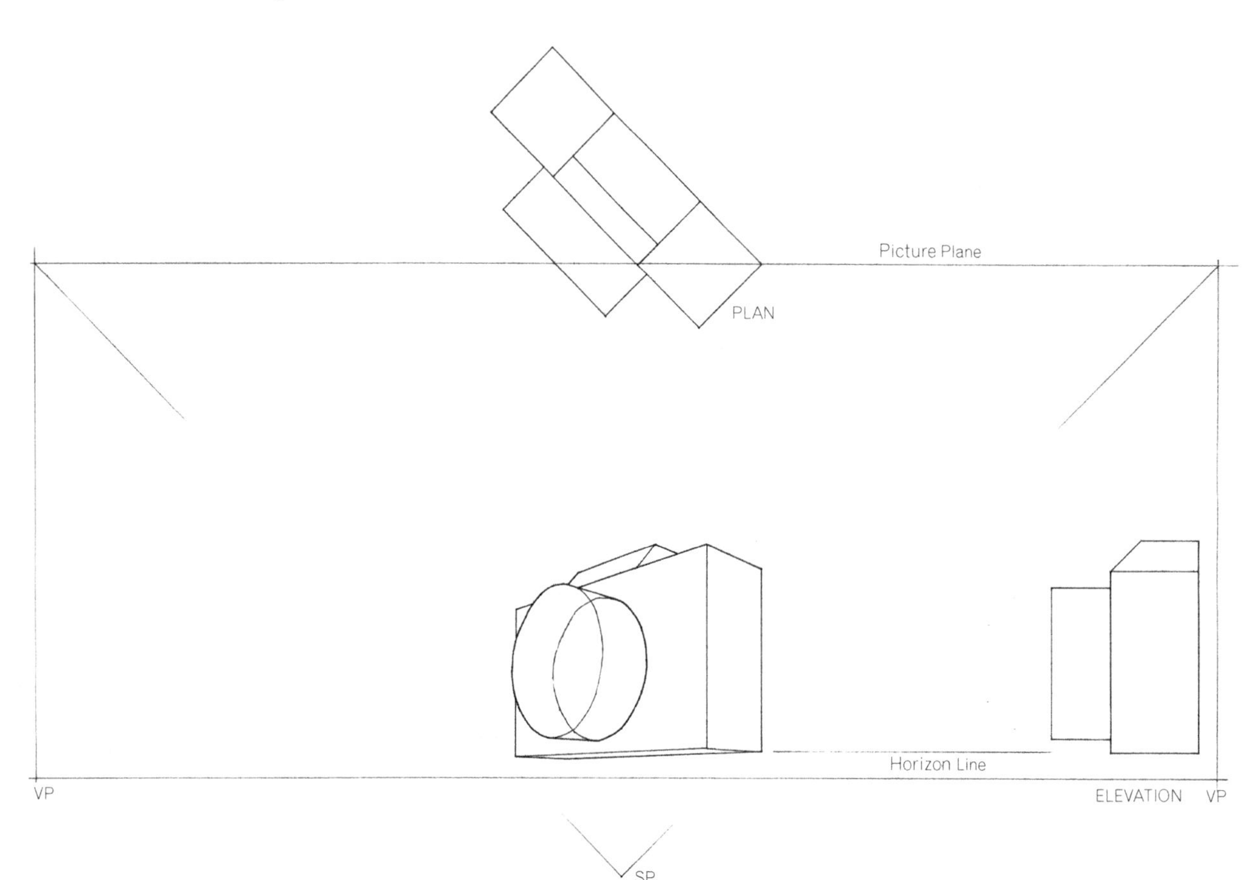

ISBN: 9780170233279

Positioning of the station point

The station point is the position the viewer is standing in relation to the object. The station point can be set up to the one side of the plan view therefore allowing the viewer to see more of one side of the object. The positioning of the station point affects the positioning of the vanishing points. This can alter the perspective dramatically. A pictorial will look more realistic the larger the distance between the vanishing points and the further the station point is from the object.

The following three pictorials are set up with the elevations, picture plane and horizon line in the same position. The only thing that is altered is the distance the station point is from the object or plan view.

The pictorial below is drawn so that the object is in proportion. Take notice of where the station point is positioned.

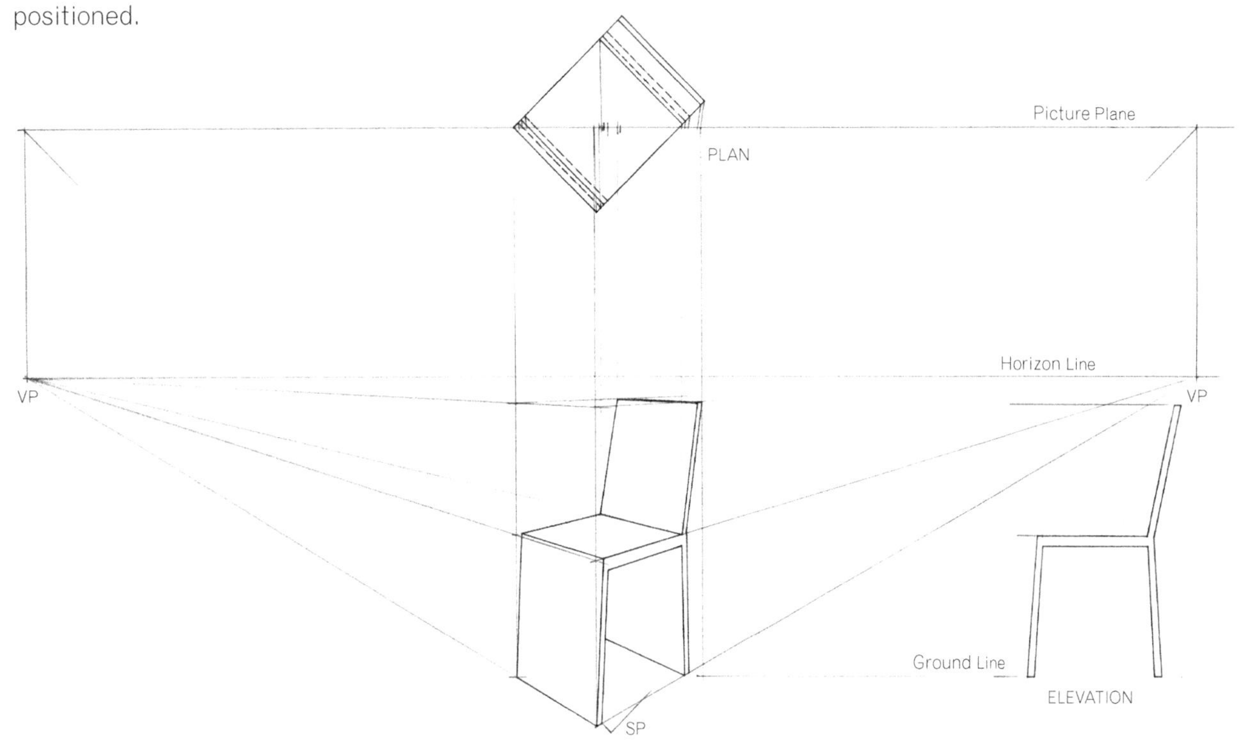

The pictorial below presents the object as a little distorted. The station point is a lot closer to the plan than in the previous pictorial.

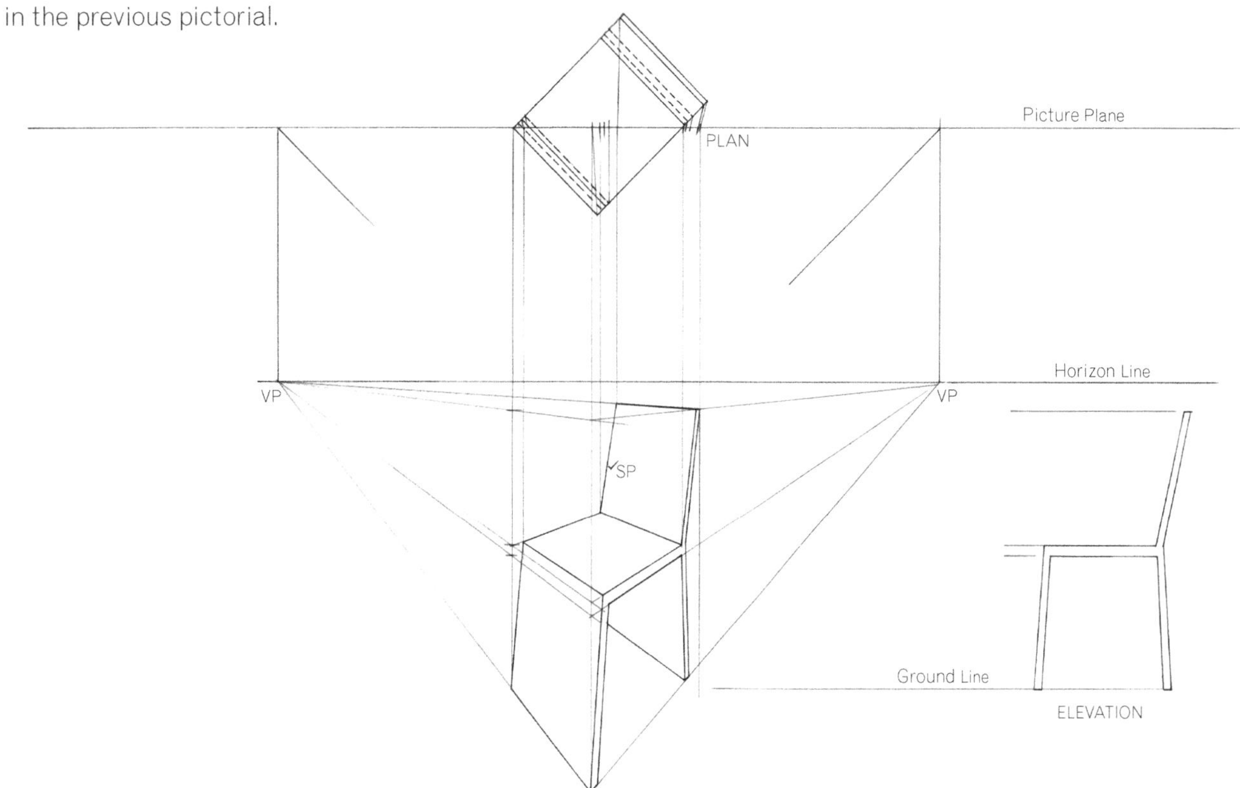

ISBN: 9780170233279

The pictorial below presents the object as very distorted. The station point is very close to the object. Notice that this has also affected how close together the vanishing points are.

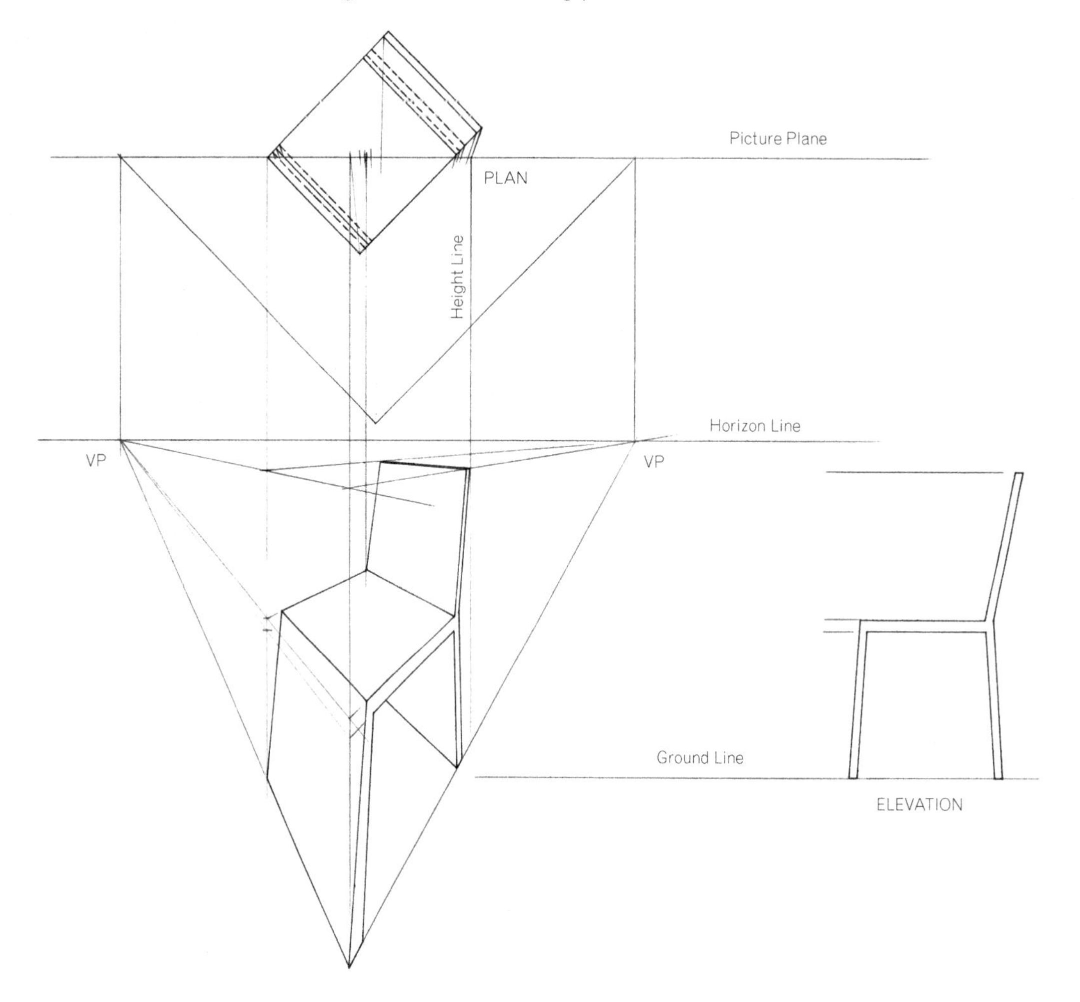

ISBN: 9780170233279

Christian Burgos — St Peter's College

The following work produced by Christian Burgos demonstrates the use of a perspective projection to construct his design for a workstation. The object is fairly difficult to construct as it includes cylinders and angles.

Picture Plane Line

Horizon Line

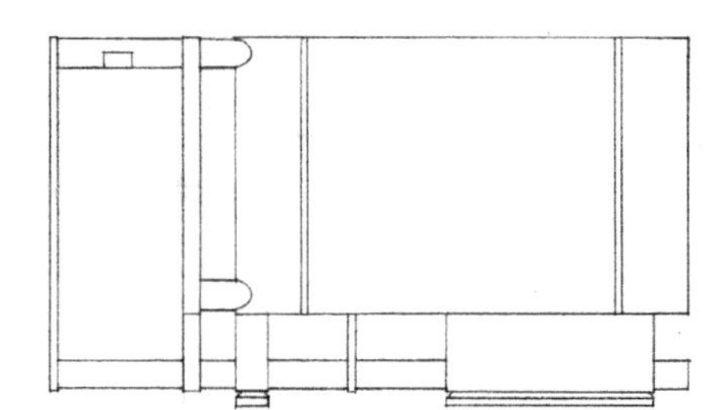

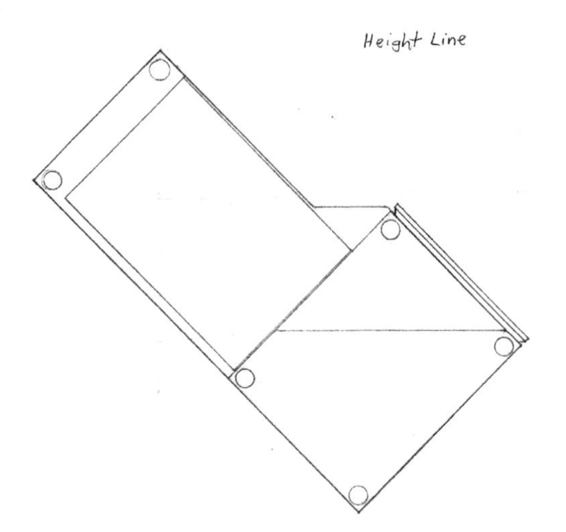

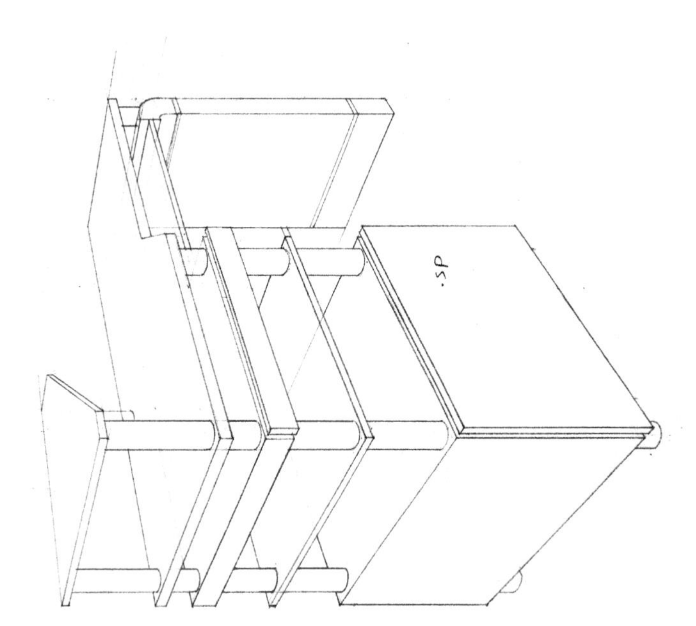

NAME: Christian Burgos
DRAWING TITLE: Mechanical Perspective
SCALE: 1:20

ISBN: 9780170233279

The following perspective views of the desk are detailed. There is a lot of plotting required. When completing drawings with this amount of detail be sure to follow the process explained methodically. It is a good idea to project the lines for each point as you need the information. For example, if you project all the height lines and depth lines at once it will be difficult to work out where they meet up. Also do not draw in all the construction lines of the SP lined up with the points on the plan as these will also confuse your drawing. Just draw the line as it crosses the picture plane. This has been shown on the following two examples.

The first example is presented with the picture plane positioned in front of the plan so the pictorial of the desk appears smaller.

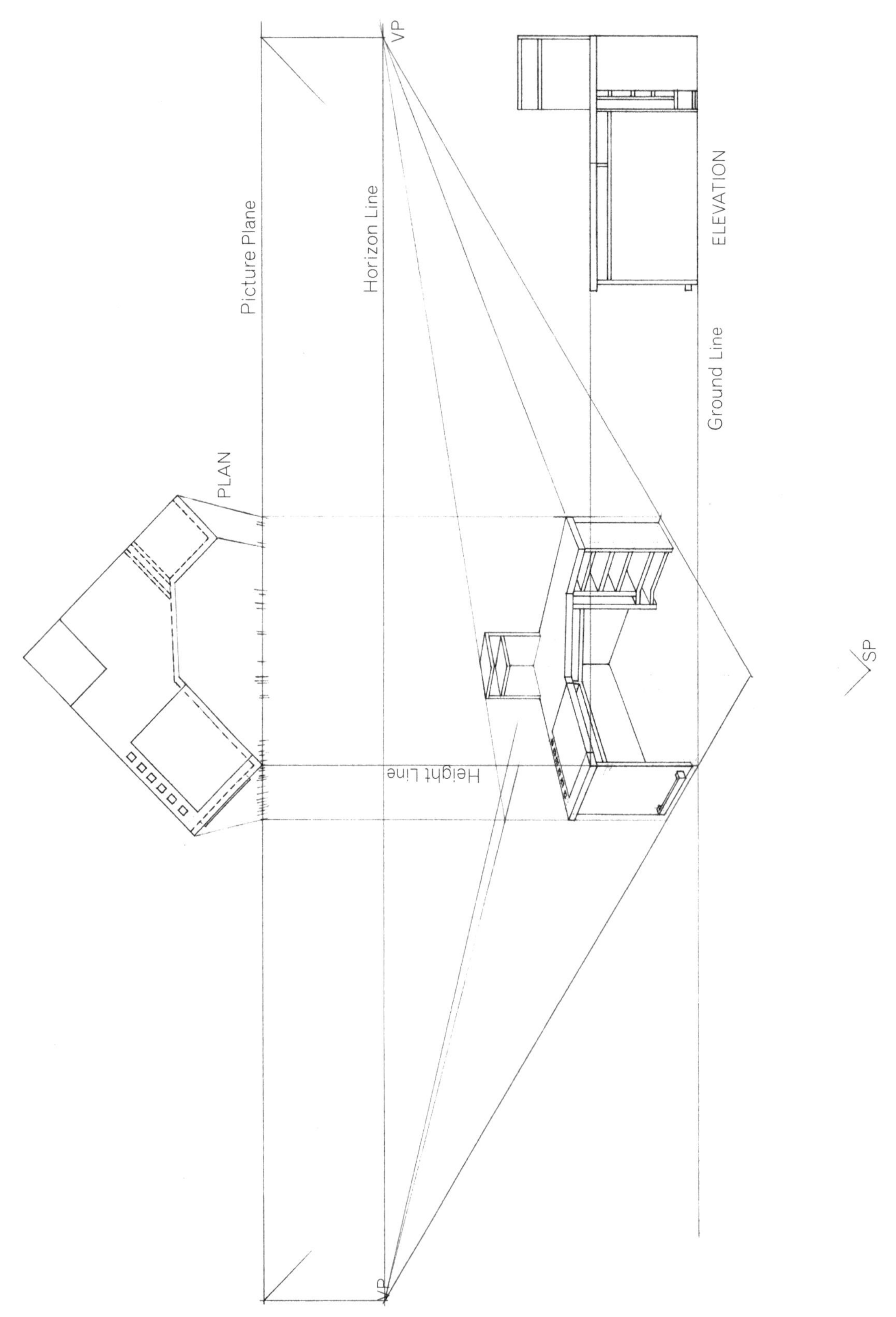

ISBN: 9780170233279

The second example is presented with the picture plane positioned behind the plan so the pictorial of the desk appears larger.

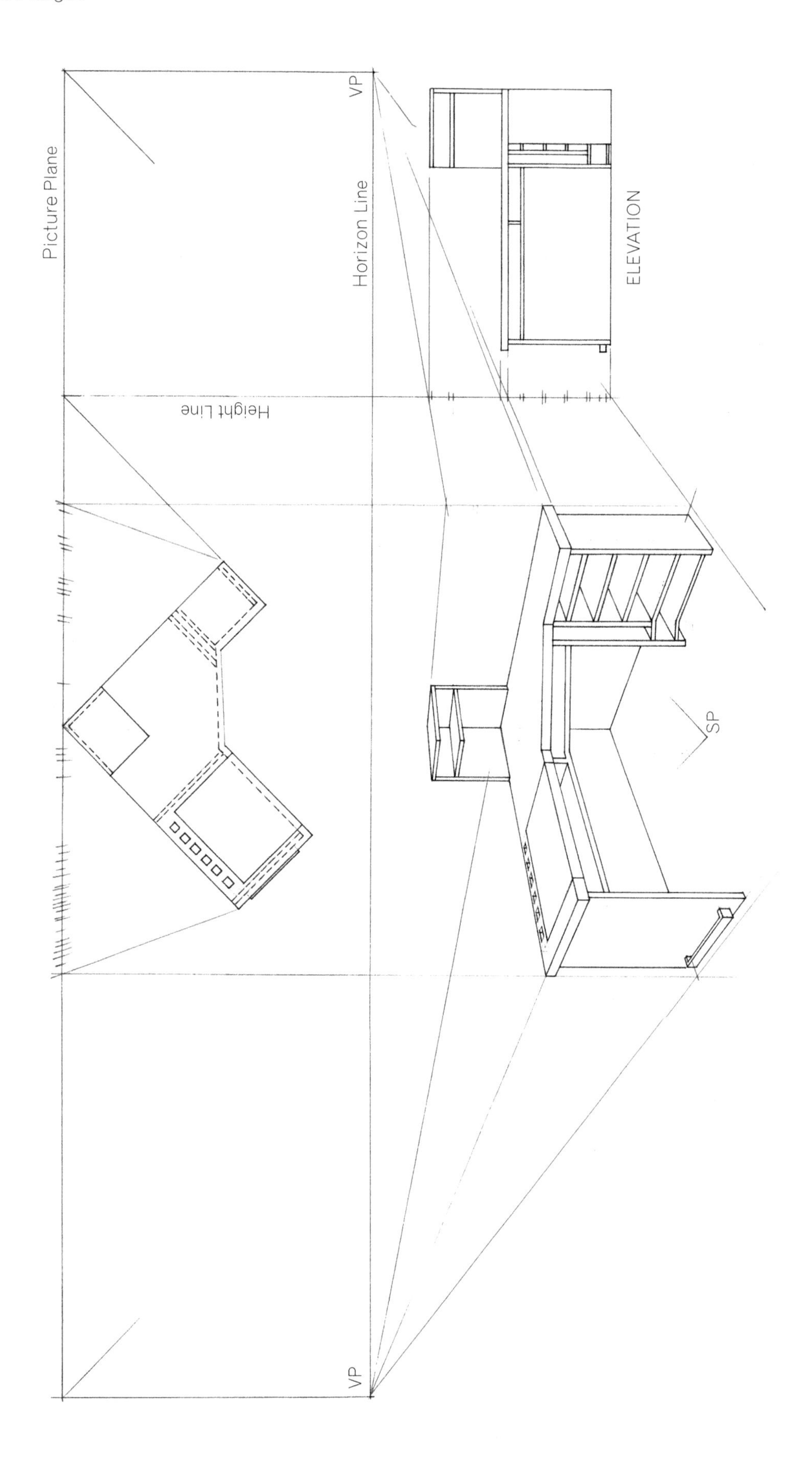

ISBN: 9780170233279

Christian Burgos — St Peter's College

The following work produced by Christian Burgos demonstrates the use of a perspective projection to construct his exterior design for a building presenting the design from two different angles. He has produced two drawings using the same floor plan but viewed from opposite directions. This gives a sound understanding of the design.

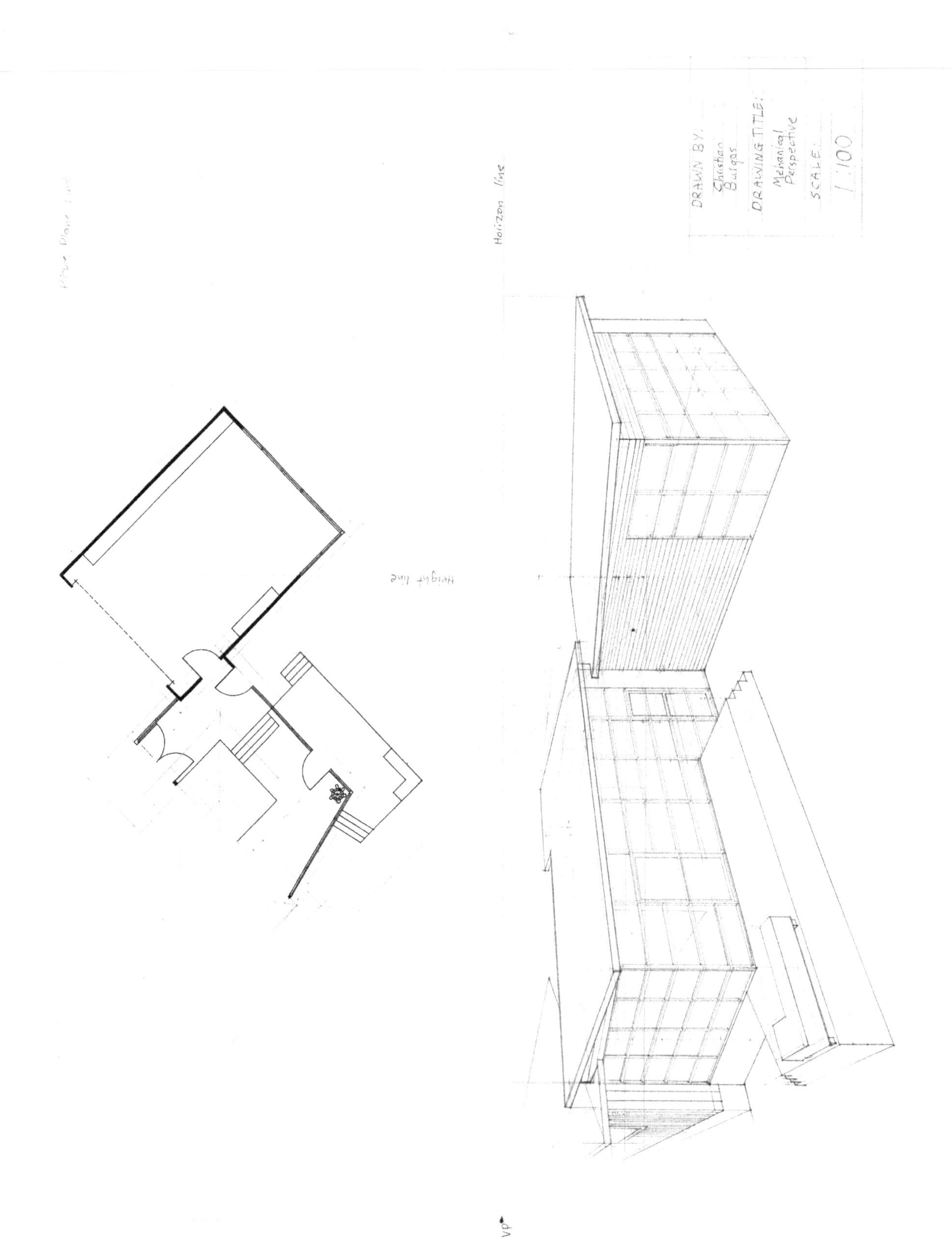

ISBN: 9780170233279

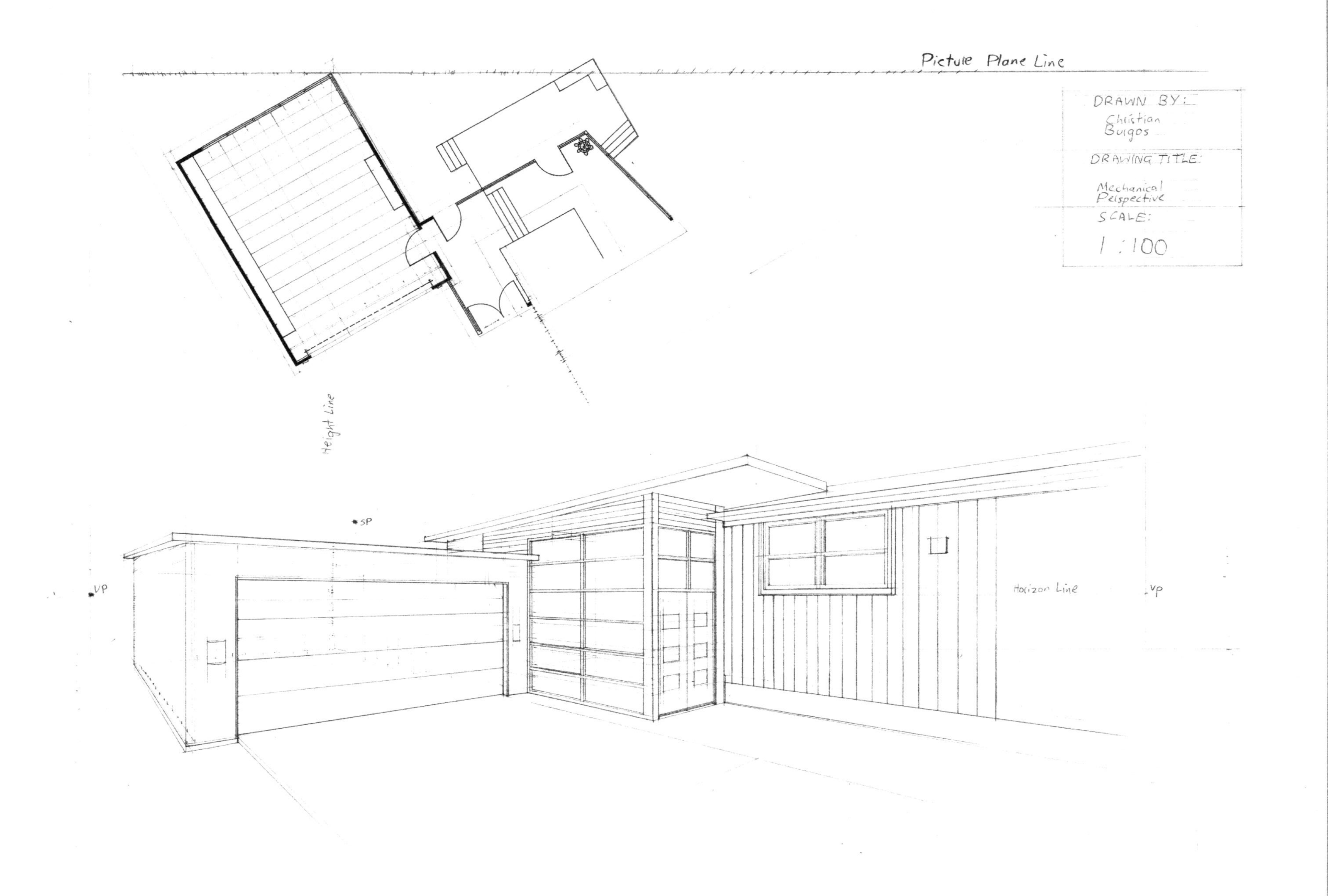

ISBN: 9780170233279

7 promotion of design work

To promote an organised body of work at Level 1 you are to select and present the features of design work to an audience. An organised body of work is your design work completed for a brief, which may include research. The presentation of work can be taken from your design work up to and including your final idea. At Level 1, you will need to demonstrate the selection and application of your visual communication techniques. Consider the use of compositional principles to strive to produce a high quality presentation that shows accuracy of layout, visual impact and precise execution of techniques.

To use visual communication techniques to compose a presentation of a design at Level 2, you are to use presentation techniques, compositional principles and modes and media to promote your design. The presentation you create is to be presented to an audience which could be for an exhibition, magazine or website. At Level 2 you must take work for your presentation from your design work up to and including your final design. Consider the use of compositional principles, modes and media and presentation techniques to strive to produce a presentation that promotes the essence of the design.

The work at both levels could be presented using the following visual communication techniques such as design sketches, instrumental drawings, models, photographs or digital media, display boards and installations depending on your selected mode of visual communication.

ISBN: 9780170233279

Compositional principles

Compositional principles are proximity, alignment, hierarchy, repetition, contrast, positive and negative space and focal point.

Proximity – the spacing between elements indicating their relationship e.g. elements that belong together should be kept close together. The spatial relationships of text communicate what information is important. Proximity is used to group elements of related information together.
Alignment – the position of blocks of text and images (left alignment, right alignment, centre alignment). Alignment is used to organise the information and space.
Hierarchy – the order in which each element (image, text) is most important.
Repetition – a repeating element of fonts, weights, lines and alignments.
Contrast – the use of opposing elements, such as colours, forms, or lines, in proximity to produce an intensified effect.
Positive and negative space – positive space refers to the shapes of objects. Negative space refers to the shapes around objects.
Focal point – the main focus of a presentation (image, text).
Typography – the design and art of characters for printing.

Size, hierarchy and position activity

The following activity could be undertaken with an image of any product or object. The example uses the image of a toy dump truck. The activity will help you to have a better understanding of the compositional principles - ***proximity, alignment, hierarchy, repetition, contrast, positive and negative space, focal point and typography.*** Photographs of the toy truck have been taken from different viewpoints, strong text which promotes this product has been selected and a range of background images that relate to dump trucks is provided.

ISBN: 9780170233279

The text, images and backgrounds can be integrated to form a promotional page for the product. There are endless ideas that can be produced for the layout of the text, focal image, detail images and backgrounds. Three examples have been shown to demonstrate the different effects of the following:

- the type of font used
- the size of a focal image and detail images
- the alignment of detail images and text
- the proximity of images and text to one another
- the negative and positive space
- how the text (heading) can be manipulated
- the hierarchy of information (images and text)
- the use of a background image
- the contrast of different colours

DUMP TRUCK

A dump truck is a truck used for transporting loose material (such as sand, gravel, or dirt) for construction. A dump truck is equipped with a hydraulically operated open-box bed hinged at the rear, the front of which can be lifted to allow the contents to be deposited on the ground behind the truck at the site of delivery.

The three presentation pages are made up of a main focal image of a close up of the truck, detail images showing the truck from different viewpoints, a background image, a heading and text describing the product. Many different ideas for presenting the product can be formed by using different styles of headings, focal images, backgrounds and layouts.

The top idea plays around with the lettering. As the product is a dump truck, this idea has been used for the heading as the d is on an angle while the rest of the lettering remains straight across the page. The large focal image is of part of the product only. The edge of the product helps to frame the page. A second image has been used as the background demonstrating the dumping action of the truck. This image is lighter than the focal image to make a clear definition between the image but still clearly displaying the function of the product. Smaller images of the truck from different angles are aligned along the side of the page. This have been framed by a grey boarder.

The middle idea has a main focal image of the the entire product. It is set up in such a way that the focal image of the truck looks like it is driving on the dirt road displayed in the background image. The background image is faded so that it does not take away from the main images while still explaining the funtion of the dump truck – to transport large quantities of dirt. The smaller images aligned along the side of the page show the product from different directions. These images are framed with a darker grey background. These images also demonstrate the function of the product. The heading and content text is aligned along the top and side of the grey background.

The bottom idea plays around with the lettering. As the product is a dump truck, this idea has been used for the heading as the text DUMP TRUCK is on an angle as if it is the dirt being dumped out of the bed of the truck. The large focal image demonstrates the function of the truck. The background is a dark grey with image demonstrating the function of the product to the right hand side. The content text is aligned along the top of the grey background.

ISBN: 9780170233279

The following pages demonstrate the process of developing a presentation – *research, selecting work, initial ideas, developing the idea and the final presentation pages.*

Research

Investigate presentation methods and evaluate the design principles used. Refer to the following terms of composition – ***alignment, hierarchy, proximity, repetition, contrast, positive/negative space and the focal point.*** To explore a range of presentation techniques you could research using magazines, the internet, reference books and advertising. Consider the following techniques used to present the ideas, how the features are promoted, principles of composition used, how the features are communicated, the integration of presentation techniques and effectiveness.

The following pages demonstrate the research for a product design and spatial design presentation.

From the research the main compositional ideas identified are:

- make sure the information, both images and text, is clear for the viewer to understand
- use mainly only three colours (including white) to present the work as this approach is simple and effective
- different viewpoints of the design are necessary to give a good understanding
- keep text to a minimum and let the images explain the design
- align the text and images and use simple, clear text.

Selection of work

You are to select the key features from your design work to present to the audience. Decide which design sketches, instrumental drawings, models, photographs or digital media you want to present.

The spatial design brief requirements for this example were to re-design the main entrance, a new garage, conservatory and decking for an existing house. The following example demonstrates the selection of work for a spatial design brief. The images selected are the following:

- Perspective sketches from the development of ideas that have been rendered using promarkers.
- 2D instrumental drawings drawn to scale – floor plan and elevations.
- Parallel and angular perspective instrumental drawings that have been scanned and rendered using Photoshop.

The following modes and media can be used when producing and selecting work for presentation:

Modes and media

Modes – digital applications, photography, image manipulation, animation, models, drawing and sketching methods.

Media – pastels, airbrush, colour pencils, collage, marker pens, paint, gouache, card and digital media.

ISBN: 9780170233279

research existing promotional layouts designs

The sorted architecture boards use a grid pattern to display the work with repetitve blocks containing photographs, drawings or text. Each block in the grid is in quite close proximity to one another with just the white space to separate the images. The presentation integrates several forms of media such as photographs, elevations and plan views. The contrast used is effective, especially the use of white on black. The main focal points are the largest photographs on each page. Each presentation board contains a lot of information both visually and written. Sorted architecture have kept colours to a minimum using mainly white, black and blue. The photographs are the most important in the hierarchy with the drawings and text quite a lot smaller on the page. Except for the headings the text is aligned to the left.

The architecture 24 website displays the idea with one main focal point. The images to the side can be enlarged by clicking on them. A grid pattern is used to display the idea with squares containing a computer model of different views of the building so just one form of media is used. Examples of houses that are complete are shown usingphotographs. This page of the website contains minimal written information. The images do the talking. Architecture 24 have kept colours to a minimum using mainly white negative spaces, green and black. Simple, clear text is used and has a center alignment. On a second page architecture 24 has used white lines on a black background.

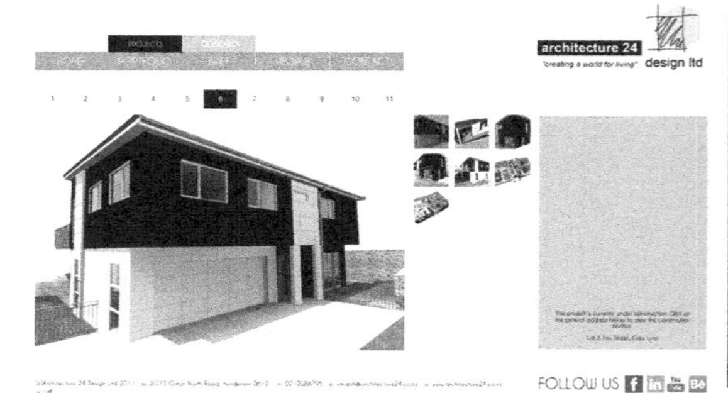

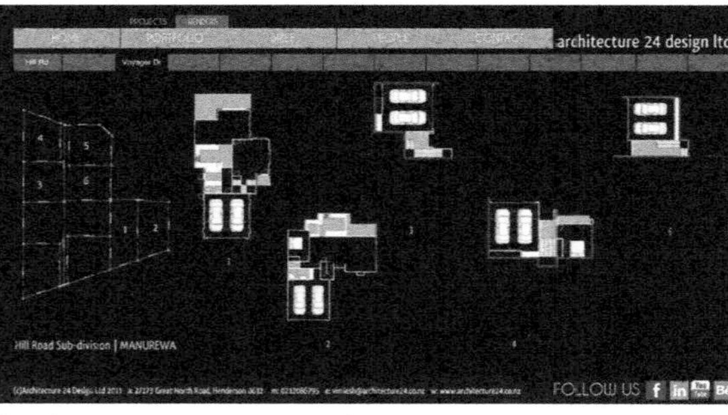

The Formway Design Studio website is a simple, elegant layout which highlights the purpose of the chair design without outlining the exact features and functions. The main focal point is the photograph of the two people using the chair in their own way. The photographs show the chair from a side view and the back and also highlights that the chair comes in two different colours. The background image highlights the movement and versatility of the chair and is integrated well with the foreground image. The text is simple and clear to the reader. All text is aligned to the left and is different sizes to highlight the most important information. The text in red outlines the function the chair has been designed for.

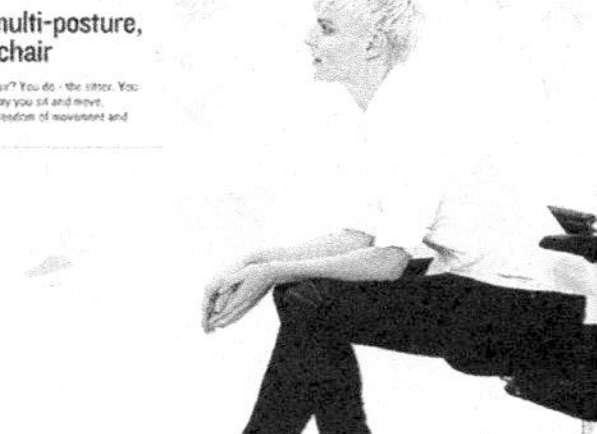

The Queens Wharf presentation includes perspective and plan view drawings. The perspective view gives a clear indication of the scale of the design. The plan views communicate the layout. The presentation allows the viewer to be well informed about the design and to also gain an understanding of what the environment will be like to use. The presentation includes quite a lot of text which is much smaller than the images. All text except for the heading has been aligned to the right. The main focal point is the perspective drawing. The use of colour has been kept to shades of grey, tan and orange on a white background. The use of minimal colours is effective to portray the information that is most important and helps link the perspective view to the plans.

Page produced by Field Landscape Architecture:
Dr Hamish Foote DocFA, MFA (1st Class Hons), BFA
C. Pete Griffiths MLA, BLA
Den Aitken BLA (Hons)

selection of work

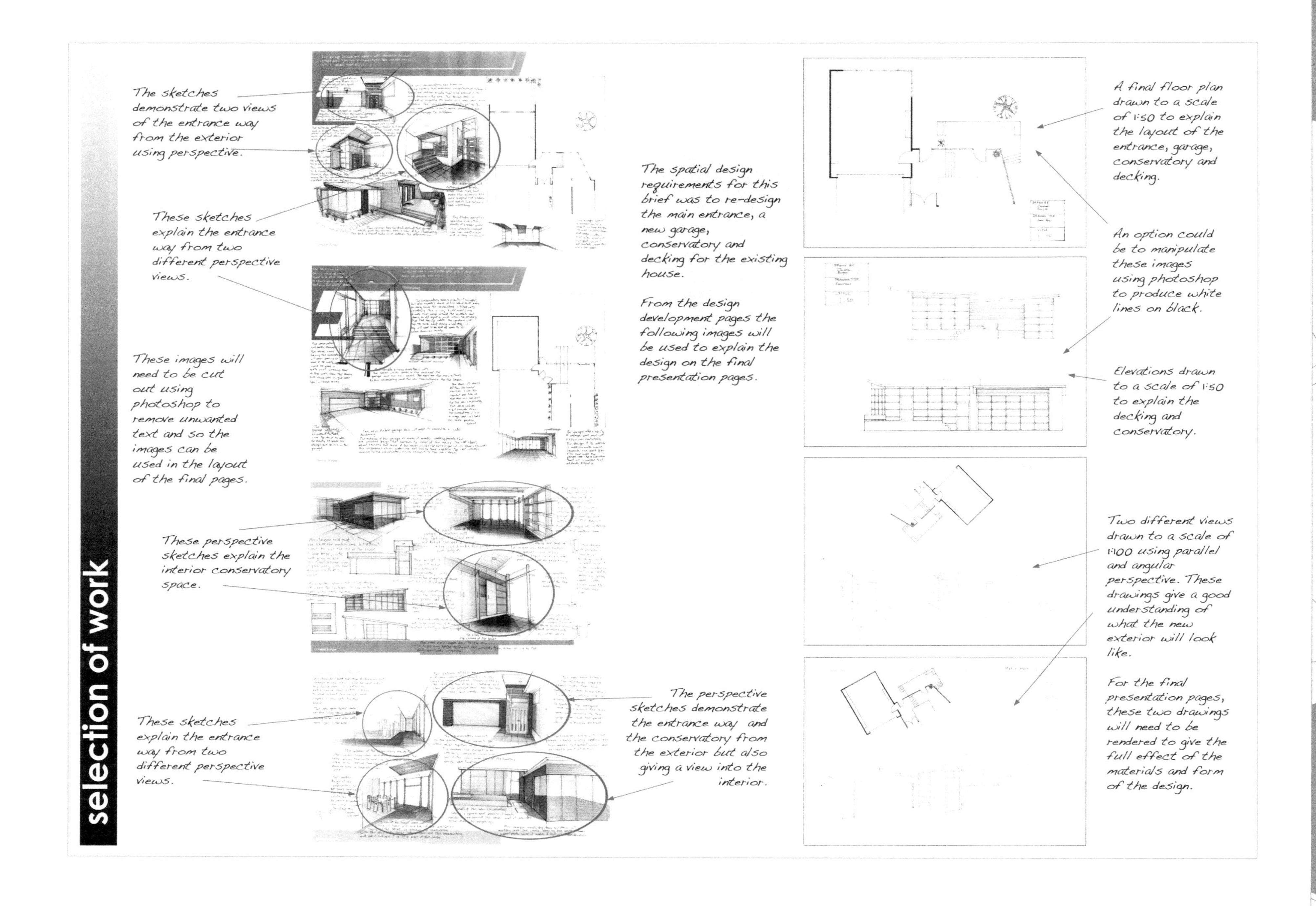

ISBN: 9780170233279

selection of work

ISBN: 9780170233279

Initial ideas

Produce different layout ideas for presenting your design sketches, instrumental drawings, models, photographs or digital media as thumbnail sketches. Consider the main layout features of your ideas, what your focus will be and the visual impact.

The following page demonstrates quick thumbnail sketches for the layout of the presentation pages.

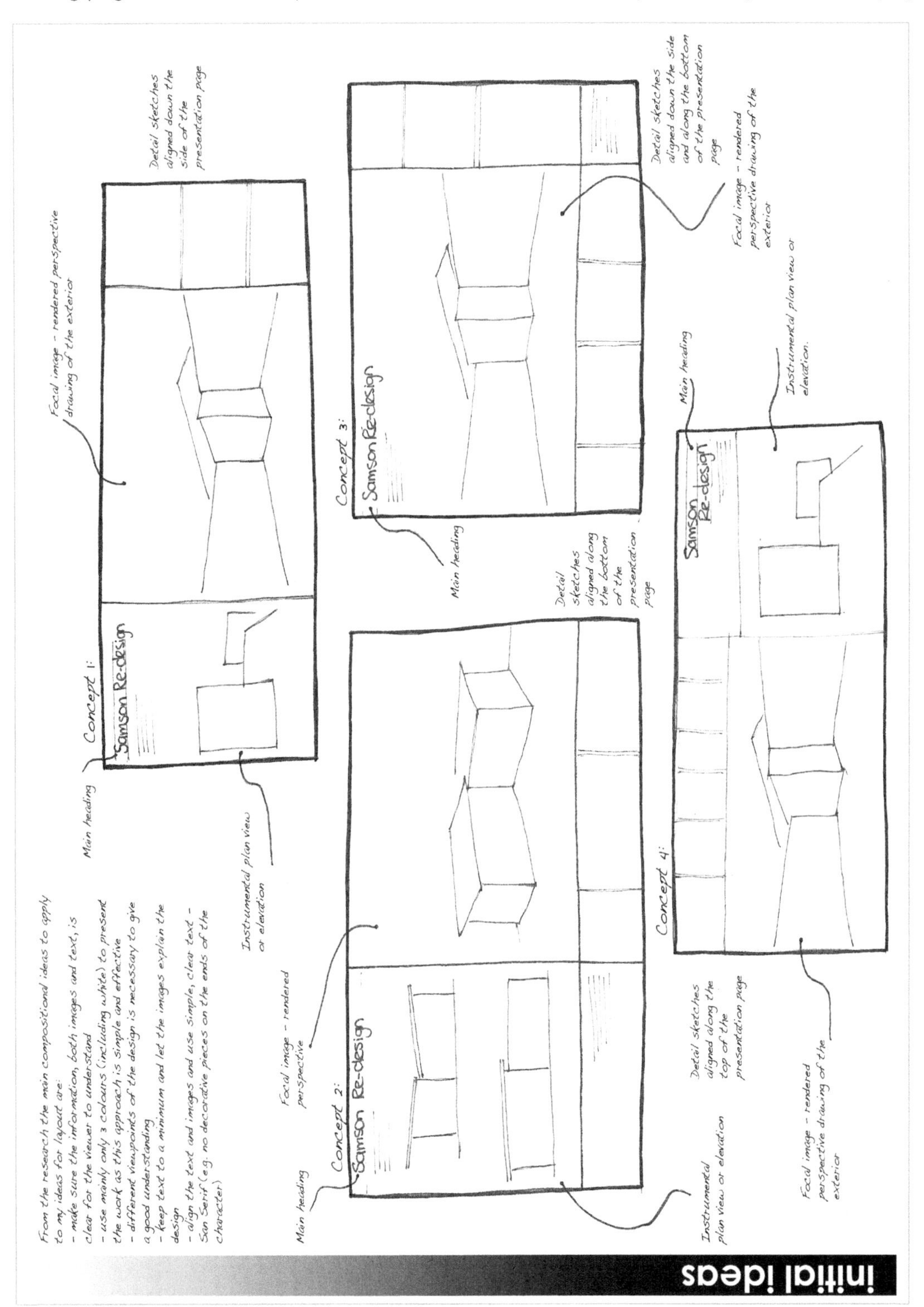

The following pages demonstrate four concepts for the layout of the presentation using the thumbnail sketches as a guide. Before laying out the images the design and instrumental drawing pages were scanned and the images cut out using Photoshop. Each image can then be moved or manipulated for the presentation layouts.

The compositional principles – ***proximity*, *alignment*, *hierarchy*, *repetition*, *contrast*, *positive and negative space*, *focal point and typography*** were considered, explored and evaluated.

ISBN: 9780170233279

initial ideas

Concept Idea 1: two page presentation

The focal point is the main images in the centre or to the side of the page. These images give a clear understanding of the overall design and are therefore the most important to highlight. The finish of the rendering is of a high quality which is also another reason for being the main focus.

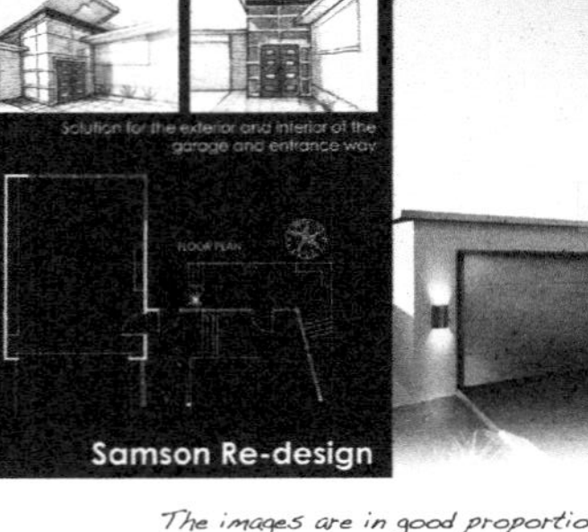

The contrast of the black against the white and colour images is effective and frames the images on the page.

The instrumentally drawn elevations and floor plan visually look effective white on black but may not be as clear to read for the viewer as black on white.

The text used – Century Gothic, is clear, simple, easy to read and visually pleasing.

The images are in good proportion to one another as the information does not look cluttered. The relevant images e.g. interior or exterior have been aligned together so that the spaces are explained from different views.

The grid pattern of the smaller aligned images down the side of the page also helps frame the page. These images show details of the interior.

Concept Idea 2: two page presentation

The instrumentally drawn elevations and floor plan are easier to read for the viewer as black lines on white paper.

The use of a colour as a background helps to frame the images on the page. The teal colour has been taken from the side of the wall of the house so that there are not too many colours introduced to the presentation.

The text used – Geneva CY, is clear, simple, easy to read and visually pleasing. The size of the lettering, upper and lower case could be explored through development.

The images are in good proportion to one another but the images along the bottom do look more cluttered than compared to the arrangement of images in Concept 1.

The grid pattern of the smaller aligned images along the bottom of the page also helps frame the page. These images show details of the interior.

ISBN: 9780170233279

initial ideas

Concept Idea 3: two page presentation

The focal point is the main images to the side of the page. These images give a clear understanding of the overall design and are therefore the most important to highlight. The finish of the rendering is of a high quality, which is also another reason for being the main focus.

The detail images are aligned along the top and around the side of the page unlike the other Concepts.

The text used – Eurostile, is clear, simple, easy to read.

The white text on grey is difficult to read.

All the instrumental 2D Drawings have been presented on the same page unlike the other Concepts. This allows for the proximity of images on the second page to be more spread out.

The main heading has been used on one of the presentation pages only which is all that is needed if the pages are being presented to the audience together.

Although each of the presentation pages for this Concept have been designed in a similar fashion there are also features that are different on each page such as the focal images and the background colour.

The background colour on each page has been taken from the focal images. The background colour is used with the same frame around the images but also gives a little difference to each page.

Concept Idea 4: two page presentation

The focal images on each page are reduced in size in this Concept which allows the presentation pages to be long and thin, although the images themselves look less dramatic.

The detail images are aligned along the top of the page, above the focal image. This seems to be less cluttered than with Concept 2.

The text used – Myanmar MN, is clear, simple, easy to read and less formal than the text used for the other Concepts.

The contrast of the colour and white on black is effective and unlike Concept 1 the instrumental floor plan and elevation is black on white.

The black background helps to frame to page. The main heading may not be needed on both pages.

ISBN: 9780170233279

Developing the design idea

Choose an idea to develop. Consider why you have chosen this idea and how the idea can be developed further. Develop and refine your idea so that the features of your design work are presented and communicated using visual communication techniques effectively.

The following pages demonstrate further development of a combination of the concept ideas.

development

From the initial ideas the main compositional ideas I would like to develop further will be:

- Concept 1 – the long thin shape of the presentation is effective, this could be further explored portrait as well as landscape.
- Concept 1 – the detail images down the side of the focal image is the best layout of all the concepts as they appear less cluttered and clearer to the viewer.
- Concept 1 – the contrast of the black against the images works well.
- Concept 2 – The teal background colour taken from the focal image looks effective so this could be an option instead of the black contrast as with Concept 1 and 4.
- Concept 3 – The presentation pages do not need the main heading on both pages like with this concept.
- Concept 4 – As with Concept 1 the long, thin shape of the presentation page is preferable.
- The white text on a dark background is effective for the main heading.
- The 2D instrumental drawings are clearer to read on Concept 2, 3 and 4 with the black lines on white paper but this does create a lot of white space.
- Although the grid of Concept 2 and 4, separate the detail images it does not make the images as clear to view as with Concept 1.

I will further develop Concept 1 but use aspects of Concepts 2, 3 and 4 for the exploration of ideas.

Samson Redesign samson redesign Samson Redesign SAMSON REDESIGN Century Gothic	Samson Redesign samson redesign Samson Redesign SAMSON REDESIGN Century Gothic
Samson Redesign samson redesign Samson Redesign SAMSON REDESIGN Geneva CY	Samson Redesign samson redesign Samson Redesign SAMSON REDESIGN Geneva CY
Samson Redesign samson redesign Samson Redesign SAMSON REDESIGN Eurostyle	Samson Redesign samson redesign Samson Redesign SAMSON REDESIGN Eurostyle
Samson Redesign samson redesign Samson Redesign SAMSON REDESIGN Myanmar MN	Samson Redesign samson redesign Samson Redesign SAMSON REDESIGN Myanmar MN

The exploration of text to the side shows the options for each of the possible fonts: Century Gothic, Geneva CY, Eurostyle and Myanmar. By changing the text from lower case to upper case and a combination of the two a different effect is achieved.

ISBN: 9780170233279

These features are explored:

- fonts for the heading text
- alignment and positioning of text
- background and text colour
- contrast of white on black or black on white
- proximity of images to one another
- use of upper and lower case text
- alignment of detail sketches of the exterior and interior
- page setup – landscape or portrait
- positioning of the focal images
- hierarchy of images

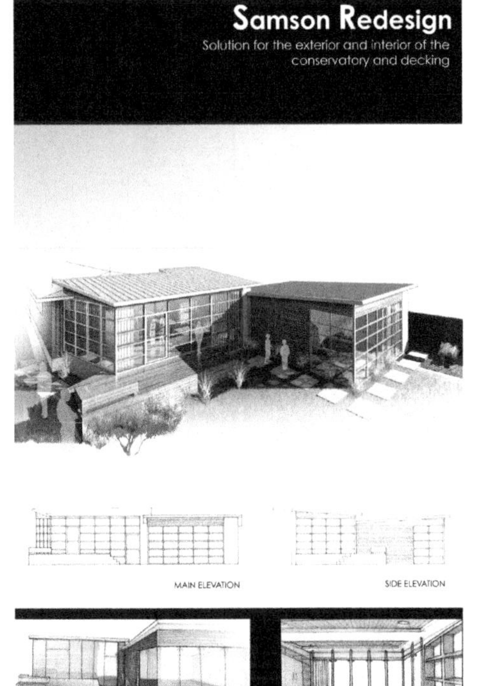

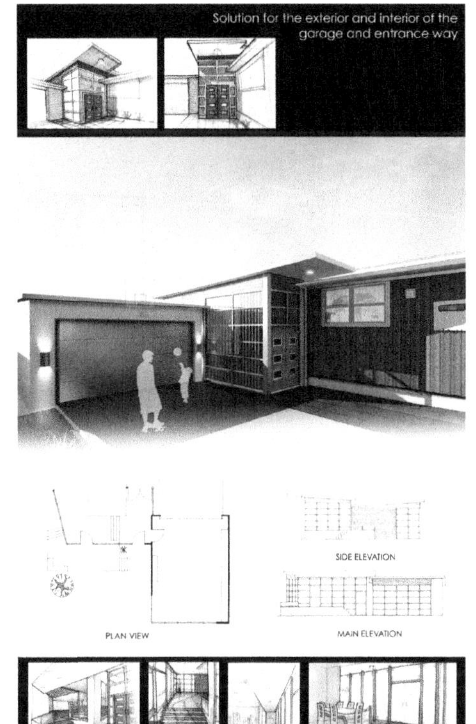

The heading is presented on one of the pages only which frees up space on the second page for more images to be included.

The layout version of the presentation pages in portrait does make the focal images (the rendered, perspective pictorials) stand out.

The instrumental drawings are very clear on the white background.

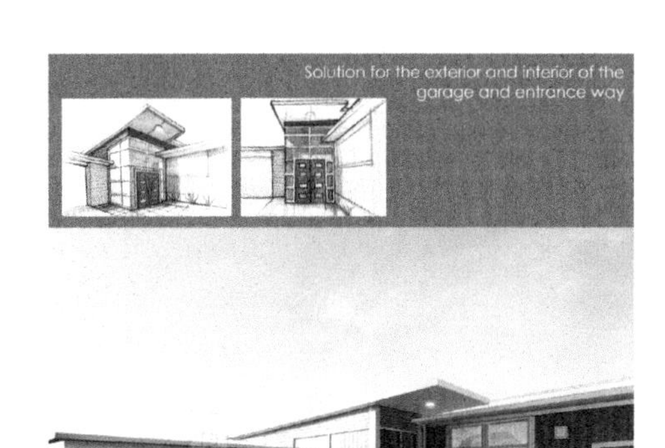

The use of black provides good contrast against the images compared to the teal, although the green is a little more subtle.

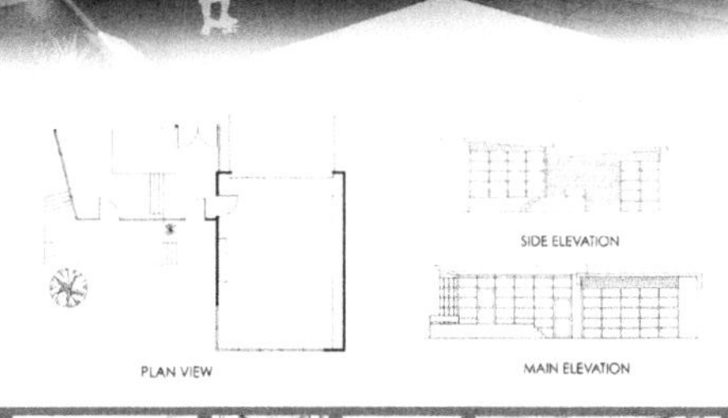

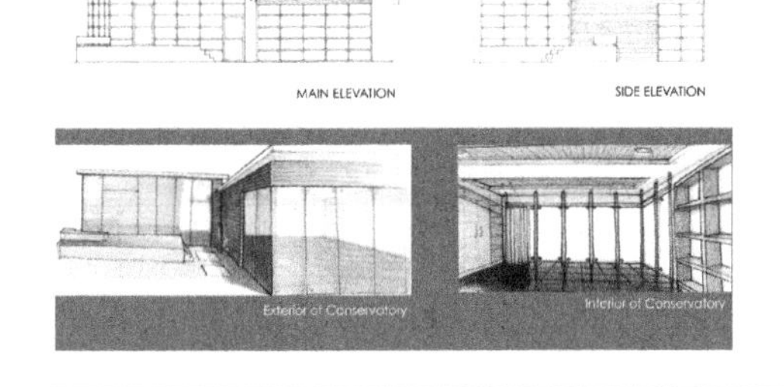

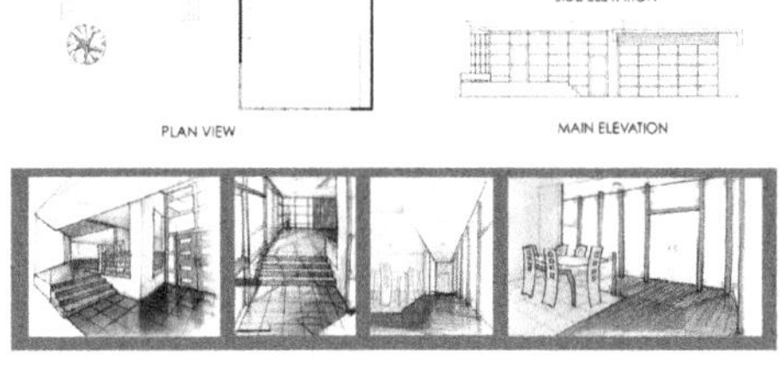

If the portrait idea is to be explored further the detail sketches at the bottom of the page would need to be adjusted. The images on presentation page 1 are too large and the details on the second presentation page could be a little larger.

ISBN: 9780170233279

development

The white back-ground on the 2D instrumental drawings does not look as effective for the landscape version of the presentation page as the portrait version.

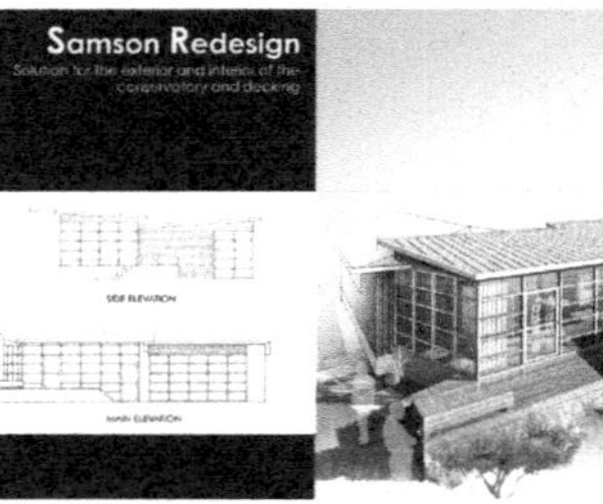

The detailed images are compiled together quite closely so are are not as clear to the viewer as they could be.

Below the images are framed by a black outline to visually connect these to the information on the left hand side of the presentation page. The images are lost with the black framing.

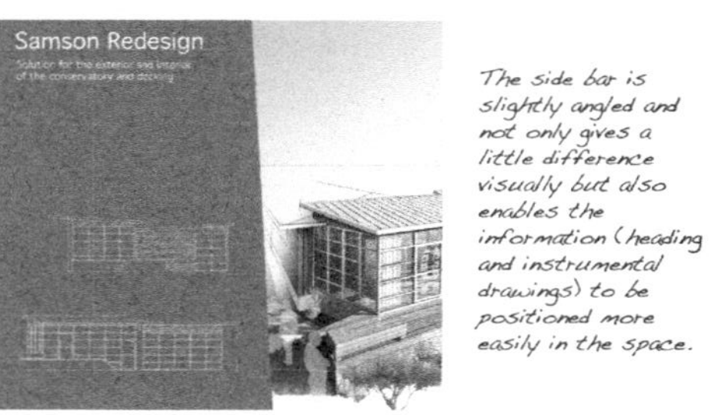

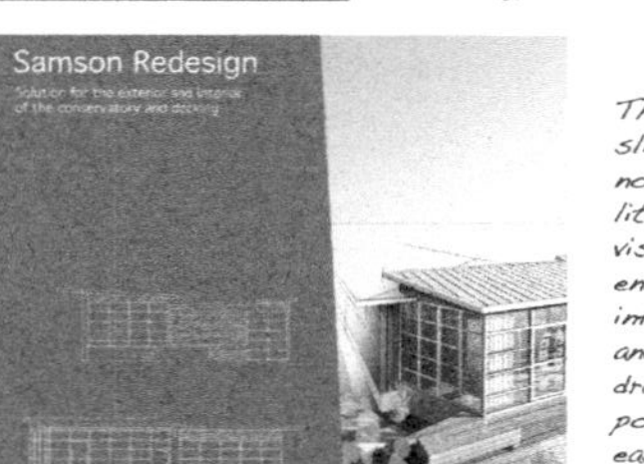

The exploration of the text for the headings gives many different options for the final presentation pages. I think the best option is this one.

The side bar is slightly angled and not only gives a little difference visually but also enables the information (heading and instrumental drawings) to be positioned more easily in the space.

The two detail sketches and the floor plan fit well into the space with the angled side bar.

The white lined floor plan on the green has the same clarity as the white lines on the black but does not have quite as much contrast.

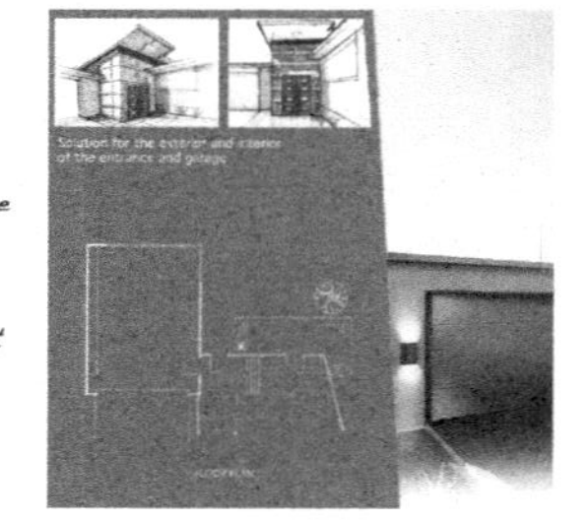

ISBN: 9780170233279

development

Below two versions of the landscape presentation pages have been shown. For the first example, the second page mimicks the first with the layout. For the second landscape version, the second page mirrors that of the first presentation page with the angled side bar on the opposite side. I prefer this layout as the pages would be presented to the audience side by side and not individually.

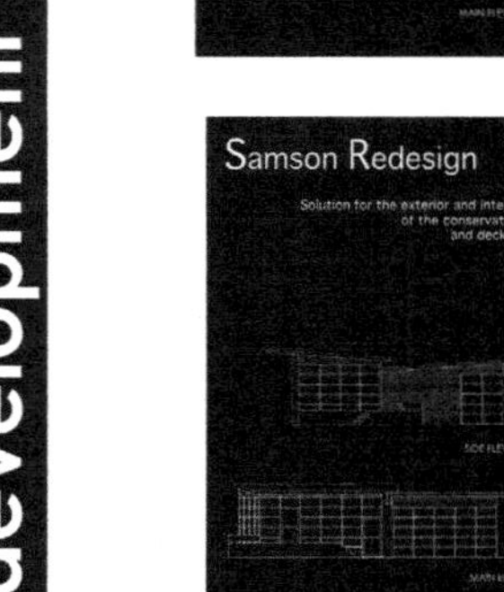

The detail images and plan view do not fit as easily on the page for the portrait version. The detail sketches have to be presented relatively small compared to the landscape version of the presentation page design.

The black boarder framing the images to the side has been reduced and the images made larger. This makes the information clearer for the viewer while also framing the page.

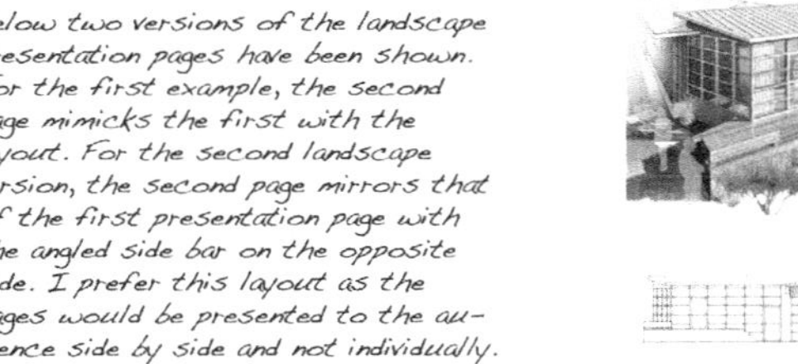

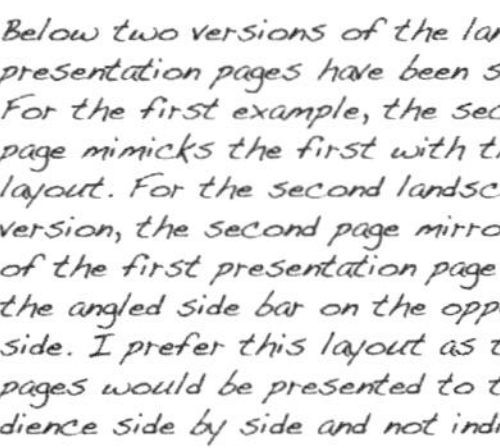

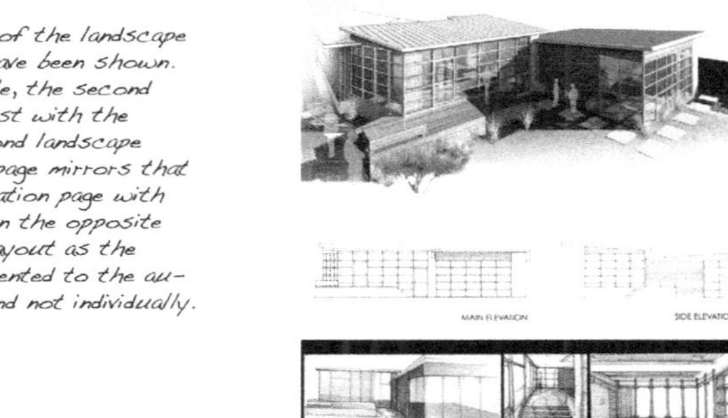

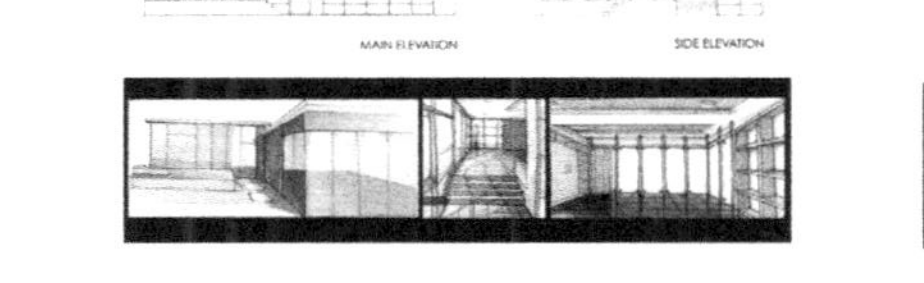

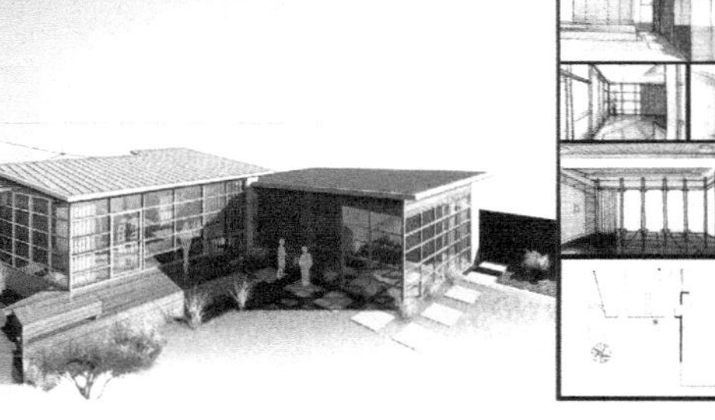

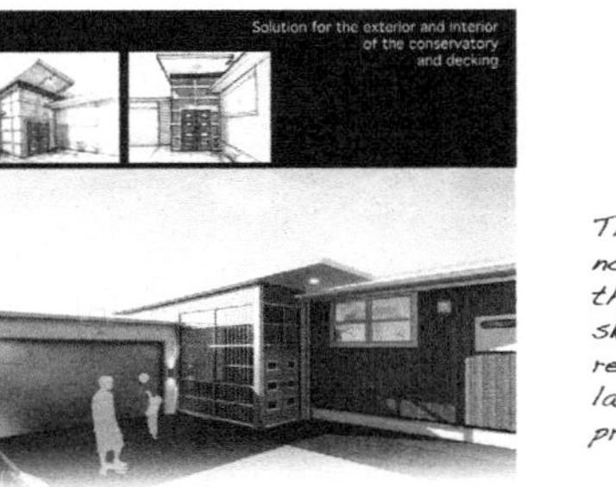

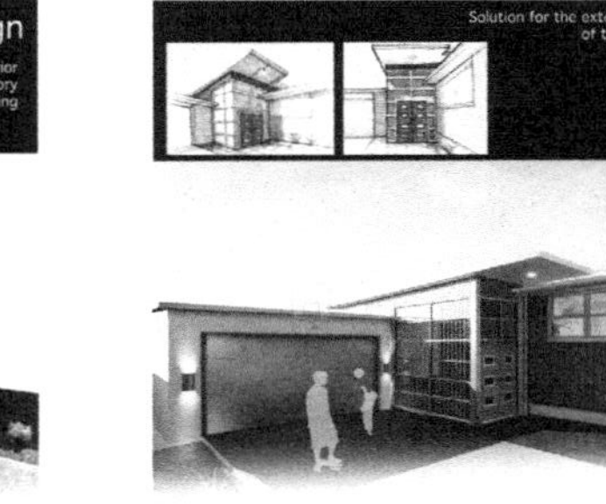

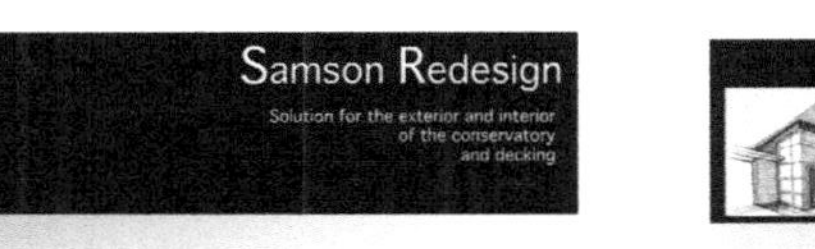

ISBN: 9780170233279

Presentation pages

Produce the presentation making sure the design sketches, instrumental drawings, models and photographs are accurate, clear and precise. The presentation needs to demonstrate cohesiveness. To achieve this integrate presentation techniques, apply compositional principles and modes and media effectively. The presentation needs to interpret and promote the essence of the design.

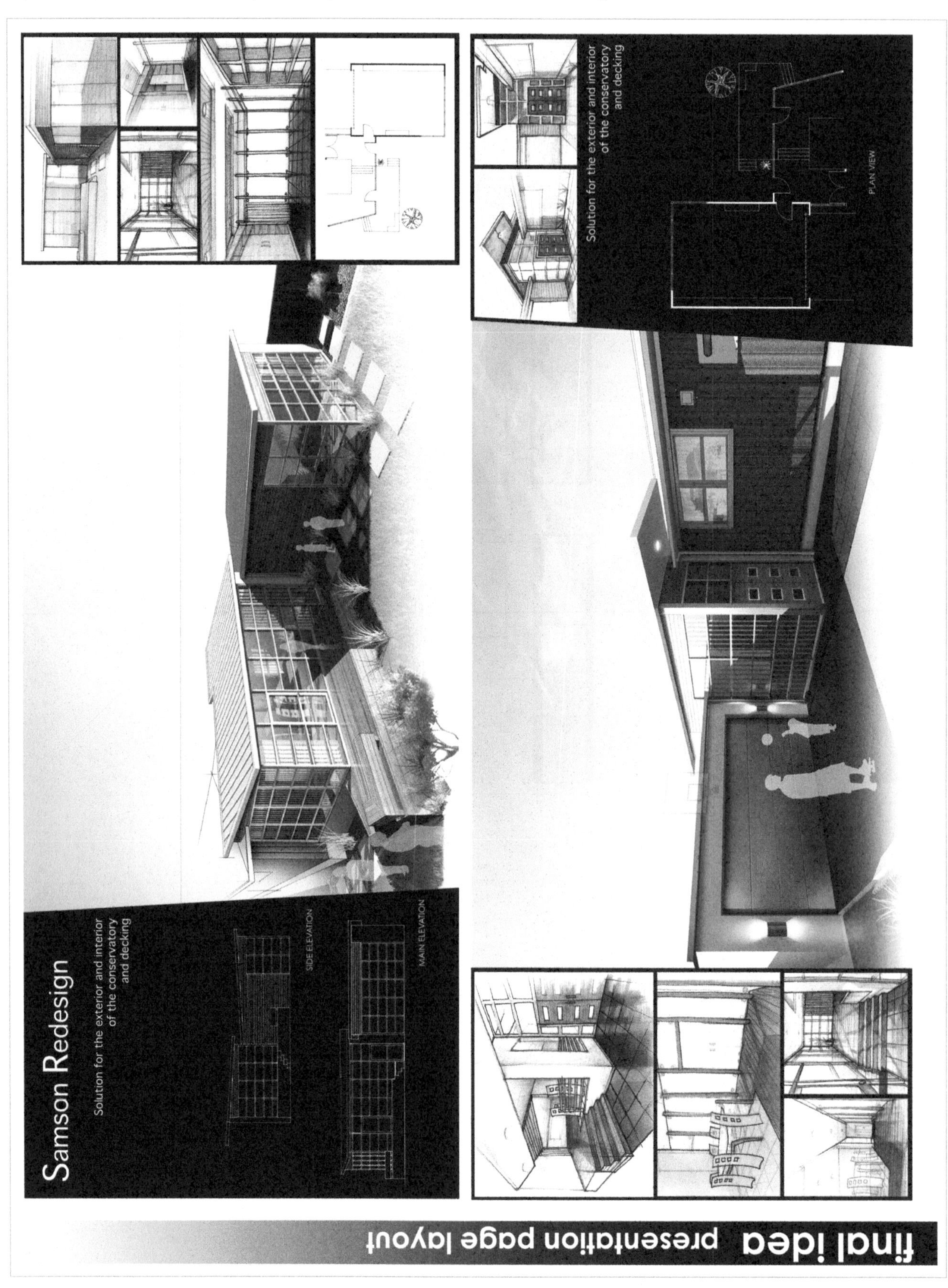

ISBN: 9780170233279

Christian Burgos — St Peter's College

The following pages produced by Christian Burgos demonstrate the presentation of the compact vehicle design. He has shown:

- *Two main focal images, which explain the product from the front and back – these images are sketches that have been scanned and rendered using Photoshop.*
- *Detail images explaining the interior of the product, the components, ergonomics and function.*
- *Simple backgrounds, which create the effect of a landscape for the vehicle. The background is different on each page but has a similar effect.*
- *Minimal text to explain the product. The headings and text on each page is slightly different with the use of colours but uses the same font.*
- *Detail images are aligned informally on each page.*

Modes that have been used are digital applications, image manipulation and sketching methods. Media that have been used are colour pencils, marker pens and digital media.

ISBN: 9780170233279

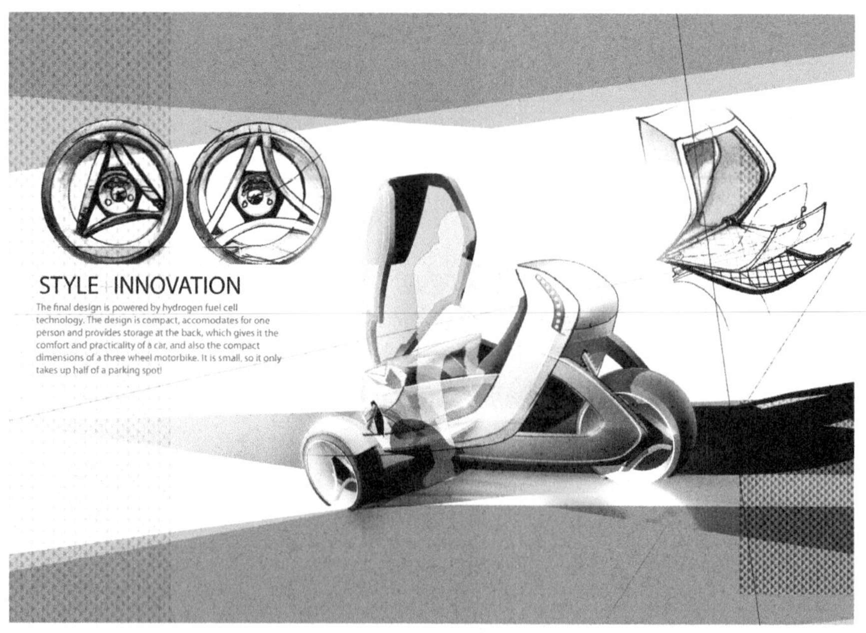

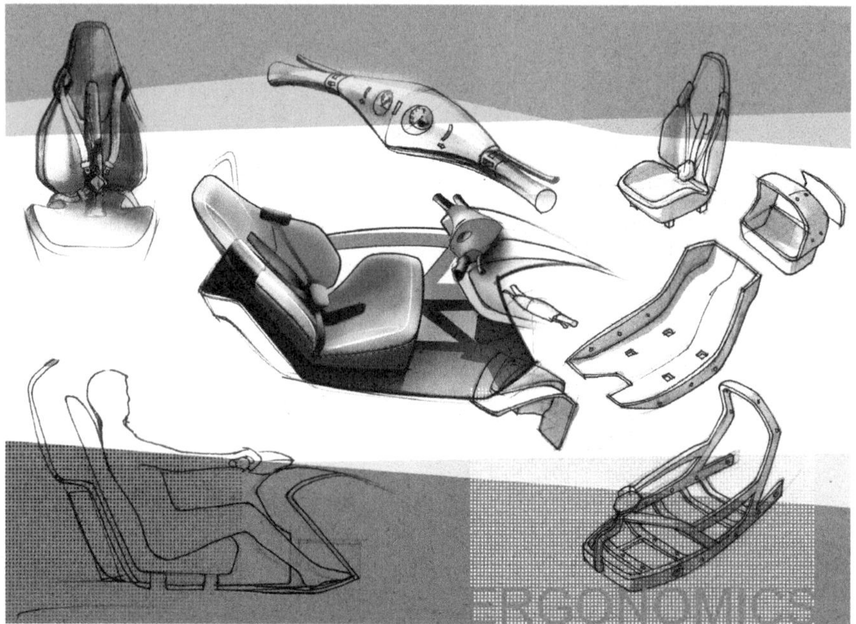

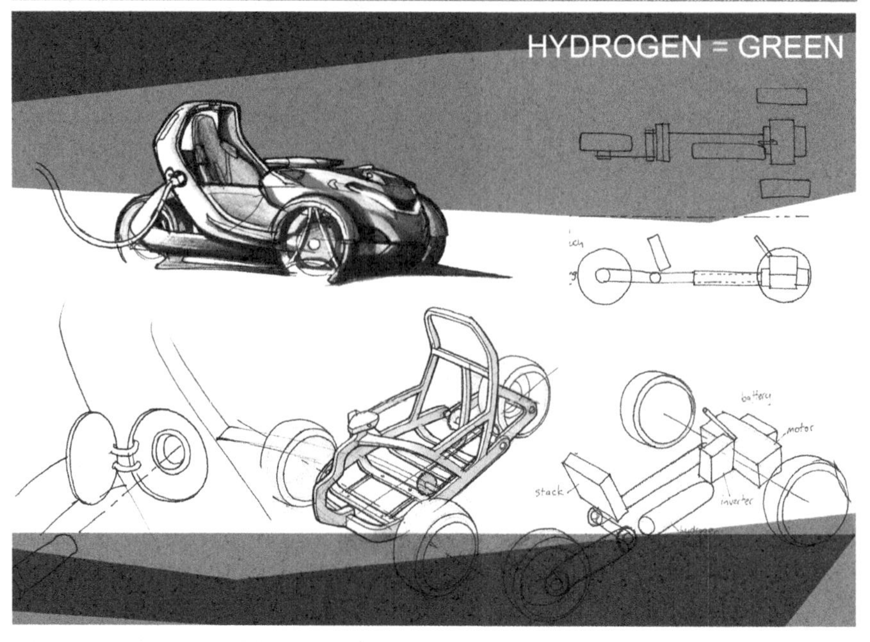

ISBN: 9780170233279

Ryan Hellier — St Peter's College

The following pages produced by Ryan Hellier demonstrate the presentation of a fireplace design. He has shown:

- *The product in the environment; the function, components and sizes.*
- *Important information has been highlighted with square boxes around those images.*
- *Minimal text to explain the product. The heading is the same on each page.*
- *The integration of different modes and media – computer generated image, instrumental drawing and sketching.*

Modes that have been used are digital applications drawing, models and sketching methods. Media that have been used are colour pencils, marker pens and digital media.

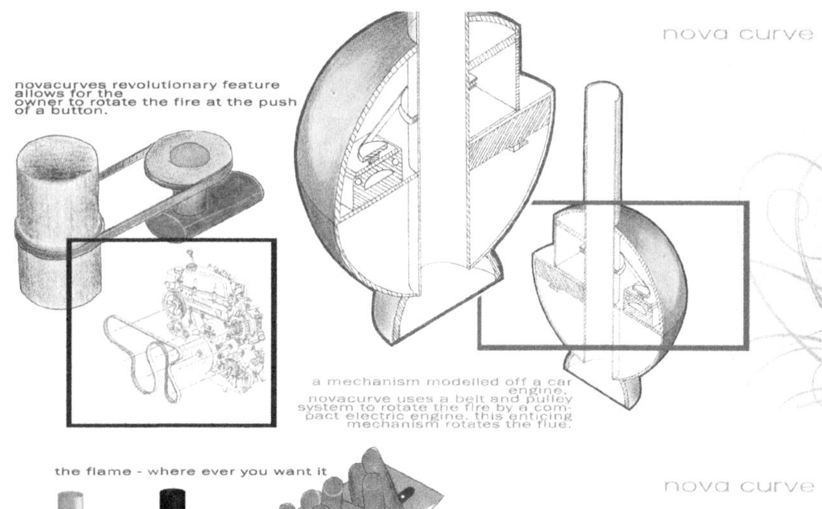

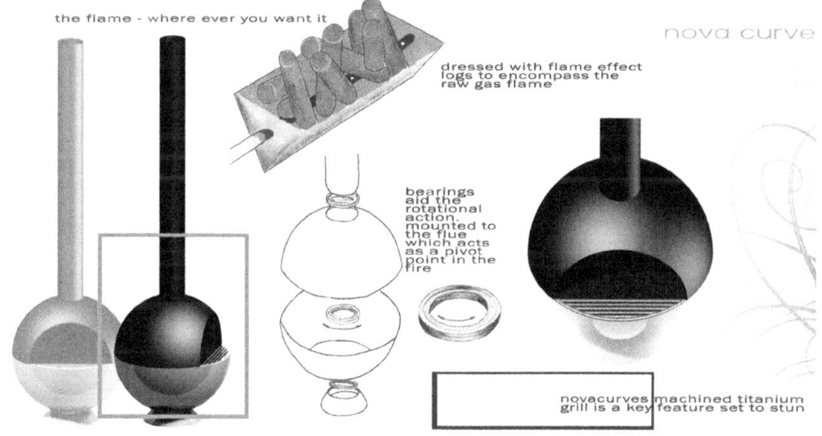

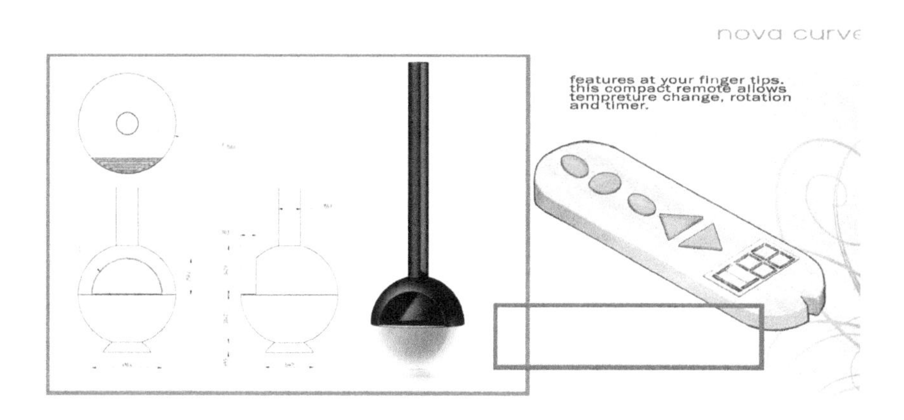

ISBN: 9780170233279

Degge Jarvie — Onslow College

The following pages produced by Degge Jarvie demonstrate the presentation of the spatial design for a cafe. The work shown here is a selection of pages he produced to present his process. Degge has shown:

- *Focal images that clearly stand out on each page highlighting the main information about aesthetics and environment.*
- *Detail images explaining the interior, structure, alternatives and function.*
- *A grid pattern, aligning the text and images, which is slightly different on each page.*
- *Headings that use grey text and the first letter of each heading is in red, which makes the word stand out on the page.*
- *Blocks of text, which explain the design principles and justification for design decisions.*

Modes that have been used are digital applications, photography, image manipulation, models, drawing and sketching methods. Media that have been used are collage, marker pens and digital media.

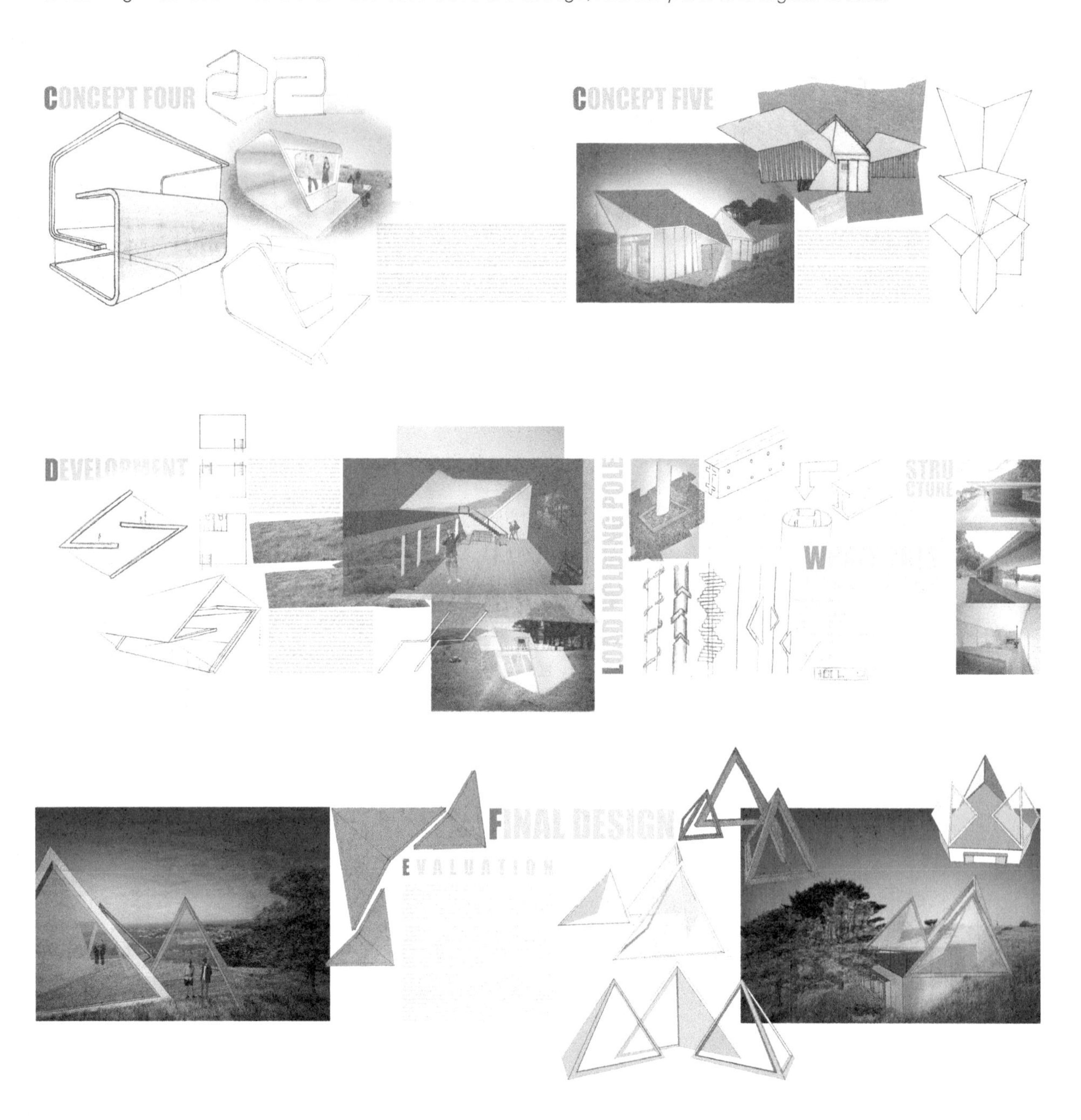

ISBN: 9780170233279

Tyler Warrington — Onslow College

The following pages produced by Tyler Warrington demonstrate the presentation of the spatial design for a cafe. The work shown here is a selection of pages he produced. Tyler has shown:

- *Focal images larger than the smaller images. The focal images are spread out on the pages.*
- *Images presented in a grid format – photographs of a card model and viewpoints of a computer model taken from many different directions.*
- *Contrast of the white model on black created on the first page and the opposite on the second page.*
- *Headings as the only text because the images explain the design. The same font is used but in different colours and sizes.*

Modes that have been used are digital applications, photography, image manipulation and models. Media that have been used are card and digital media.

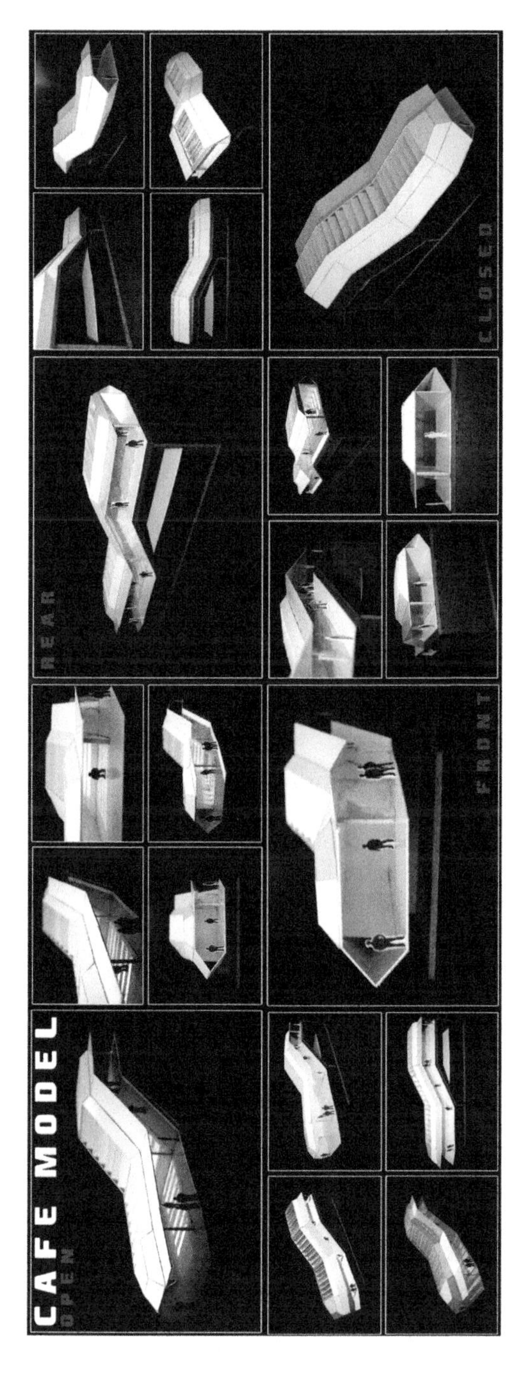

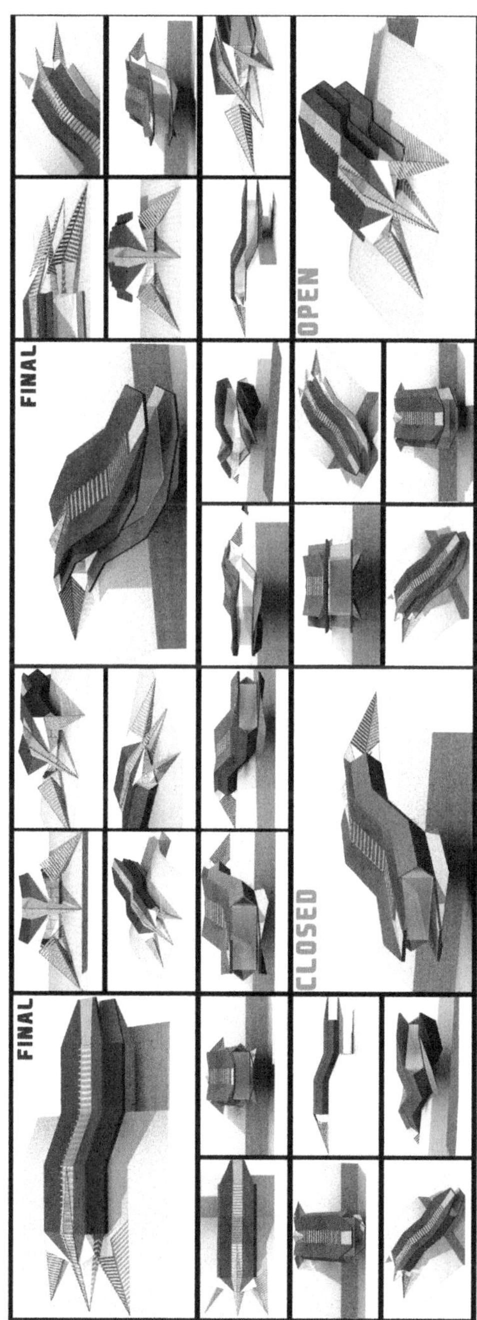

ISBN: 9780170233279

Jake Sanders — Onslow College

The following pages produced by Jake Sanders demonstrate the presentation of the spatial design for a cafe. The work shown here is a selection of pages he produced to present his process. Tyler has shown:

- *Large focal images on each page from different viewpoints of the café exterior.*
- *Detail images of the interior and elevations on one presentation page.*
- *The café in its environment from different viewpoints.*
- *The background of the environment on all pages but faded with a slight change of colour on three of the pages to help the foreground images stand out.*
- *Headings presented in the same layout and text on each page.*
- *Informative text on the Energy & Sustainability page to explain the function.*

Modes that have been used are digital applications, image manipulation and models. Media that have been used are digital media.

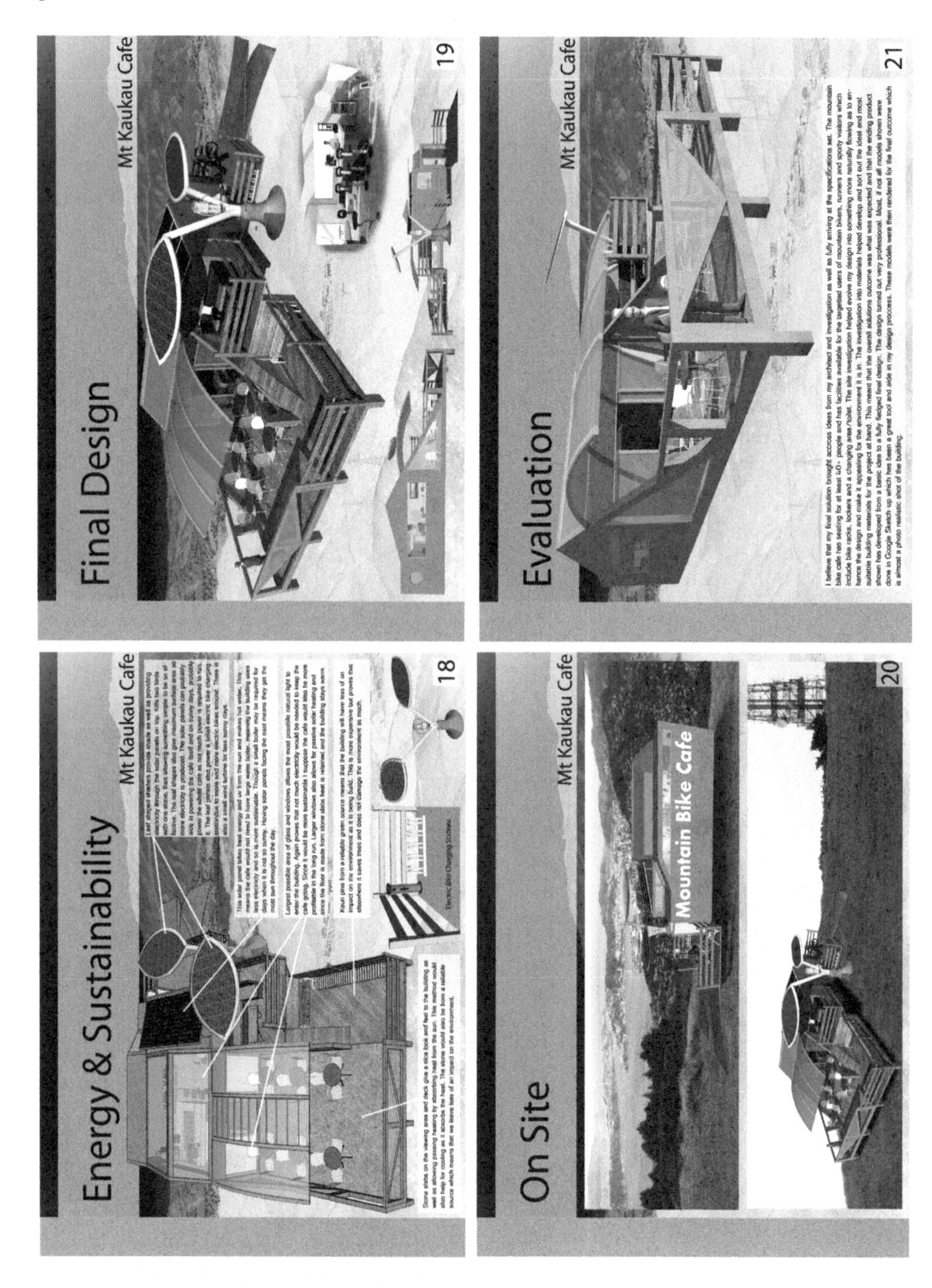

ISBN: 9780170233279

Sheldon Carr — Onslow College

The following work produced by Sheldon Carr demonstrates the presentation of research not his own design work. Interesting compositional principles have been used. The pages have been divided up using a ripped paper effect. Sheldon continues this effect through the development of his design ideas. Contrast is created between the off-white and black spaces. Images and text are aligned on the page with no focal image.

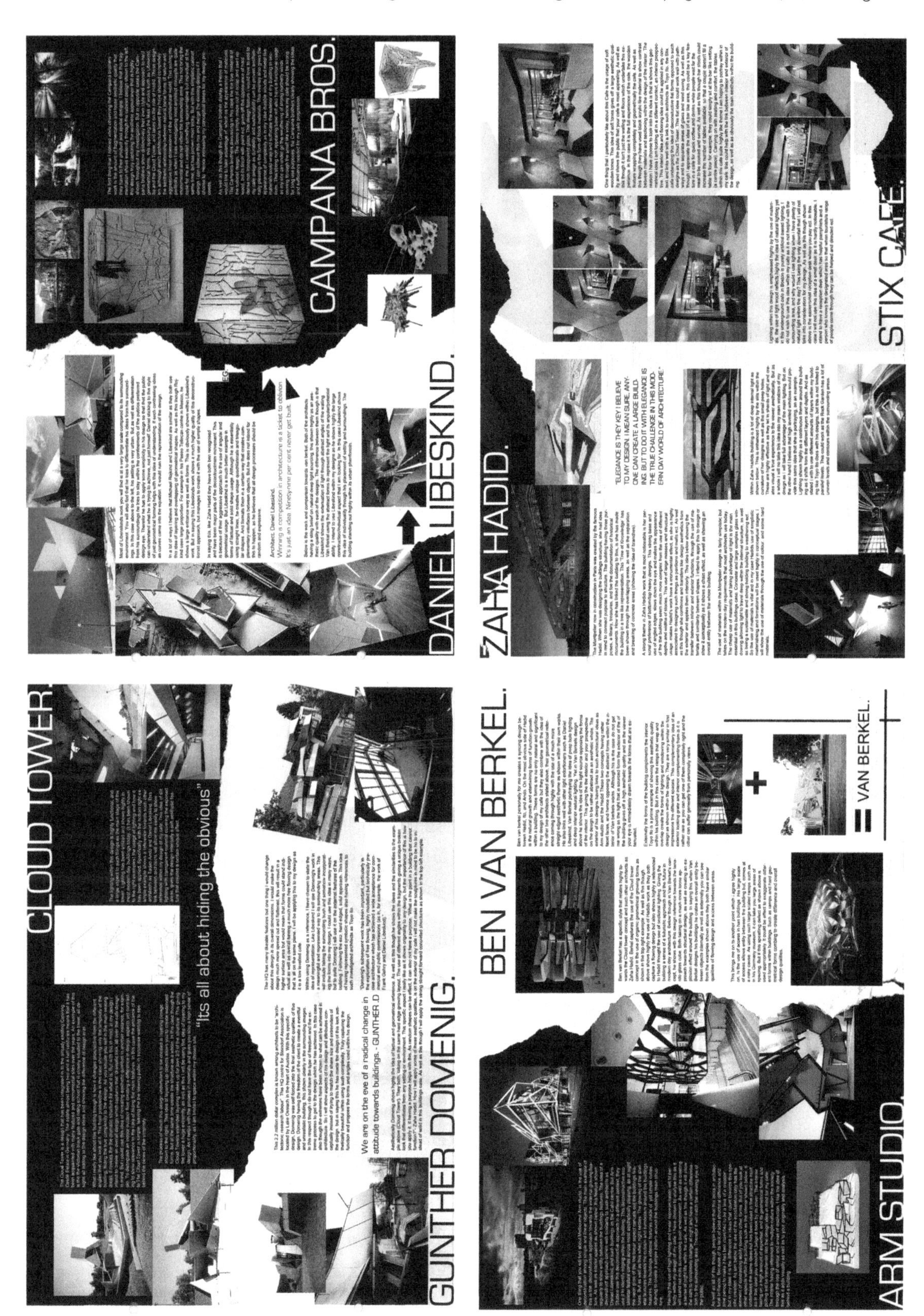

ISBN: 9780170233279

Aesthetics — what a product looks like

Alignment — left alignment, right alignment, centre alignment

Apex — the top point of a pyramid

Arc — a curve scribed with a compass

Auxiliary — an extra view on an orthographic projection

Balance — the appearance of a product

Brief — a design problem to be resolved, including specifications

Casting — a manufacturing process for metal

Centre line — line made up of long and short dashes

Collage — composition of materials

Component — part of a product

Compositional principles — proximity, alignment, hierarchy, repetition, contrast, positive and negative space and focal point

Concept — ideas for a design solution

Construction — light lines used to sketch or draw an object

Contrast — the use of opposing elements

Conventions — rules or standards for presenting parts of a drawing

Crating — a technique for constructing a sketch or drawing

Cutting plane — where an object is cut in one view and shown in another

Design process — the stages of drawing and modelling to solve a problem

Design elements — derived from the design principles e.g. shape, durability

Design principles — aesthetics and function

Development — exploring, reviewing and refining your ideas

Diameter — the width across a circle

Diametric — drawing (paraline), two angles are equal

Digital media — electronic media

Dimensioning — the measurements on a sketch or drawing

Drawing — to use instruments

Durability — the strength of an object

Elevation — view of one side of an object

Equilateral — applies to a shape with equal sides

Ergonome — a scaled replica of a person which has moving parts

ISBN: 9780170233279

Ergonomics — consideration of the human form
Evaluation — analysis and justification about a design
Exploded sketch — a sketch that has the parts pulled apart (no rulers)
Exploded instrumental pictorial — a drawing that has the parts pulled apart (rulers are used)
Eye level line — the height of the viewers eyes, used in perspective sketching
Floor plan — a bird's eye view of a building or room
Focal point — the main focus of text or image
Form — three dimensional object
Forming — a manufacturing process, e.g. plywood forming
Freehand sketching — no use of instruments
Function — how a product works
Geometric construction — the construction using instruments of shapes and solids
Geometric shape — two dimensional – square, circle, hexagon etc
Geometric solid — three dimensional – prism, pyramid, cone, cylinder
Graphic media — the use of different media are markers, pencil, paint etc
Graphic modes — instrumental drawing, freehand sketching, digital media, mockups and models
Hatching — lines to indicate where the material has been cut through for a section sketch or drawing.
Hidden detail — short dashed lines to indicate what is hidden on a particular view
Hierarchy — the order in which each element is most important
Highlight — where light is reflected on an object and is presented on a rendering
Horizon line — the eye level line in perspective
HP — label on an orthographic projection, (Horizontal Plane)
Idea generation — visual communication techniques used to generate ideas
Initial Ideas — a range of first ideas using sketches and/or mockups
Injection moulding — a manufacturing process for plastic
Instruments — set squares, t-squares, compass
Instrumental drawing — drawing using instruments
Instrumental pictorial — isometric, oblique, planometric, trimetric, diametric
Isometric — drawing (paraline) using 30/60 degree set square
Justification — evaluations that justify your design decisions
Leader line — lines on an orthographic projection used for dimensioning to indicate where the measurement starts and finishes
LHE — Left Hand Elevation on an orthographic projection
Light source — where the light is coming from on a rendering to indicate where the light and dark areas are
Manufacturing process — processes to make mass produced products
Material value — timber, metal, plastic, concrete etc
Metal bending — a manufacturing process used for sheet and rod metal
Mockup — a model to test out ideas quickly
Model — a well finished model to present a final idea
Mould — a solid that is used to form other materials
Multi-view — three or more views for an orthographic projection
Negative space — the shapes around objects
Oblique — drawing (paraline) using a 45 degree set square
Orthographic symbol — a symbol on an orthographic projection
One point perspective — the vanishing lines go to one point
Paper architecture — 3D constructions to experiment with form
Parallel line development — a surface development, construction lines are projected parallel
Pattern — an arrangement of repeated parts
Perpendicular — a line which is at 90 degrees to a line or surface
Perspective — a pictorial sketching method, there are two methods 1pt and 2pt
Pictorial — three dimensional sketching and drawing
Plan view — a birds eye view of an object
Plane — a two dimensional surface
Planometric — drawing method (paraline) using a plan as the base.
Positive space — the shapes of objects
Prism — a geometric solid with parallel sides
Product design — the design of objects and artifacts

ISBN: 9780170233279

Projection — lines and points taken from one view and plotted on another

Proportion — the relation of a part of an object to another

Proxlmlty — the spacing between elements

Pyramid — a geometric solid with sides that go to an apex

Radial line development - a surface development, construction lines are projected on an arc

Radius — half the diameter of a circle

Reference lines — lines that are made up of a long dash and then a short dash. Views on a working drawing are projected through a reference line

Research — investigation of information that relates to the design brief

RHE — Right Hand Elevation

Rhythm — pattern or repetition

Scale — a drawing or model produced in proportion to the real life object

Section view — a view where you can see the interior of an object

Shading — dark and light areas on an object to give a sketch or drawing a three dimensional effect

Shadow — a shaded area where light does not get to

Shape — hexagon, square, circle, triangle, octagon

Sketch — a freehand drawing, no instruments should be used

Sketch models — manipulation of card or other materials to generate ideas

Spatial design — the design of exterior and interior spaces

Surface development — a solid drawn folded out flat

Symmetrical — two halves that are mirror image of each other

Texture — the look and feel of a surface

Third angle orthographic projection — a working drawing of an object that has a plan and elevations

Three dimensional — form showing the width, height and depth of an object

Title block — contains information about a drawing

Tonal change — the change in light and dark areas

Trimetric — drawing (paraline) based on three scales

True view — a view that is seen face on

Truncated — cutting a prism or pyramid on an angle

Two dimensional — shape showing the width and height only

Two point perspective — lines go to two vanishing points

Typography — the design and art of characters for printing

Vanishing point — the point on the horizon in a perspective drawing where the lines vanish

Working drawing — scale drawings that are used for reading measurements for the manufacture of a product

ISBN: 9780170233279